HOLOGRAPHIC IMAGING

HOLOGRAPHIC IMAGING

Stephen A. Benton
V. Michael Bove, Jr.

Illustration and design by Elizabeth Connors-Chen

Additional material by William Farmer, Michael Halle, Mark Holzbach, Michael Klug, Mark Lucente, Ravikanth Pappu, Wendy Plesniak, Pierre St.-Hilaire, John Underkoffler

WILEY-INTERSCIENCE

A JOHN WILEY & SONS, INC., PUBLICATION

Published by John Wiley & Sons, Inc., Hoboken, New Jersey.
Published simultaneously in Canada.

For general information on our other products and services or for technical support. please contact our Customer Care Department within the United States at (800) 762-2974, outside the United States at (317) 572-3993 or fax (317)572-4002.

Wiley also publishes its books in a variety of electronics formats. Some content that appears in print may not be available in electronic format. For information about Wiley products, visit our web site at www.wiley.com.

Library of Congress Cataloging-in-Publication Data:

Benton, Stephen A.
Holographic imaging / by Stephen A. Benton and V. Michael Bove; Illustration and Design by Elizabeth Connors-Chen.
p. cm.
ISBN 978-0-470-06806-9 (cloth)
1. Holography. I. Bove, Michael. II. Title.
TA1540.B46 2007
621.36'75—dc22 2007022429

10 9 8 7 6 5 4 3 2 1

For the Benton Family Farm
and Teakettle Farm
and all their inhabitants, great and small

Contents

Foreword: Holography

Charles M. Vest

Why would the world need another book on holography, a mature science and technology for which most of the seminal work was done in the 1960s and 1970s? The answer is that this book is inspired and largely written by a true master teacher—the late Steve Benton. It covers virtually every aspect of the field—from fundamentals to "real world" issues—in a consistent, wonderfully accessible manner. Amazingly, this volume is equally attractive as an introduction for newcomers to the field or as a reference work for seasoned professionals. Artists, hobbyists, scientists, and engineers will be equally at ease with it.

Optical holography as we know it today was developed largely by three great pioneers—Emmett Leith of the University of Michigan, Yuri Denisyuk of Russia's Ioffe Institute, and Steve Benton of MIT. They established a fundamentally different way of working with light. Their discoveries and inventions brought us three-dimensional imaging, a variety of ways of processing optically encoded information, and new means of visualizing and measuring various physical phenomena. Steve Benton was not only an important and pioneering contributor to these developments, but he also broadened the communities of interest in holography, expended the conceptual base to a more general view of three-dimensional imaging, and worked to bring holography from the age of film to the digital age.

Holography's importance is attested to by its continuing ability to inspire and excite those who encounter it for the first time. Few if any other scientific concepts or technologies of the second half of the twentieth century are so readily accessible in their basic form to those with little formal scientific or engineering training. Young children continue to marvel at floating three-dimensional images they encounter in science museums, school science projects, and even souvenir or jewelry shops. Chances are very good that for many readers this book too will open a new world—or at least a new way of looking at the world. Readers will be able to start their journey at a relatively simple descriptive level and pursue it as far down the path of depth of understanding and art of practice as far as they wish.

Steve Benton is most widely known as the inventor of the rainbow hologram that utilizes the full optical spectrum to create three-dimensional, full-color images. From this work he also created the embossed holograms that today are ubiquitous on credit cards, and other security and identification products. These things are explained clearly in this book as a part of a wide-ranging treatment of the subject.

This book is at once the product of, and homage to, a great teacher and scientific "man for all seasons." Steve Benton's inventiveness, enthusiasm, and *joie de vivre* are seen in the nature and quality of his writing, but even more so in the commitment of his students and colleagues to bring this volume to completion following his untimely death.

Charles M. Vest is professor of mechanical engineering and president emeritus of the Massachusetts Institute of Technology. He is a member of the U.S. President's Council of Advisors on Science and Technology and the National Academy of Engineering, and is the author of Holographic Interferometry, *John Wiley & Sons, 1979, and of* Pursuing the Endless Frontier: Essays on MIT and the Role of Research Universities, *MIT Press, 2005. He is a director of IBM and DuPont and has received 10 honorary doctorates.*

Foreword: Nerd Pride

Nicholas Negroponte

In the 1970s, art served a function similar to athletics at MIT. Both were seen as a relief from stress, fun to do, socially engaging. But they were not part of the Institute's serious business of doing science and creating technology. Instead, art's purpose on campus was extracurricular and ancillary. If the students learned a little about art, it was thought, they'd be better rounded individuals. Great artists who then taught at MIT—I am thinking of Minor White and Richard Leacock—were brought to the Institute in large measure as a counterweight to the general geekiness; call it aesthetics for nerds.

Steve Benton changed all that. He was a bred-in-the-bone scientist, a brilliant physicist who proudly wore a "Nerd Power" pocket protector. His work in optics was so highly esteemed that Steve became the first, and so far only, Media Lab faculty member to jump two rungs of the MIT promotional ladder at once. His commitment to the arts was equally profound and well illustrated by his eventual directorship of the Center for Advanced Visual Studies (CAVS), founded by Gyorgy Kepes in 1968, which Steve headed from 1996 until his death in 2003.

Kepes, the last of the Bauhaus originals, was a great philosopher of art and technology. When Kepes started CAVS—I know because I was there—he had in mind a place like the Center for Advanced Studies at Princeton, the academic home of Albert Einstein and John von Neumann. Little did Kepes know that a thinker of their magisterial ilk would someday head *his* program.

It is well known that the Media Lab was born within MIT's small School of Architecture and Planning, not the much larger School of Engineering, a more logical choice at first glance. This decision held several advantages. One was to keep us below anybody's radar about science and technology, which gave the lab the chance to break all the rules, gain momentum, and establish itself before anybody took notice. Another benefit was the natural *salon des refuses* provided by arts and design. It was more socially and academically acceptable to have iconoclastic science and idiosyncratic engineering in our branch of academia. For this reason, the Media Lab lived happily and undisturbed on the lunatic fringe, because nobody noticed—in the beginning.

Less well known is that the Media Lab's degree program grew out of the Department of Architecture's Masters of Science in Visual Studies, which I headed before Steve came to MIT. This program was so broad, it even included electronic music. Go figure. It also included photography, which as a discipline at MIT was going through a difficult period following the death of its founder, Minor White.

My own campaign for the Media Lab to achieve primacy of place, instead of serving as an occupational therapy clinic, took a major and credible turn for the best when I proposed we convert photography to holography, bring in the world leader in that field, and use holography as an archetype for the future of arts at MIT. At the time, people at the Institute thought it was unlikely we could attract Benton away from Polaroid, where he not only worked, but was also the direct protégé of Dr. Land himself. Fortunately, Jerry Wiesner (the 13th President of MIT and co-founder of the Media Lab) and Dr. Land (whom Jerry knew as Din) were such close friends that this idea was discussed openly and received Land's immediate blessing. Thus began Steve Benton's quarter century at MIT, three years before we moved into the new I. M. Pei building on Ames Street. Shortly afterwards he became the Academic Head of the Media Lab and created a robust PhD program.

What followed were two-and-a-half decades of remarkable work—I am talking about his own research—at the intersection of both art and science, the kind Kepes wrote about. Engineering deadlocks were broken by card-carrying artists. Some of our geeks provided artistic expressions of lasting effect on the art world. The symbiosis went deeper than any before it. For evidence, just consider the pages that follow.

When MIT recruited its 15th President, it found none other than a holographer at the University of Michigan, where Charles Vest was then Provost. During his presidency, Chuck, as he is called, taught only one class a year at MIT; that was in Benton's course, proudly at the Media Lab. While their specializations within holography were different, their scientific interests overlapped. Chuck held Steve in the highest admiration, as he movingly recalls in his introduction that precedes this one

When Steve fell ill, Chuck and I decided together that we should hold an international symposium in Benton's honor. Of course, the invitees included the world's foremost holographers, including Emmett Leith and Yuri Denisyuk. After a small amount of planning, we moved the date up by six months to accommodate Steve's worsening prognosis. Even on short notice, it was easy to get these busy people to come from around the world. Alas, it was not soon enough.

Thirty-six hours before the meeting, Steve died. Some of the more distant participants had already started their trips—they wouldn't know until the day of the symposium that it had become a memorial.

Jeannie Benton asked to be the first speaker. She quickly turned commiseration into celebration, breathing life into the solemn event, giving everybody both energy and goosebumps. One result of the symposium is this compendium in Steve's honor. It documents remarkable work and real attitude. What it cannot provide as easily, but you will find between the lines, is family. Steve's family and students were indistinguishable. This was the hallmark of his teaching and research and explains the perfect attendance at his memorial. It was an opportunity that none would miss if possible.

Steve was born on December 1st. So was I. So was Neil Gershenfeld, also a senior faculty member at the Media Lab. We used to wish each other happy birthday and joke among ourselves that being born on December 1st was the key to tenure. Now all three of us have departed the lab; Steve with sad finality. Yet sadness hardly is his legacy. Steve wasn't only a gifted scientist and man of parts. As his widow reminded a silent auditorium that day, Steve also exemplified "demo or die," the Lab's cheeky take-off on "publish or perish." He always demo'ed and now had died. He was a cherished friend, colleague and example to us all, and we miss him.

Nicholas Negroponte co-founded the MIT Media Lab with Jerome B. Weisner, starting in 1980. Thereafter he served as its first Director until 2000, at which time he became Chairman. He was a founder of and columnist for WiReD *Magazine, which led to his New York Times best seller,* Being Digital. *He is currently on leave from MIT and is the founding Chairman of the One Laptop per Child non-profit Association and the 2B1 Foundation, that work together to bring $100 laptops to children in the developing world.*

Guide to Color Plates

Betsy Connors-Chen

Only a small number of images, spanning almost forty years of holographic imaging, are reproduced in this book. These represent research started in the late sixties by Dr. Benton at the Polaroid Corporation, continuing with the founding of the Spatial Imaging Group in 1983 at MIT, and including some work done by former students of Steve's. All of the students and affiliates listed at the end of this section contributed to the work in the Spatial Imaging Group and to the family atmosphere that Steve promoted in the lab. Along with Steve, these people advanced the research and imaging from 1983-2003; this list also includes members of the Media Lab's Object-Based Media Group who continue to carry the Holo-Video research forward. We have tried our best to confirm the attributions here, and apologize in advance for any errors we may have made.

These color plates reflect the group's history, techniques, and technology through a diverse set of technologies including transmission and reflection, single and full color, real-object and computer generated holograms; they represent a range of visual applications including scientific and architectural visualizations, product design work, and artistic expression. We also include a few non-MIT holograms and images of historical interest.

Color plate 1

1. *Dr. Dennis Gabor,* by R. Rhinehaart, McDonnell Douglas Electronics Corp., 1971, laser transmission hologram, Courtesy of the MIT Museum [MOH-1979.01], photo by B. Connors-Chen

2. *Emmett Leith and Juris Upatnieks*, 1964, Courtesy of the Bentley Historical Library, University of Michigan, Collection: Emmett N. Leith papers [photo box 1]

3. Stephen A. Benton with Holo-Video Mark II, 1997, photo by A. Blount

4. *Yurii Denisyuk,* by Ana Marie Nicholson, 1978, white light reflection hologram, Courtesy of the MIT Museum [MOH-1993.47.186], photo by B. Connors-Chen

Color plate 2

1. *Train*, by Emmett Leith and Juris Upatnieks, 1964, one of the first off-axis laser transmission holograms, Courtesy of the MIT Museum [MOH-1983.37], photo by B. Connors-Chen

2. Untitled [Chess Set], by Stephen A. Benton, 1968, Polaroid Corporation, first white light transmission "rainbow" hologram, Courtesy of the MIT Museum [MOH-1979.66], photo by B. Connors-Chen

3. *Holographic Filament,* by Yurii Denisyuk, 1958-1962, first white light reflection hologram, Courtesy of the MIT Museum [MOH-1993.47.180], photo by B. Connors-Chen

Color plate 3

1. *Rind II*, by Stephen A. Benton, 1977, Polaroid Corporation, (Herb Mingace, Will Walters) white light transmission hologram, Courtesy of the Benton Foundation Collection, photo by B. Connors-Chen

2. *Chess Pieces,* by Stephen A. Benton, 1979, Polaroid Corporation, (Herb Mingace, Will Walters), full color, white light transmission hologram, Courtesy of the Benton Foundation Collection, photo by B. Connors-Chen

3. *Tricia*, by Stephen A. Benton, 1980, Polaroid Corporation, (Jean Marc Fournier, Herb Mingace), full color (black/white) transmission holographic stereogram, Courtesy of the Benton Foundation Collection

Color plate 4

1. *Martian Grove,* by MIT Media Lab Spatial Imaging Group, 1984, (Stephen A. Benton, Lynn Fulkerson, Jennifer Hall, Mike Teitel, Julie Walker), full color transmission, computer generated, holographic stereogram, Courtesy of the Benton Foundation Collection

2. *Leonardo's Vision*, by MIT Spatial Imaging Group, 1985, (Stephen A. Benton, Herb Mingace, Bill Molteni, Mike Teitel, Julie Walker), full color transmission, computer generated, holographic stereogram, Courtesy of the Benton Foundation Collection

3. *Robie House,* by MIT Media Lab Spatial Imaging Group, 1986, (Stephen A. Benton, Cliff Brett, David Chen, Mark Holzbach, Peter Jurgensen, Eric Krantz, data by Lana Miranda), full color transmission, computer generated, holographic stereogram, Courtesy of the Benton Foundation Collection

Color plate 5

1. *Flower*, by MIT Media Lab Spatial Imaging Group, 1988, (Stephen A. Benton, Sabrina Birner), edge-lit hologram, Courtesy of the Benton Foundation Collection

2. *Stephen A. Benton with Boston Camaro Alcove Hologram,* by MIT Media Lab Spatial Imaging Group, 1986, (Stephen A. Benton, Mark Holzbach, Michael Klug, Eric Krantz, Mike Teitel, data provided by the GM Design Group), computer generated, alcove transmission hologram, Courtesy of the Benton Foundation Collection

3. *Teacup*, by MIT Media Lab Spatial Imaging Group, 1993, (Stephen A. Benton, Wendy Plesniak, Michael Halle, Michael Klug), one-step full-color reflection Ultragram on DuPont photopolymer, Courtesy of the Benton Foundation Collection, photo by B. Connors-Chen

4. *Cadillac Wheel*, by MIT Media Lab Spatial Imaging Group, 1990, (Stephen A. Benton, Michael Halle, John Underkoffler, Michael Klug), two-step reflection Ultragram, Courtesy of the Benton Foundation Collection

Color plate 6

1. *Still Life*, by MIT Media Lab Spatial Imaging Group, 1988, (Stephen A. Benton, Wendy Plesniak, Michael Klug), full color transmission, computer generated, holographic stereogram, Courtesy of the Benton Foundation Collection, photo by M. Klug

2. *Honda Acura NSX*, by MIT Media Lab Spatial Imaging Group, 1993, (Stephen A. Benton, Wendy Plesniak, Michael Klug, data and images provided by Honda), full color reflection holographic stereogram, Courtesy of the Benton Foundation Collection

3. *Photoelectron Tumor Treatment Medical Image Stereogram*, by MIT Media Lab Spatial Imaging Group, 1991, (Stephen A. Benton, Michael Halle, Michael Klug, with Ron Kikinis, Ferenc Jolesz *et al.*, Surgical Planning Lab, Brigham and Women's Hospital and Photoelectron Corporation), Courtesy of the Benton Foundation Collection, photo by B. Connors-Chen

Color plate 7

1. *World's Largest Hologram*, by Zebra Imaging, 1997, full-scale Ford P 2000 Concept Car Data, 18' × 6', 27 tiles, full-color, full-parallax, reflection holographic stereogram, Zebra Imaging, Inc., Austin TX, photo by Zebra Imaging, Inc.

2. *Holo-Video Flowerpot*, (Mark I display), by MIT Media Lab Spatial Imaging Group, 1990, (Stephen A. Benton, Mark Lucente, Pierre St.-Hilaire, John Underkoffler, Hiroshi Yoshikawa), photo by M. Lucente

3. *Holo-Video Volkswagen*, (Mark I display), by MIT Media Lab Spatial Imaging Group, 1991, (Stephen A. Benton, Mark Lucente, Pierre St.-Hilaire, John Underkoffler, Hiroshi Yoshikawa), photo by Pierre St.-Hilaire

4. *Holo-Video Tumor Treatment Medical Image*, (Mark I display), by MIT Media Lab Spatial Imaging Group, 1991, (Stephen A. Benton, Michael Halle, Mark Lucente, Pierre St.-Hilaire, data provided by Photoelectron Corporation and Surgical Planning Lab, Brigham and Women's Hospital), photo by Pierre St.-Hilaire

Color plate 8

1. Holo-Video Mark I three color set-up, by MIT Media Lab Spatial Imaging Group, 1991

2. *Holo-Video Honda EPX,* (Mark II display), by MIT Media Lab Spatial Imaging Group, 1992, (Stephen Benton, Mark Lucente Wendy Plesniak, Carlton Sparrell, Pierre St.-Hilaire), Photo by Pierre St.-Hilaire

3. Holo-Video Mark II 18-channel light modulator, by MIT Media Lab Spatial Imaging Group, 1992, photo by V. M. Bove, Jr.

4. *Holo-Video RIP Lincoln Cube,* (Mark II display), by MIT Media Lab Object-Based Media Group, 2005, (Wendy Plesniak, Tyeler Quentmeyer, James Barabas, V. Michael Bove, Jr.), photo by V. M. Bove, Jr.

5. Holo-Video Mark III light modulator, by MIT Media Lab Object-Based Media Group, 2007, (Daniel Smalley, Quinn Smithwick, V. Michael Bove, Jr.), photo by V. M. Bove, Jr.

Polaroid group, 1967-1998

Jeanne Benton
Betsy Connors-Chen
Jean Marc Fournier
William Molteni
Herb Mingace
William Walters

MIT students, 1982-2007

James Barabas
Sabrina Birner
Paul Christie
Betsy Connors-Chen
Oliver Cossairt
William Farmer
Amy Fisch
Lynn Fulkerson
Michael Halle
Michele Henrion
Samuel Hill
Mark Holzbach
Mary Lou Jepsen
Arno Klein
Michael Klug
Joel Kollin
Eric Krantz
Mark Lucente
Ryder Nesbitt

Ravikanth Pappu
William Parker
Elroy Pearson
Wendy Plesniak
Tyeler Quentmeyer
Pierre St.-Hilaire
Daniel Smalley
John Sutter
Michael Teitel
John Underkoffler
Julie Walker
John Watlington
Aaron Weber

MIT research staff

Jeff Kulick
Thomas Nwodoh
Carlton Sparrell
Steve Smith

Visiting researchers/postdoctoral fellows

Paul Hubel
Nobuhiro Kihara
Christian Moller
Quinn Smithwick
Akira Shirakura
Ichiro Tamitani
Hiroshi Yoshikawa

Introduction: Why Holographic Imaging?

About This Volume

At the time of this book's final preparation for publication (2007), both commercial and consumer photography have nearly completed a remarkably rapid transition from chemically based processes to digital, electronic technology. Before that change could happen, there had to exist inexpensive and high-quality electronic image capture, digital image processing, "soft copy" display, and hard copy printing.

Holography is just beginning to undergo the same transformation. Because, as we will see shortly, capturing a scene holographically requires recording the directions of light rays in addition to their intensities and colors, much more information is involved than in an ordinary photo—and systems for capturing and then dealing with that much information electronically are not yet fully developed. In particular, there really isn't yet such a thing as a practical electronic holographic "camera," so many of the achievements in electronic holography have been in the service of imagery that already exists in 3-D digital form, such as computer graphics models and volumetric medical scans.

A consequence of the digitization of photography was a sort of "darkroom democratization," in that suddenly everyone with a camera and access to a personal computer could do expressive things that formerly required training, patience, expensive specialized equipment, and overcoming the (understandable) fear of splashing around with chemicals in the dark. Internet connectivity enabled publishing these expressive images to unlimited audiences with almost no delay or cost. Holography as traditionally practiced has involved even more patience, more expensive and unusual equipment, and longer amounts of time in the dark—sometimes with even nastier chemicals—so making the process electronic and thus similarly accessible to more people certainly sounds like a good idea!

Because the move to electronic holography seems not only desirable but also inevitable, and because the contributors to this book have been among the pioneers in that area of research, we will (especially in the later chapters) look into both the theoretical and the practical issues in making the transition happen—all the way to holographic television! But first we will embark on an exploration—both historical and technological—of "traditional" holography. The ideas, conceptual approaches, and math tools we learn along the way will be just as applicable in the not-too-distant future when office-supply stores will stock supplies for holographic printers.

We intend to make this book accessible and useful for readers with a broad range of backgrounds. As a result, we have to strike a mathematical balance: we're going to try to be reasonably mathematically rigorous, but we'll rely on trigonometry and algebra as much as we can, and avoid where possible the use of complex numbers, vectors, and multivariate calculus. Thus although our equations may look a little different from those in some other texts, and our proofs may take a few lines longer, we haven't "dumbed them down," and we hope our more mathematically sophisticated readers will find our approach of concentrating on the physical phenomena instead of the mathematics more intuitive than the more usual way of going about these things. On occasion, we'll do the math both ways, if each approach can illustrate something helpful.

At the time of Steve Benton's untimely passing, he had for several years been working on expanding into a book the lecture notes from his popular MIT class Holographic Imaging. Michael Bove's research group at the MIT Media Laboratory had for over a decade been collaborating with Steve's group on electronic 3-D displays, and he agreed to finish the task, as well as to extend the reach of the material into advanced areas not covered by the course. Several of Steve's former graduate students have given their time to help with the latter part of the project, which is only appropriate as they have been among the internationally recognized leaders in pushing the boundaries of holography both during their MIT years and afterwards. Michael Halle, Julie Walker Parker, and Betsy Connors-Chen spent many hours working with Steve on organizational and layout concepts in the early stages of this book's development, and their efforts were extremely important in helping this volume take shape. Betsy deserves particular recognition for bringing coherence and clarity to a collection of diagrams whose original versions spanned over twenty years of PC graphics software, and for curating the archive of photos from which the color plates were selected. If we haven't explicitly listed authors' names on a chapter, readers can assume Benton with some Bove mixed in (seamlessly, we hope).

Steve Benton had a deep faith in holography not just as a fascinating scientific phenomenon or an involving craft practiced by a community of skilled artisans, but as an inevitable step in the evolution of visual communication, and he passed that faith on to those of us who have worked to bring this book to completion. We hope it further passes along to our readers.

The Window View Upon Reality

For centuries, popular culture has speculated on the future of visual communication, and has imagined that, as a matter of course, the resulting images would be three dimensional—that they would accurately render sensations of depth, locations, and spatial relationships.[i] One can only imagine the collective sense of betrayal when conventional photography turned out to be flat! Only a few years after the spread of photography, the public embraced stereoscopic photography, a feeble imitation of the glorious imaging expected from the inventors of their day. Since then, ever better methods for "perfect 3-D" have emerged from decade to decade, each promising more realistic and satisfying imaging than the last. Just when the ultimate limitations of traditional optical methods (such as lenticular photographs) seemed to be all too obvious, a completely new technique emerged in the early 1960s, one that promised an incredibly high quality of depth, detail, and tonal gradation; it was called "holography." Although it was invented in 1947 as a complex solution to a specific problem in electron microscopy, holography actually presented a solution to a fundamental question of wave recording and reconstructing—so fundamental that it eventually won the Nobel Prize in Physics for its inventor, Prof. Dennis Gabor (in 1971, after the advent of the laser had made the impact of holography visually obvious).

Unlike photography (and painting, drawing, printing, etc.), holography enables "steering" light in a way that reconstructs the directions of light rays coming from a 3-D scene. That additional degree of freedom (or of fidelity, if you prefer to think of it that way) is what makes a hologram the most complete and visually satisfying 2-D record of a 3-D scene we know how to make, as it works with the strongest perceptual cue by which our eyes and brains interpret depth. The ability to produce a thin piece of material that causes light to go in controllable directions (by means of diffraction) is such a useful feature that holographic processes find many valuable applications other than just making attractive pictures. But this book will largely concentrate on the three-dimensional "window view upon reality" that Gabriel Lippmann predicted (another Nobel Prize winner in Physics, and the inventor of a 3-D technique called "integral photography").[ii]

References

i. Two of Benton's favorite examples from the *Gulliver's Travels* school of early science fiction are:

from the *Fables of Fénelon*, Fénelon, F. (F. de Salignac de la Mothe) (this piece is probably from around 1699): "Water was placed in great basins of silver or gold, and the object to be painted was placed in front of that basin. After a while the water froze and became a glass mirror, on which an ineffaceable image remained." (Of course, like a mirror image, it was three dimensional!)

from *Giphantie,* Tiphaigne de la Roche, C.-F. (1760): The chief of a remote African tribe takes Giphantie into his home, where the sea can be seen through a window. Giphantie, amazed (so far from the shoreline), rushes to the window and bumps his head on something. He reports: "That window, that vast horizon, those black clouds, that raging sea, all were but a picture..." (Again, obviously three dimensional!) He goes on to describe the picture-making process: "The elemental spirits have composed a subtle matter, very viscous and quick to dry, by means of which a picture is formed in the twinkling of an eye. They coat a piece of canvas with this material and hold in front of the object that they wish to paint. It is then carried away to some dark place. An hour later, the impression is dry, and you have a picture. The correctness of the drawing, the truth of the expression, the stronger or weaker strokes, the gradation of the shades, the rules of perspective, all this we leave to nature, who with a sure and never-erring hand, draws upon our canvases which deceive the eye." (Change a few words and it sounds a lot like holography itself!)

ii. Lippmann, G. (1908). "Épreuves Réversibles. Photographies Intégrales," *Comptes Rendus*, **146**, pp. 446-451.

CHAPTER 1

Holograms and Perception

Provoking Spatial Perceptions

Any discussion of three-dimensional images properly begins with a discussion of human vision, and the mechanisms by which we perceive spatial relationships, including shape, position, distance, and motion through space. These can be roughly grouped into three types, depending on whether they are driven by (1) single-eyed (monocular) vision or (2) properly-combined two-eyed (binocular) vision, and by whether (3) they are stimulated by static or moving images (or perhaps the motion of the observer) in various combinations. A thorough discussion goes beyond the scope of this book, although we will revisit the topic in later explorations of the design of holographic images. Many references are available that explore these issues in detail (*e.g.*, Okoshi (1976),[i] Patterson & Martin (1992)[ii]).

	static	dynamic
monocular	overlap perspective focus, etc.	kinetic depth effect motion parallax
binocular	convergence fusion edge effects	

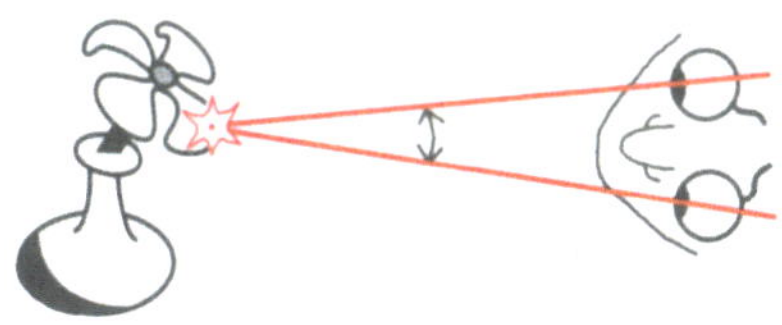

For our purposes, we will concentrate on the triangulation of point sources by binocular vision as the primary stimulus, or "cue," for spatial vision. Implicit in this are other cues arising from motion of one eye from side to side, which makes a kind of "temporal triangulation" possible, although the sliding of near objects over far objects also seems to be an important cue (time-varying occlusion correlated with observer motion). The eyes separately fixate on an image point (bringing its image onto the retina's *fovea*, the small central area of its most acute vision), and the angle of convergence between the eyes is sensed via muscular *proprioception* (a fancy way of saying that the brain knows the positions of the muscles that move your eyes). By combining these stimuli with knowledge (derived from experience) of the inter-ocular or inter-pupillary distance, the brain can make a fairly accurate estimate of the distance to a point. A mathematically equivalent approach is to say that the two eyes receive slightly differing 2-D views of a three-dimensional scene, which are fused to produce a single perception (without double vision in most cases), and that it is the "effort of fusion" that produces the impression of distance. As important as these binocular cues are, they are readily outvoted by simple monocular cues, especially by overlap (occlusion, opacity) cues. That one object's image terminates at the boundary of another is very convincing evidence that it is behind the other, and being hidden by it, in spite of possibly conflicting binocular cues. We will see this for ourselves in the study of *pseudoscopic* holographic images to come!

This simplified view makes it possible to say that it is only necessary to reproduce the *directions* in which light is traveling in order to produce a three-dimensional image. And it is this capability of holography that distinguishes it from other forms of photography. It must also provide the other depth cues, such as surface shading and occlusion, but those will follow naturally. First we will concentrate on the directions of light waves reaching the eyes through different parts of the hologram.

Optical Information

What do we mean by the "direction of the light" and its reproduction? Well, what do we mean by "light" at all? It has often been said that holography is photography in light so "coherent" (*e.g.* laser light) that it becomes useful to describe it as a *wave* phenomenon. But we are much more familiar with the drawing of *rays*, which are the imaginary trajectories of imaginary particles (photons) traveling through the air. If particles of light, or their corresponding rays, are emitted by a point source of light and reach the pupil of an eye, that eye must rotate so that its optical axis (or "line of sight") is aligned with the ray in order to focus the light onto the fovea. Which is to say that the eye's optical axis must pass through the point source in order to see it clearly. If we think of the point source instead as emitting spherical *waves* of light, the eye must still rotate so that the center of the lens is perpendicular to the wavefront so as to focus the waves' source on the fovea, which is again the same as saying that the "line of sight" must pass through the point source for good vision. So the task is independent of whether we consider light to be represented by rays or by waves; even so, we will worry quite a bit about what representation to use.

Light as Waves and Rays

You have probably heard repeatedly of the particle/wave duality properties of radiation. Sometimes, light behaves like a stream of particles; sometimes it behaves like a collection of waves. In fact, it is neither. We are like blind people feeling an elephant for the first time: what we think it is depends on where we grab it, and we may never quite grasp the entire concept. As both the geneticist J. B. S. Haldane and the astrophysicist Sir Arthur Stanley Eddington reputedly said, "Nature is not only stranger than we think, it is stranger than we *can* think!" Light is neither particles nor waves, and quantum mechanics has proposed a hybrid probabilistic model that is being argued even today. For all of the purposes of this book, it will suffice to adopt a simple wave model of light (*i.e.*, we will use a "classical" analysis). It will also suffice in most cases to represent these waves by their perpendiculars or normals at the areas of interest. These normals look a lot like rays! And they should, because the energy of a wave flows perpendicularly to the wavefront (in all but some crystalline materials). Thus we can use ray-like drawings, which are convenient, as long as we understand that we are talking indirectly about waves, or at least the directions of the wavefronts. And it is the directions of the wavefronts received by our two eyes that are compared to give rise to an impression of distance, so these graphical "rays" are enough (for now, at least—we will elaborate on this question in subsequent chapters, especially Chapter 8).

Capturing the Directions of Rays

We can now consider the basic problem of three-dimensional

imaging to be the recording and reproduction of the directions of the light rays that strike some surface between the scene and the viewer. If we can reproduce the directions and relative strengths of all the rays accurately, then looking at this magical surface should be like looking through a window: we should see a three-dimensional image of the scene floating behind it with perfect realism, just as it would have looked if we saw the scene itself. We have created a "window with a memory."

A few other things become clear at this point, by the way. The image is *not* floating in thin air—we can see it only if we look *through* the window, and not if we look *around* it. The world's best known "hologram," the Princess Leia projection from *Star Wars*, is pure science fiction and Hollywood special effects: there are no known physical processes that could produce such an image from a projector off to one side—there has to be some optical element in the line of sight, somewhere. Of course, George Lucas produced this effect by superposition, but it has come to represent what most people mean by "hologram" (as in "Look out, he's got a *hologram*!" in *Total Recall*, etc.). Likewise, the Haunted Mansion at Disney World and Disneyland employs no true holograms, but a combination of magician's tricks that have been known for over a century, especially the "Pepper's Ghost" in the ballroom scene.[iii] In 2006 a Japanese research team from Keio University, the National Institute of Advanced Industrial Science and Technology (AIST), and Burton Inc. received a great deal of press coverage for a volumetric display system that uses a focused infrared pulsed laser to induce tiny glowing plasma discharges at points in free air; although this system is truly 3-D and is based on lasers, it isn't holographic (and it lacks the ability to generate proper occlusion clues, as the glowing spots glow in all directions, so it's possible to see the back of a 3-D object through the front). It is important to remember that there are really two definitions of holography in our culture: "wavefront recording and reconstruction by interference and diffraction" (the technical field we are about to study) and "the psychologically ultimate three-dimensional imaging medium of the future" (what most people think we are working on!).

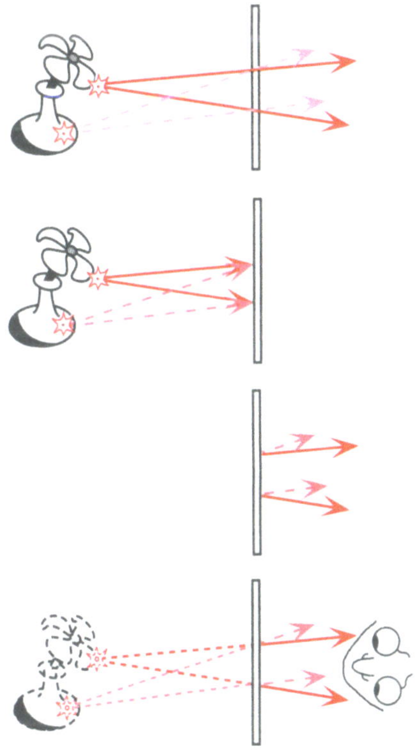

Back to reality: the problem with our proposed ray-direction recording and playback scheme is that there is no known material that is sensitive to the direction of light—only its energy (or wave amplitude), which triggers an individual micro-crystal, molecule, or electronic structure, can be sensed. This is not to say that no such material could ever exist; we just can't obtain one at the moment. We know that a pane of ordinary glass briefly "traps" light passing through it, and releases it very shortly afterward, which accounts for the delay in propagation that we ascribe to its *index of refraction*. At least we might someday hope for a time-delay window with delay times measured in hours instead of attoseconds!

In the meantime, optical inventors have come up with a succession of techniques for approximating the variation of ray direction between the two eyes, starting with Wheatstone's stereoscope in 1838. Stereoscopes sample and reproduce the ray direction variation very coarsely—only twice! Most users prefer 3-D technologies that do not require them to use viewing aids, such as stereoscopes or spectacles; this has given rise to the class of *autostereoscopic* displays, of which holography is the most recent and the most spectacularly realistic.

Classical Optical Techniques

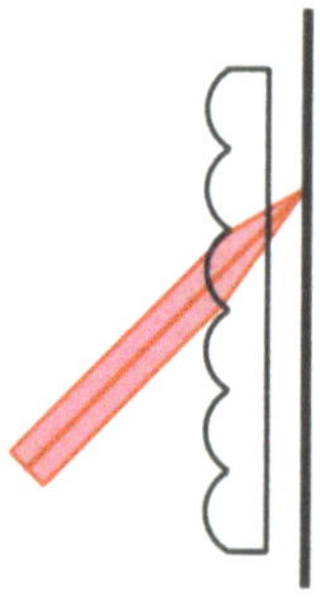

This is not the place for a detailed catalog of viewer-aided and autostereoscopic display technologies—Okoshi's book offers a fairly complete account of that history. The technology that comes closest to anticipating the visual impact of holography is Lippmann's integral photography, which places an array of small spherical lenses in front of a photographic film layer, the so-called "fly's eye lens array." The smaller the lenslets, the finer the sampling of the variation of light ray direction becomes, but the less accurate the reproduction of that direction, due to diffraction by the small diameter of the lenslets. Lippmann's proposal had some problems: as he first described the method, it produces an image of reversed depth (pseudoscopic); this was initially overlooked as no experimental tests of the technique were undertaken for several years. In the 1950s, Roger de Montebello perfected a second-generation technique that corrected several of these problems, but he also found severe limits on the image depth that could be provided without blurring.

Holographic Direction Recording

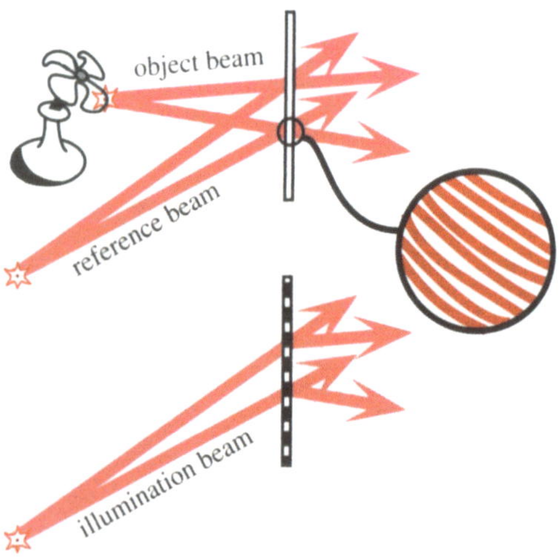

Photographic holography typically uses conventional photographic recording materials, ultra-fine grained versions of the same silver-halide emulsions that are used for black-and-white photography (the volume of the grains is about 1/30,000 of the usual, producing an equivalent ASA rating of about 0.001!). Which is to say that these materials are not sensitive to light direction either—holography records the direction information only indirectly. A second spread-out beam of light also exposes the film, overlapping the first at a carefully pre-arranged angle. That second, or *reference*, beam has to be coherent with the information, or *object*, beam; it has to have the same frequency, and be locked in phase with the object beam. In practice, that means that they both have to come from the same laser (ordinary light is nowhere near coherent enough). Where they overlap, a characteristic "picket fence"-like interference pattern is formed which is imprinted on the film. The larger the angular difference between the beams, the finer the pattern becomes (it is very fine indeed, usually more than one thousand dark and light line pairs per millimeter). A 3-D scene consists of many points at different locations, and their waves impinge on the film at different angles; each of these produces its own interference pattern, creating superimposed picket-fence patterns of different rotations and spacings.

Later, when the exposed and developed film (now the *hologram*) is illuminated with laser light at the same (reference beam) angle, the picket-fence-like pattern diffracts some of the light, with finer patterns deflecting it through greater angles. If everything works out as expected, the diffracted angle will equal the object beam angle, and we will have reconstructed the direction of the object beam at that point. It goes far beyond that simple fact, though. What emerges from the hologram is a perfect replica of the entire wave reflected by the object (plus some other waves). A viewer looking at the hologram does indeed see that "window view" of a 3-D image of the object, just as it looked during the exposure!

Of course we have to prove all these assertions, and wrestle with the limitations on their validity—that is what the rest of the book is about. And we have to understand how we make holograms that we can view in ordinary white light, which is when some of this starts taking on practical utility. But this should give you a general sense of what we are trying to do, and how.

Origins of Holography

Dennis Gabor was a German-trained electrical engineer, born in Budapest, Hungary, and interned in England during World War II. While there, he worked on a three-dimensional movie projection system in London, and later on electron microscope imaging for the British Thomson-Houston company in Rugby, England. The magnetic lenses of electron microscopes are imperfect for fundamental reasons—they distort the shape of the spherical electron waves coming from point-like objects. Gabor hoped to record that wave shape in the electron microscope, and then correct it with optical waves created by specially ground lenses, but to do this he had to be able to record wavefront shape as well as amplitude/intensity, the wave's *phase* or *local direction* in our terms. People had been struggling with this problem for years, and it was considered unsolvable until a key idea came to Gabor while he was waiting for a tennis court one Sunday afternoon. When Gabor published his two-beam recording method in 1948, it was dismissed by most "experts" until they took a close look at his example photographs—something obviously worked! But the requirements that the object and reference beam be coherent limited Gabor's "holography" (inspired by the Greek for "whole" and "message," *holos* and *graphos*) to very small objects. Gabor had not even thought about holography as a three-dimensional imaging technology until he saw the results at the University of Michigan in the early 1960s.

Emmett Leith and Juris Upatnieks were electrical engineers at the University of Michigan's Willow Run Laboratories, near Ann Arbor. During the 1950s, they were working on a highly secret radar technique that allowed images of nearly photographic resolution to be generated by combining data from along a long flight path—the Project Michigan side-looking radar system. The key to the technique was an optical image-processing system that illuminated a long strip of radar data film with light from a mercury arc, focused it through a series of exotic lenses, and produced an incredibly detailed image. Slowly, Leith realized that he had rediscovered Gabor's concepts of holography, but in a much more general context. In 1962, low-power helium-neon lasers began to become commercially available, and Willow Run was one of the first labs to have one to experiment with. After verifying its usefulness for the side-looking radar project, Leith and Upatnieks started extending their ideas to the recording of three-dimensional table-top scenes. First they studied back-lit scenes, and by 1964 they had made holograms of front-lit objects—most notably a brass model of a steam locomotive that one of the machinists at the lab had made. They showed these holograms at the Fall 1964 meeting of the Optical Society of America, where a long line of scientists waited patiently in the hallway to glimpse this amazing sight. This triggered the long and tumultuous history of holographic imaging, which Leith and Upatnieks dubbed "wavefront reconstruction photography." Many artifacts from these early stages of holography research are now at the MIT Museum in Cambridge, Massachusetts, joining the collection of the Museum of Holography that is housed there.

We encourage readers with an interest in the historical development of holography to consult the very comprehensive book by Johnston (2006).[iv]

Application Areas

Although it is three-dimensional imaging that jumps into most people's minds when you say "holograms," the fact is that most of the applications of holography have been in other fields. Three-dimensional photography has been the beautiful "love child" of holography (until quite recently), while other applications did the work and earned the money that kept most of the research going. Readers with a detailed interest in these non-imaging applications may wish to refer to Hariharan (1996),[v] and Ludman, *et al.* (2002).[vi] To simplify things, it is useful to categorize the applications of holography into five groups:

1. Holographic optical elements (HOEs)

Holograms can deflect and focus light just as prisms, lenses, and mirrors do (for one color at a time), but they are much lighter and more compact, and usually cheaper to make. Some folks call them "diffractive optical elements," which may be more accurate. Suffice to say that wherever laser light is used, a HOE is now a serious candidate to replace a conventional optical element, such as in supermarket scanners, CD/DVD players, automotive and aircraft heads-up displays, and so forth. These elements can, of course, be made by exposing a holographic recording material, but a common mass-production method called *binary optics* fabricates them using chipmaking techniques such as photolithography and micromachining.[vii]

2. Optical computing

There is a small but devoted cult within the computer science community that believes that photons will someday replace electrons for high-speed highly-parallelized processing of data. There are already a few installations where this is beginning to come true. Within that domain, there are several tasks that holograms can do with some unique attributes. Because the thickness of a recording material can be accessed in a particularly efficient way by holographic readout, very high storage densities can be reached (*e.g.* around a gigabyte per square inch of film, or 10^{12} bits in a cubic centimeter of crystal). Also, holographic storage holds the promise of *associative addressing*: illuminating the hologram with a small part of an image that it has seen before can produce a weak image of the rest of the image. A high-volume associative memory (or content-addressable memory) would have important uses in artificial intelligence computing, for example.

3. Optical metrology and microscopy
Because a hologram can produce an incredibly accurate replica of a wavefront recorded at another place at another time, the images it produces can be measured with great precision. A room-sized nuclear containment vessel can be recorded in a laser flash, for example, and its image then examined at leisure at a distant and non-hazardous laboratory for cracked metal parts, corrosion, and so forth. The same property can also be used for very small subjects. In *ordinary* photography, the higher the resolution that is needed, the shallower the depth of field that can be focused. Thus a normal microscope can't be used to make direct measurements on moving three-dimensional arrangements of small things, but a holography enables capturing a snapshot of the entire volume which can then be examined later. The nuclear physics team at MIT built a holographic recording system for a giant-size bubble chamber (3 meters deep) used in the search for the tau lepton. Holography allows 30 micron bubbles to be tracked throughout the depth of the chamber.

4. Non-destructive testing (NDT)
Likewise, two optical wavefronts can be compared with high accuracy, even though they were recorded or observed at very different times, and with the object under very different conditions. Because the phase of the wavefront changes very rapidly with very small object motions, the interference pattern formed between two holographic recordings of a scene are very sensitive to small changes. Only a hundred millionths of a millimeter of object motion will change its image from light to dark—this can be caused by mechanical stress, or by the effect of a defect hidden deep in the structure of the object. Most aircraft tires are retreaded many times, and for many years all these recaps were required to be checked by holographic interferometry (holographic non-destructive testing), which was the only sector of holography making any money at the time!

5. Three-dimensional imaging (display holography)
In spite of being concerned mainly with "pretty pictures," display holography has had a major impact on all the fields mentioned above. Ultimately, they all have similar concerns about making bright and clear holograms, but display holographers attacked these problems first, and in peculiarly inventive and unorthodox ways (most of them didn't know any better). Their improvements in manufactured materials and processing chemistry and techniques were taken up by the industrial labs with some reluctance, but worked so well that this "trickle-up technology" has become an important part of the field as a whole. But with their focus on holographic imaging for many reasons (fine art, museum display, security devices, advertising, portraiture, and so forth), the display

holography community still seems somewhat separated from the other sectors of the field. Since the development of mass production techniques for white-light holograms, a whole new set of technologies have come into the mixture, and the field is changing rapidly these days. Much of our holography research at the MIT Media Laboratory has explored making synthetic holograms of computational "objects," to make better understanding of their spatial organization possible, in spite of their complexity. Early application areas include computer-aided design, medical imaging, and scientific visualization, as you will see toward the end of the book.

Styles of Analysis

Just as people use holography for many different purposes, they use many different styles of analysis to understand and control the technique. Physicists tend to use three-dimensional analyses based on Green's functions, which can be hard to visualize, and don't hook into optical design thinking very well. Electrical engineers have made many contributions to the field by looking at the volume as a series of flat and parallel planes. The light "signals" on one plane are related to those on another by fairly simple (for them, anyway) integral transforms, or convolutions of impulse functions. For our purposes, it is much easier to concentrate on just one two-dimensional surface, the x-z plane, perpendicular to the hologram plane. The sources and rays of interest will be restricted to this plane (mostly) and light will travel in the $+z$ direction (mostly). We will find that limiting ourselves to a single plane is what makes "shop math" (algebra, plane geometry, trigonometry) really useful. Things that we learn by limiting the geometry to the x-z plane will cultivate many practical insights that can be generalized later on, if we feel so inclined. Actually, we will have to let the rays travel a little ways out of the x-z plane to discuss focusing properly, especially to talk about *astigmatism* (forewarned is forearmed!). Those of you who have already had some electrodynamics may well be skeptical of such a simplified approach, but we have many optical components to fold into our story, and the authors predict that you will be grateful for this point of view. And we will show you how to generalize the approach to the full x-y-z space before we are done, we promise.

References

i. Okoshi, T. (1976). *Three-Dimensional Imaging Techniques*, Academic Press, New York.

ii. Patterson, R. and W. L. Martin, (1992). "Human Stereopsis," *Human Factors*, **34**, 6, pp. 669–692.

iii. Reflection-based "Pepper's Ghost" illusions originated with John Henry Pepper (1821–1900) at the Royal Polytechnic Institute in London in 1862. The technique is similar to that used to create "heads-up" displays in airplane cockpits and on auto windshields.

iv. Johnston, S. F. (2006). *Holographic Visions,* Oxford University Press, Oxford, UK.

v. Hariharan, P. (1996). *Optical Holography: Principles, Techniques, and*

Applications, Cambridge University Press, Cambridge, UK.

vi. Ludman, J., H. J. Caulfield, and J. Riccobono (2002). *Holography for the New Millennium,* Springer-Verlag, New York.

vii. Goodman, J. W. (2005). *Introduction to Fourier Optics,* Roberts & Co., Englewood, CO, section 7.3.

CHAPTER 2

Light as Waves

Light

What we think of as "light" is actually a ripple-like disturbance of combined electrical and magnetic fields (in the so-called "classical" or non-quantum-mechanical approximation). As such, most serious optics books dutifully begin with a discussion of Maxwell's equations, which can also be widely found on T-shirts around the physics and electrical engineering departments of many college campuses.[i,ii,iii] The electric and magnetic fields are vectors, **E** and **H**, respectively, indicating the directions between lower and higher electrical and magnetic potentials. Everything follows from these mathematical elaborations by James Clerk Maxwell (1831–1879) of observations by Michael Faraday (1791–1867), that there is a coupling of the spatial variations of one of the fields (denoted by the "curl" or "div" of its vector) and the time variation of the other field, and vice versa (the first two equations). Either field may be stimulated—by a temporal variation of charge density in one case, and of current in the other—giving rise to a wave that immediately couples one to the other. Together, the electric and magnetic fields propagate away from the source like ripples in a pond with characteristic shapes that depend on how the disturbance was started, a manifestation of the *wave equation* that can be derived from Maxwell's equations. In this chapter, we will first consider some aspects of the shapes of the waves, then their time variations, and finally some underlying aspects of the electromagnetic waves themselves.

$$\nabla \cdot \mathbf{D} = \rho$$
$$\nabla \cdot \mathbf{B} = 0$$
$$\nabla \times \mathbf{E} + \frac{\partial \mathbf{B}}{\partial t} = 0$$
$$\nabla \times \mathbf{H} - \frac{\partial \mathbf{D}}{dt} = \mathbf{J}$$

$$\left(\frac{\partial^2}{\partial x^2} + \frac{\partial^2}{\partial y^2} + \frac{\partial^2}{\partial z^2}\right) E_x = \varepsilon_0 \mu_0 \frac{\partial^2}{\partial t^2} E_x$$
$$\left(\frac{\partial^2}{\partial x^2} + \frac{\partial^2}{\partial y^2} + \frac{\partial^2}{\partial z^2}\right) E_y = \varepsilon_0 \mu_0 \frac{\partial^2}{\partial t^2} E_y$$
$$\left(\frac{\partial^2}{\partial x^2} + \frac{\partial^2}{\partial y^2} + \frac{\partial^2}{\partial z^2}\right) E_z = \varepsilon_0 \mu_0 \frac{\partial^2}{\partial t^2} E_z$$

Wave Shapes

The term "wave" really refers to "a self-propagating disturbance" such that a disturbance at some location, such as from a pebble dropped into a pond, produces a disturbance somewhere else at a later time, without any molecules of water actually traveling from the first place to the second. Physicists often refer to those familiar ripples in a pond when talking about waves, and use ripple tanks to illustrate their thoughts, but water-wave propagation is actually quite a complex problem, even in two dimensions. We will be concerned instead with light waves propagating in three-dimensional space, such as from the point-like focus of a laser beam. There are three simple shapes of light waves that will cover most of the cases we will have to deal with.

$$c = \frac{1}{\sqrt{\varepsilon_0 \mu_0}} \equiv 299{,}792{,}458 \text{ m/s } \left(3 \times 10^8\right)$$
$$= 186{,}282 \text{ miles/second}$$

$$speed = \frac{c}{n}$$

$$n = 1.33 \text{ for water}$$
$$= 1.50 \text{ for glass, plastic}$$
$$= 1.000294\ f\left(T^\circ\right) \text{ for air}$$

1. Spherical waves

If the wave source is a spark-like disturbance at an idealized point in space, say at $(x, y, z) = (0, 0, 0)$, then the resulting pulsed electrical and magnetic disturbances will spread out like an inflating spherical balloon, with the radius of the sphere increasing linearly with time at a rate we call the "speed of light," which is determined by the properties of the medium (typically air, which is close enough to a vacuum for most purposes). The speed of light in a medium is given by one over the square root of the product of the dielectric constant ε_0 and magnetic permittivity μ_0 of the medium, and is equal to 3×10^8 meters/sec in a vacuum. In denser media, such as air, water, or glass, the dielectric constant increases and the waves slow down. The ratio of the speed in a vacuum to the speed in a particular medium is called the *refractive index* of that medium, about which we will hear more later. For the moment, let's consider waves in a uniform medium with a refractive index of unity (space, or air).

We have to be a little careful about the definition of the term "wavefront!" For a spark source, we can think of the wave as defined by a small interval in time when the electric field is non-zero, a "spike" in other words, at a single point in space. Maxwell's equations say that the resulting pulse-like disturbance will move outward from that point at the same speed in all directions, forming an expanding sphere. The pulse intensity cannot actually be the same in all directions, but let's imagine for the moment that it is (we are usually interested in only a fairly small range of angles anyway, where it can be virtually constant). As the wave spreads out, its amplitude drops as one over the distance (this will provide "conservation of energy," to be discussed later on), but the "spike" stays a "spike" as it moves outward. We will think of the locations in space where we could observe the spike at time t as describing a sphere of radius r, where

$$r = \sqrt{x^2 + y^2 + z^2} = ct \qquad (1)$$

c is the speed of light, mentioned above, and t is the time since the spark. This sphere is what we will call the *wavefront*. The disturbance moves quickly outward, always moving perpendicularly to the wavefront at every location, so that the radius of curvature of the spherical wavefront increases as the wave moves outward, and is the same everywhere on the wavefront.

What about "rays?" We sometimes think of a point of light as emitting imaginary particles outward, which travel at a constant speed, their trajectories described by straight lines called "rays." Our emphasis in this book will be on the description of light as a wave phenomenon instead, in which the relative time properties of the light energy headed in various directions becomes very important, which information the ray description generally loses. Nevertheless, we can draw arrows perpendicular to the spherical wavefront at any location and get a good prediction of where the energy will be found an instant later (except in *birefringent* crystals, where the light's behavior depends on its polarization); it is these arrows that we will refer to as "rays" when we sketch what is going on in an experiment.

Sketches are important to optickers and holographers alike, and become problematical because we have only a two-dimensional paper surface to sketch them upon! Usually, these sketches will represent *slices* through a 3-D volume, although we will also attempt isometric-like views of a scene (usually a layout of optical components), which is a *projection* through a 3-D volume—quite a different kettle of fish (the differences will usually be clear from the context). In most of what we do, light will be traveling from left to right (considered to be good optical design practice), and we will adopt the z-axis as the horizontal axis, with the x-axis pointing upwards.

So, we can attempt a sketch of a spherical wave as seen at a single instant of time as a "snapshot" of a slice-view of the spike, which simply looks like a circle.

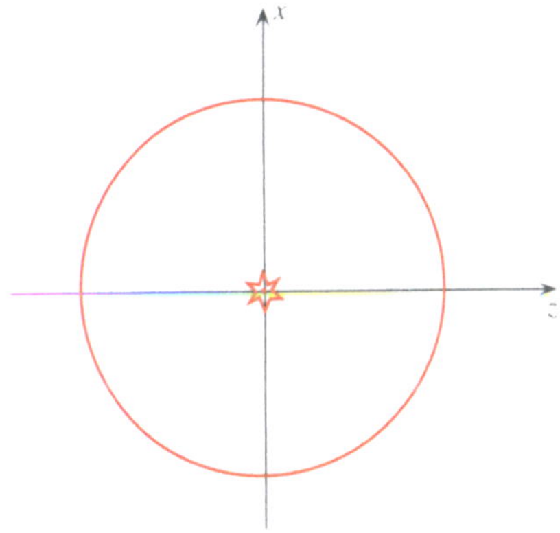

If we take a sequence of such "snapshots" at equally-spaced intervals of time, we get a series of concentric rings, also equally spaced in radius.

The direction of energy propagation is everywhere perpendicular to the surface of the spherical wave, so the wavefront reproduces itself an instant later as a sphere of larger radius.

Mathematically, if we describe the source "pulse" as some function, $p(t)$, at the center, then the pulse arrives at a radius r after a delay time given by r/c, and falls off in strength as $1/r$. This can be expressed as an electric field of strength $E(r)$ given by

$$E(r,t) = \frac{1}{r}\, p\left(t - \frac{r}{c}\right) \tag{2}$$

or, in terms of the (x, y, z) coordinates

$$E(x,y,z,t) = \frac{1}{\sqrt{x^2 + y^2 + z^2}}\, p\left(t - \frac{\sqrt{x^2 + y^2 + z^2}}{c}\right) \tag{3}$$

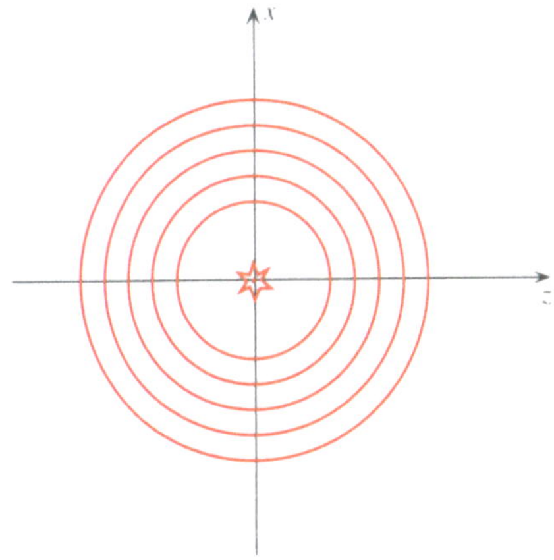

2. Plane waves

After a spherical wavefront has propagated for a very long distance, its wavefronts become effectively flat or "planar" over the area of interest to us (which is to say that its radius of curvature has become nearly infinite), so we refer to them as "plane" waves. For example, the light from a nearby star—other than our own sun, which is too close to seem to be a point-like source—arrives as a plane wave (of course, if we changed locations in the galaxy, we would find that the angle of the plane wave would change, and thus that the wave really is spherical). Thus plane waves are really an abstraction, but physicists are very fond of them for simplifying analyses, and we have to take them into account as an interesting limiting case of a spherical wave. Because the source location, which would ordinarily define the center of our optical coordinate system, is a long ways away, we refer instead to the local inclination of the wavefront as observed at the new center of our experimental coordinate system. We describe the plane wavefront as having an angle, θ, measured between the wavefront normal and the horizontal or z-axis.

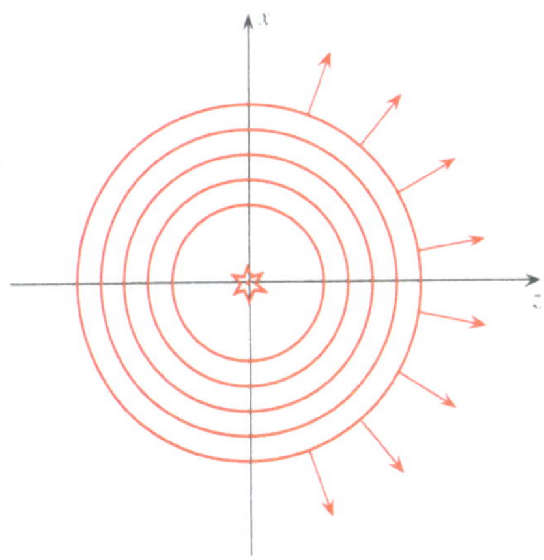

The location of this wavefront at time $t = 0$, shown first, is given by the x, z locations of equal electrical voltage,

$$\begin{aligned} x &= -z/\tan\theta, \text{or} \\ x\sin\theta + z\cos\theta &= 0 \end{aligned} \tag{4}$$

A short time, Δt, later, the wavefront will have moved a distance $d = c \cdot \Delta t$ perpendicularly to its surface, and the equation for the wavefront becomes

$$x\sin\theta + z\cos\theta = c\,\Delta t \tag{5}$$

Mathematically, then, if the source pulse has the form $p(t)$ at the $(x, y, z) = (0, 0, 0)$ point, then the field seen at any other location (x, y, z) is retarded by d/c, where d is the shortest distance between

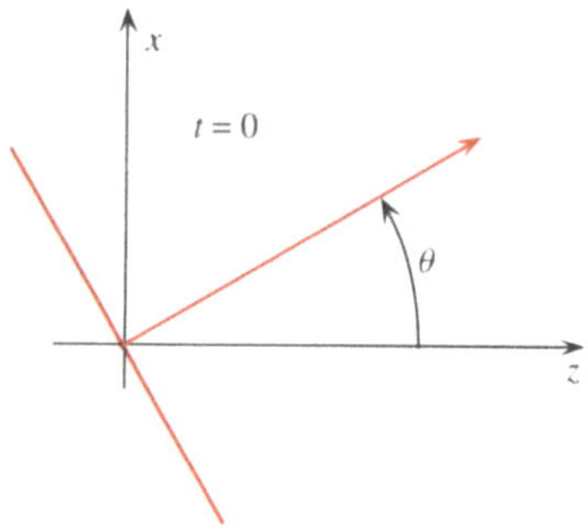

the origin and the wavefront passing through the observation point (the perpendicular distance from the origin to the wavefront)

$$d = x\sin\theta + z\cos\theta \tag{6}$$

The magnitude of the pulse does not diminish because the wavefront is no longer "spreading out" as it would for a spherical wave; the wave amplitude is constant as it moves ahead:

$$E(x,y,z,t) = p\left(t - \frac{x\sin\theta + z\cos\theta}{c}\right) \tag{7}$$

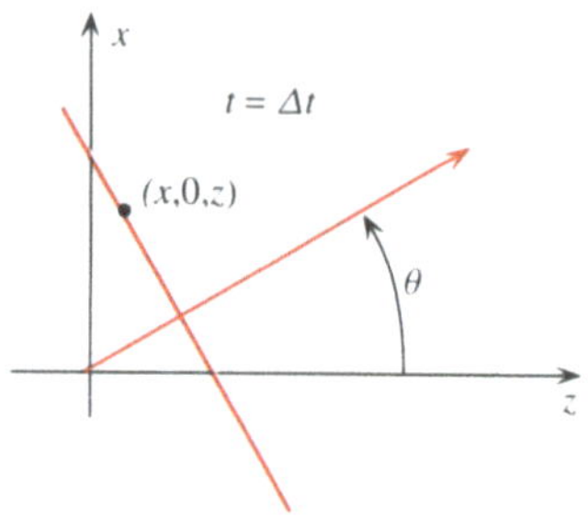

3. Cylindrical waves

Later on, we will encounter waves that have different curvatures in different directions, called *astigmatic* waves. A simple first case of one is a cylindrical wave, which we can think of as propagating from a surge of current in an infinitely long wire; let's say the wire is stretched in the y-direction. The wavefront will lift off of the wire as a cylindrical tube, and propagate outward as a tube of constantly increasing radius equal to the speed of light times the propagation time. At some distance from the wire, let's say one meter, the wavefront will be curved around the wire in one direction, but not curved in the other. These are hard to sketch clearly, but an *isometric-style* projected view would look like the illustration in the margin. Mathematically, it would have a representation like

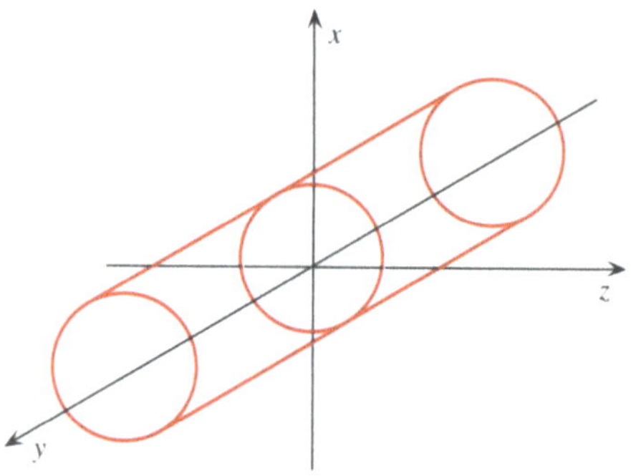

$$E(x,y,z,t) = \frac{1}{\left(x^2+z^2\right)^{\frac{1}{4}}}\, p\left(t - \frac{\sqrt{x^2+z^2}}{c}\right) \tag{8}$$

Light as Repetitive Waves

So far, we have considered a single pulse-like wave propagating through 3-D space, but visible light is a repetitive wave, which is what makes holography possible too! In the case of light, the pulses are smoothed out so that the electric and magnetic fields are smoothly varying functions of time. If we stood at a particular point in space and measured the electric field of a wave passing by (spherical or plane), we might observe a voltage as seen here:

Mathematically, this is described by the trigonometric sine function, with time as its argument. Every T seconds (we call T the "period" of the wave), the argument increases by 2π or "full circle" (360°) and the voltage pattern repeats:

$$E(t) = E_0 \sin\left(2\pi \frac{t}{T}\right) \tag{9}$$

The sine function derives its name from the sinuous "look" of the curve, which describes the x-coordinate of a point on the rim of a wheel as it turns through 360° or 2π radians.

You might ask "Why are the waves sinusoidal, instead of saw-toothed or triangular?" The answer is, approximately, that the waves are generated by electrons oscillating at the ends of "springs" that represent the change of energy as the electron's orbit moves nearer to and farther from the nucleus. The actual process quickly gets into quantum complexities that we don't have time to deal with here! Similarly, our eyes respond only to the sinusoidal components, because the sensing structures are resonantly tuned. This all turns out to be handy, because the techniques of mathematical physics have largely been developed around sinusoidal signals since the days of Fourier, the extraordinary late-18th/early-19th century French mathematician and physicist.

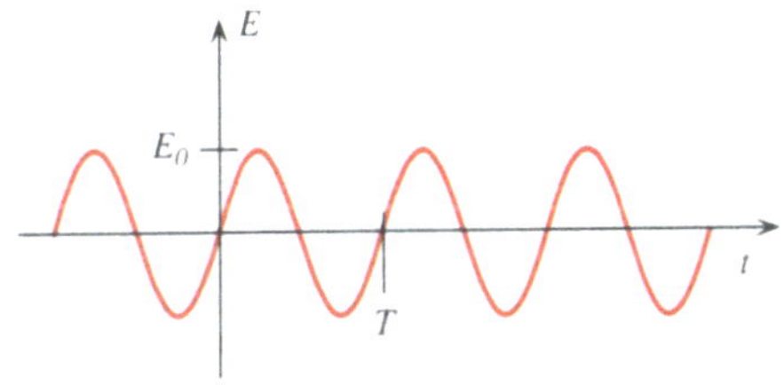

As we move the observation point further from the source, the receipt of the wave is delayed a little by the extra propagation time, which causes an apparent shift in the sinusoidal wave by some angle, which we call the *phase* of the wave, and about which we will say much more later on. The strength of the wave also drops off a little, as the $1/r$-law dictates.

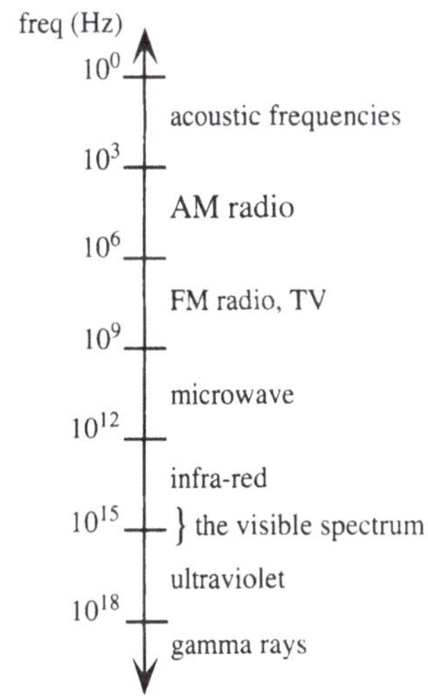

The rate of repetition is the only thing that separates light waves from radio, television, microwave, and gamma waves! The mathematical physics of all these varieties of electromagnetic waves are the same, but their practical and physiological properties are quite different. We think of the various waves as being arrayed in terms of their "frequencies," measured in cycles per second (called Hertz, Hz). Their "period" is what we have already seen as T, measured in seconds (or microseconds, or in the case of light, attoseconds). Their frequency is given by ν (the Greek letter "nu", in cycles per second), where $\nu = 1/T$. Electrical engineers like to speak in terms of the "radian frequency," $\omega = 2\pi\nu = 2\pi/T$ ("omega," in radians/second), but we will speak strictly of the "natural frequency," ν, in these discussions.

The electromagnetic spectrum is described in most physics books, and we will outline it only briefly here. Suffice to say that the principles of holography apply to all frequencies of waves, not just visible light.

Of the entire electromagnetic spectrum, only a tiny sliver, less than a two-to-one ratio of frequencies (compared to the nine octaves of the audio spectrum), serves to evoke a response in the human eye that we call "seeing." Within this visible part of the spectrum, different regions evoke quite different sensations, which we distinguish by the term "colors." For unknown reasons, optickers like to describe the visible spectrum in terms of the wavelengths in a vacuum of the radiations that are involved. These wavelengths vary between 400 and 700 nanometers (nm, 10^{-9} m), and it is the extreme shortness of these wavelengths that accounts for many of the practical problems of making holograms.

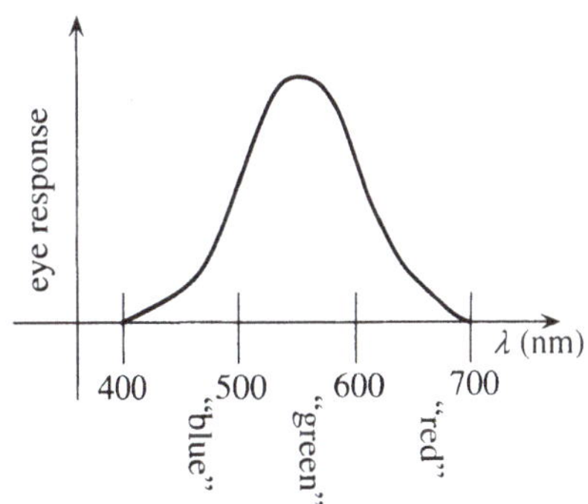

The sensation of color produced by light of various wavelengths (when viewed as an isolated spot in a dark surround) varies in a fairly reliable way as the wavelength varies from long to short. The color names of "red, orange, yellow, green, blue, and violet" and so forth are associated with various regions of the spectrum for that reason. We will simplify matters by referring only to the "red", "green," and "blue" areas, which will serve as additive color primaries.

Light as Sinusoidal Waves

Now, mingling the wave shapes from our discussion of pulsed waves, and the sinusoidal repetitiveness of ordinary light, we can come up with a combined description of light in a form that can readily be manipulated in mathematical terms. Again we refer to illustrative sketches as capturing a "snapshot" of the wave, but this time the concentric circles represent the successively emitted peaks of the repetitive sinusoidal waves (not a succession of snapshots, as before). The separation of the circles is the distance that the wave travels in T seconds, one cycle of the vibration, and is called the "wavelength," designated by λ (the Greek letter "lambda"), so that $\lambda = cT = c/\nu$. Then we can write, for a *spherical* wave:

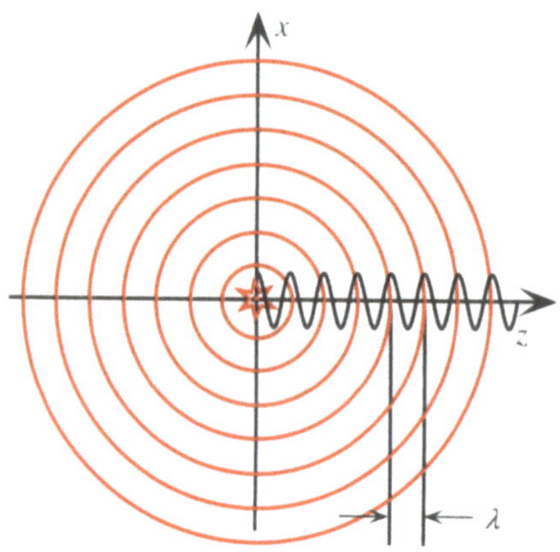

$$E(r,t) = \frac{E_0}{r}\sin\left(2\pi\frac{\left(t-\frac{r}{c}\right)}{T}\right) = \frac{E_0}{r}\sin\left(2\pi\left(\nu t-\frac{r}{\lambda}\right)\right) = \frac{E_0}{r}\sin\left(2\pi\nu t-\frac{2\pi}{\lambda}r\right) \tag{10}$$

When we go on to consider plane waves, the situation is not much different: just plug in the new form of the pulsing function into the same old wave shape formula, and the new function results:

$$E(x,y,z,t) = E_0\sin\left(2\pi\frac{t-\frac{x\sin\theta+z\cos\theta}{c}}{T}\right) = E_0\sin\left(2\pi\left(\nu t-\frac{x\sin\theta+z\cos\theta}{\lambda}\right)\right) \tag{11}$$

Coherence in Waves

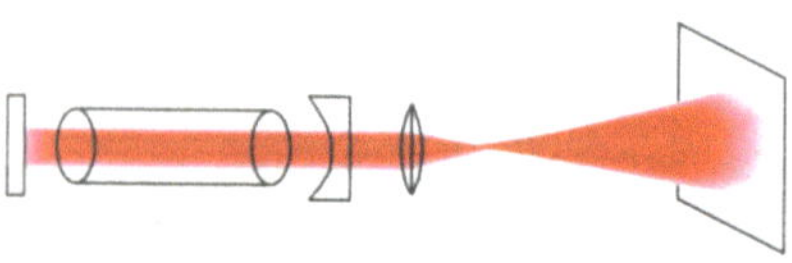

Our simple model of laser light assumes that it emerges from an ideal point source (the focus of a microscope lens, *e.g.*, as shown in the sketch). Within the diverging beam are the concentric spherical wavefronts, invisible to the eye. We also assume that this light has a perfectly well-defined and unvaryingly constant frequency. But both of these assumptions simplify the behavior of real lasers in ways that we should at least acknowledge—before continuing to ignore them for the most part! The term used to describe these properties of light is their *coherence*, and it has two "dimensions," the *spatial* coherence, which describes the departure from ideal point-source-like behavior, and the *temporal* coherence, which describes the departure from ideal single-stable-frequency behavior. Both of these follow from the physics of resonant laser cavities and light-amplifying media, which allow several oscillating modes along the direction of the resonator and from side to side.

Spatial coherence: point sources

Laser cavities can, if nothing is done about it, resonate in a wide variety of modes, each with a slightly different frequency and spatial distribution [iv]. Analogously, mechanical structures also vibrate in different spatial modes (*e.g.*, the sound of a drumhead depends on where you strike it)—these vibrational modes of objects are easily seen with holographic interferometry. We usually distinguish between the various lateral or side-to-side modes, and the various longitudinal or along-the-cavity modes. The lowest-order or preferred longitudinal mode is the so-called TEM_{00} mode ("t-e-m-zero-zero"), which produces a nice bell-shaped output beam, called a *Gaussian* beam after the German mathematician Carl Friedrich Gauss (1777–1855), who first described the exponential function involved.

Other low-order lateral modes produce donut-shaped beams, two-lobed beams, and four-leaf-clover-shaped beams. If they are present, then the spot focused by a microscope objective will be larger than expected from a single zero-order mode. However, almost all lasers used today are "single mode" type, producing only the Gaussian beam profile. But if the laser system becomes overheated, or mechanically distorts for any reason, it can easily produce other low-order modes that will degrade its operation for holographic purposes. The main problem caused by the other modes is that their frequencies are significantly different than the lowest-order mode, which decreases the coherence length of the laser light, discussed below.

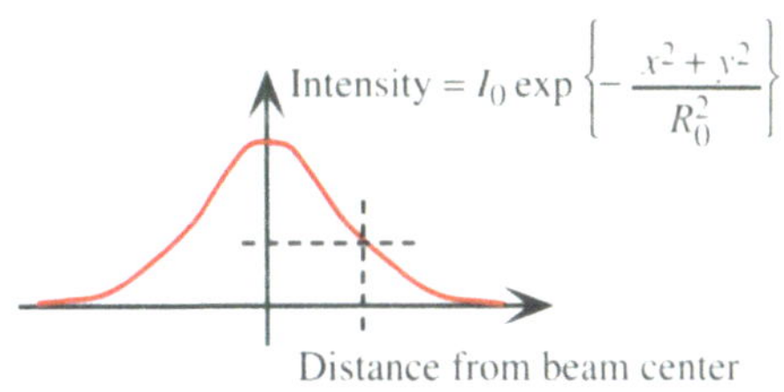

Temporal coherence: monochromaticity

Usually, "monochromatic" means that something is seen by the eye as a single color. When talking about lasers, though, monochromatic has a more specific meaning: single frequency. Even when a laser is operating "multi-mode," all of the output beams look the same color! You may know that resonators such as organ pipes and violin strings can be overblown or excited so as to produce overtones or higher harmonics, usually integer multiples of the lowest allowed or fundamental frequency (*e.g.* for which the string is a half wavelength long). Typical gas laser resonators are a million half-wavelengths long, and are operating at extremely high harmonics of the basic frequency, given by $f_{cavity} = c/2L$ (in the range of a hundred megahertz). The laser's amplification medium is capable of providing gain over a fairly wide range of frequencies, depending on exactly what the material and conditions are. Thus the resonator can be simultaneously operating at several nearby harmonics of the basic cavity frequency. The combination of these modes appears like a single output signal that is fluctuating in amplitude and frequency very rapidly, returning to the same frequency every round-trip cavity time (one over the fundamental cavity frequency).

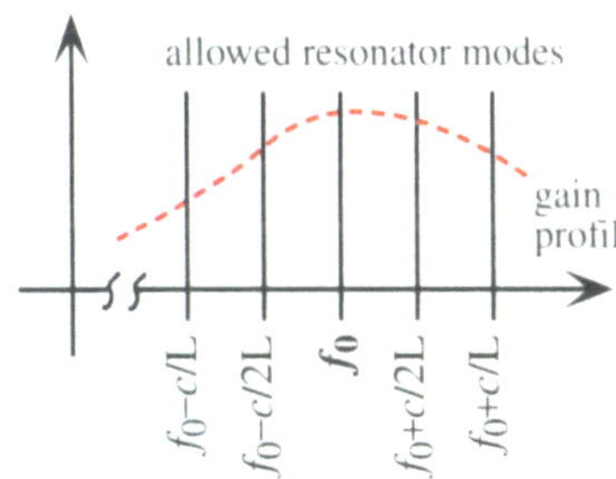

Because the output frequency is fluctuating so widely, only light that emerges from the laser at nearly the same instant can interfere with itself—light that comes out at a little later time will produce an unsteady interference pattern that will average to zero contrast over a very short exposure time. The acceptable delay between light samples is usually expressed as the *coherence length* of the laser, the distance that light travels before the frequency changes so drastically as to destroy the interference pattern. For typical helium-neon lasers, the coherence length is somewhere between 100 and 150 mm (four to six inches). A holographic image of a scene will gradually lose brightness for components deeper than 50 to 75 mm from the object point whose path length has been carefully matched to that of the overlapping reference beam. Semiconductor diode lasers (as used in

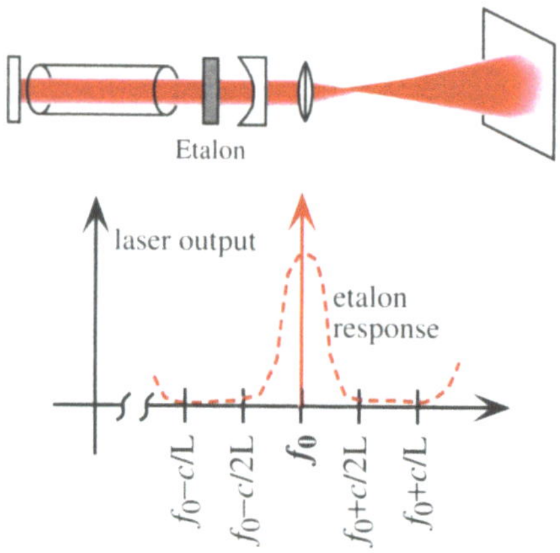

laser pointers, DVD players, etc.) have much shorter cavities (a fraction of a millimeter) than gas lasers, so the harmonics of the fundamental cavity frequency are spaced farther apart than in a gas laser; however the emission spectrum of the semiconductor material is fairly broad so the cavity can still potentially support many modes. As it happens these lasers often have coherence lengths at least as good as helium-neon lasers, though the mechanism by which they do that (effectively, the way in which one mode becomes dominant) is beyond the scope of this book.

The coherence length of a gas laser can be dramatically increased by the use of an *etalon* in the laser cavity. This is typically a piece of very carefully polished glass with partially reflecting coatings on each surface. Because it is only 10 mm or so thick, its cavity frequency is quite high, and its harmonics are deliberately separated by more than the width of the laser medium's gain bandwidth. If an etalon harmonic can be aligned with the central resonance of the cavity, only one output frequency will be allowed, and it can have roughly 50% of the power previously put out in the collection of frequencies. This produces true single-frequency operation, and the coherence length can become many hundreds of meters. However, the system is still vulnerable to mechanical vibrations, which alter the separation of the main cavity mirrors, and thermal drift (which does the same thing). Thus, although manufacturers cite some amazing coherence lengths, they have to be measured over the time of a holographic exposure to be useful predictors, and can be much shorter in practice. These days, almost all medium- and large-frame lasers for holography include etalons (single-frequency operation makes life *so* much easier!), but we will still have to worry about it with helium-neon lasers.

Laser speckle

Another quality of laser light that you have perhaps noticed is the gritty or sandy appearance of the surfaces that it illuminates. We call this grit "laser speckle." It is an interference phenomenon that arises from the microscopic randomness of surfaces that look to our eyes like flat and smooth surfaces, and as such we can't say much about it before we start looking at interference in more detail. Even then, the statistical techniques required go beyond the scope of this introductory book.[v] But we can at least start cataloging some interesting properties of laser speckle, so that we can know what to look for:

1. Laser speckle is always in focus; that is, its contrast is high no matter where our eyes are focused.
2. Laser speckle follows our motion. Or rather, it seems to stand still at whatever plane our eyes are focused at. Speckle can be a useful way of checking the accuracy of your eyeglass prescription!
3. The size of the speckles increases if the diameter of the pupil decreases—such as by looking through a pinhole.

If you play around with lasers you will gain plenty of experience with laser speckle. Interestingly, many people become so accustomed to it that they stop noticing it!

E&M Nature of the Waves

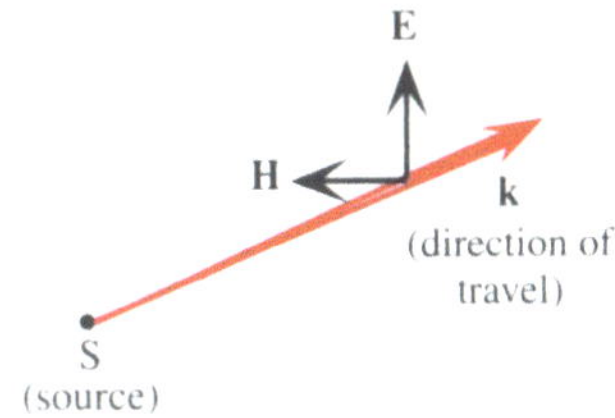

Now we have to deal with some of the realities of the electromagnetic nature of these waves. Firstly, the electric field is a vector quantity, so we should designate it as a bold-face variable, $\mathbf{E}(x, y, z, t)$, and the vector's direction is always perpendicular (or "transverse") to the direction of propagation (except in some crystals). The magnetic vector is also transverse, and also perpendicular to the electric vector. Often, the electric vector vibrates up and down, or at some angle, so that its end point traces out a straight line. Such light is called "linearly polarized." In other cases, the tip of the electric vector may sweep out a circle or ellipse, and the light is called "circularly polarized" or "elliptically polarized." Light from incandescent bodies, such as the sun or electric lamps, varies its polarization state every few femtoseconds, and is called "unpolarized." But laser light is usually very well polarized, and is usually linearly polarized. The direction depends on the orientation of the Brewster windows for a gas laser, and is customarily vertical. Maxwell's equations require light to be a transverse vibration, which means that no point source can radiate equally in all directions; there have to be some directions of no radiation (for the same reason that you can't comb a hairy basketball flat—there must always be some "cowlicks").

Polarization will come to be fairly important—two reasons come to mind:

1. The strength of the reflection of light from a glass surface depends on the polarization of the light (unless the light is coming in perpendicular to the surface), and
2. Only the parallel polarization components of two waves can combine to produce interference patterns (which we discuss in Chapter 3). Perpendicularly polarized beams (or rather, *orthogonally* polarized beams, in the general elliptical polarization case) cannot interfere at all.

Intensity (Irradiance)

When it comes time for a light beam to do some work, such as expose a piece of photo film, we have to consider where the necessary energy comes from. In virtually all cases, it is the electric field that does the work; the magnetic field is just "along for the ride." Electrical engineers know that the power absorbed by a resistive load is proportional to the average of the square of the electrical voltage across the load, divided by the resistance of the load. Similarly for optical power, which we usually measure in watts per square centimeter and call the "irradiance" or the "intensity" of the light (the latter being an obsolete term in the metric system, but still very commonly used)—it is proportional to the time average of the square of the electric field amplitude. Radiometry and photometry are baffling topics, as are many forms of accounting; they account for "what happened to the photons, or the lumens, that came out of the laser?" Suffice to say that if a uniform light beam has an electric field of the form

$$E(x,y,t) = A\ \sin(2\pi\nu t)\ \text{volts/meter} \tag{12}$$

then its average squared value will be

$$\left\langle |E(x,y,t)|^2 \right\rangle_{\substack{\text{time}\\ \text{average}}} = \frac{A^2}{2} \tag{13}$$

and it will deliver an irradiance of

$$\begin{aligned} I(x,y) &= (\varepsilon_0 c)\left\langle |E(x,y,t)|^2 \right\rangle_{\substack{\text{time}\\ \text{average}}} = \left(\frac{\varepsilon_0 c}{2}\right)A^2 \\ &= 2.65\times10^{-3} E_{\text{rms}}^2\ \text{watts/meter}^2 \end{aligned} \tag{14}$$

SPACE \ TIME	per second	total time
per sq. cm	intensity (mW/cm^2)	exposure (ergs/cm^2)
total area	power (mW)	energy (Joules)

in the MKS system of metric units. Full sunlight provides about one kilowatt per square meter, from which you can estimate its peak electric field.

Non-linear detection

All detectors (photocells, photo film, photodiodes, etc.) produce a current (electrons per second) that is proportional to the power in a light beam (which is proportional to the number of photons per second). The sensitivity may vary over the electromagnetic spectrum, but the linear electrical output is always proportional to the square of the optical input (the light amplitude). Most optical engineers have thought of irradiance as the linear input variable, but for coherent-light optickers, the amplitude of the wave is the important linear variable. It is the "square-law detection" (*i.e.*, non-linear detection) of this signal that causes many of the effects that seem so strange about coherent optics!

Intensity, power, energy, and brightness

Holographers often meter their beams, and it is important to understand what the various units of measurement are, and what they mean. Also, it is prudent to start thinking about the safe use of lasers, and this also requires understanding the various measures of laser light, and how they might affect a recording film or your eyes. There is nothing dangerous about using lasers in this ways we will be discussing in upcoming chapters, but they are certainly capable of being misused with unhappy results.

There are different terms to describe whether we are measuring a light beam over a small area within the beam, and are interested in its energy density, or over the beam's entire area, and are interested in its total "flow." Similarly, we have to distinguish between a measurement of a rate of flow at a particular instant, or the cumulative flow over the entire length of a pulse or of an exposure time. The chart to the side notes the various terms. We will walk through them one by one.

The power of continuous-wave lasers, such as laser pointers or He-Ne lab lasers, is typically rated in milliwatts (perhaps between 1 and 10). However, if the beam is spread out with a lens, the "heat" felt by our hands will be proportional to the "intensity" or "irradiance," the power per unit area. And if this is totaled up over time, we will determine the total amount of "cooking" each small area of our hands have suffered, their "exposure" in milliwatt-seconds per unit area, also called milliJoules per square centimeter (photo film sensitivity is typically measured in ergs/cm^2, a CGS unit; an erg is 1/10,000th of a milliwatt-second). We add that small lasers are too weak to feel with your hand (go ahead, try it with a laser pointer!), and that a typical flashlight emits about ten to a hundred times as much power (which is, of course, more spread out than a laser beam).

If you are dealing with a pulsed laser, such as a ruby laser, you will instead be told its total energy output per pulse, in Joules (watt-seconds). A one Joule laser is pretty big, and puts out as much light in a few tens of nanoseconds as a 10 mW He-Ne does in a minute and a half. If you divide the Joules by the number of square centimeters of the spread-out beam (and multiply by 10,000,000 to go from Joules to ergs), you will get the exposure of a piece of film put there. The danger of pulsed lasers comes from the very high instantaneous power of the beam at its peak, which may cause explosive damage to surfaces. A one-Joule laser with a 30 nanosecond-wide pulse reaches a peak power of 33 *billion* milliwatts. You would feel, hear, and remember that one! We won't be discussing the use of pulsed lasers in this book.

We will come back to these ideas when we make measurements for holographic exposures, which will involve overlapping beams, but the concepts will be the same. This discussion has been in terms of the "thermal" or radiometric power of a laser beam; a whole other set of units and measurements is used to describe perceived brightness, which is measured in lumens and is a function of the wavelength of the light as well as its radiometric power, since your eyes don't have the same sensitivity to all wavelengths. The conversion from power P in watts to brightness F in lumens for a monochromatic light source (which is the kind we're discussing here) is

$$F = 683\, P\, V(\lambda) \tag{15}$$

where $V(\lambda)$ is what is known as the *CIE Eye Response Curve* and has a peak value of 1.0 at 555 nm (green) and falls to zero at the red and violet ends of the visible spectrum. Your eyes are a lot more sensitive to the light coming from a typical green laser than to a red or a blue one, which is one reason green laser pointers are so fashionable. $V(\lambda)$ is not an analytic function but rather is defined numerically, based on the average of measurements on a number of human observers. On the right we show the function's value every 25 nm across the visible spectrum; tables of its values are available much more finely sampled in other references.

λ (nm)	$V(\lambda)$
400	.000396
425	.00730
450	.0380
475	.113
500	.323
525	.793
550	.995
575	.915
600	.631
625	.321
650	.107
675	.0232
700	.00410

Conclusions

We have skimmed through a lot of optics to find the mathematical descriptions of spherical and plane sinusoidal waves, which will serve us in good stead in the chapters just to come. You should make sure that you follow the logic that leads to the terms in the parentheses so far, as they will soon mutate into still further and more complex forms! Once those are under control, we may not often worry about the formalities of describing waves in detail, unless we are interested in the details of holographic optical element design. Likewise, there are lots of details about measuring the intensities of optical beams that we should know about, but only a few calculations that we will make over and over again. Nevertheless, as holography takes on new and different forms, there are likely to be times when we have to worry about measuring beams based on fundamental principles. For example, we have ignored the effect of exposure angle on the necessary dose for a holographic material—this is acceptably accurate for typical angles, but might require re-examination if one is talking about edge-lit holograms that involve very large beam angles.

References

i. Stratton, J. A. (1941). *Electromagnetic Theory,* McGraw-Hill, New York.
ii. Lorraine, P. and D. Corson (1970). *Electromagnetic Fields and Waves,* W. H. Freeman, San Francisco.
iii. Haus, H. A. (1984). *Waves and Fields in Optoelectronics,* Prentice-Hall, Englewood Cliffs, NJ.
iv. O'Shea, D. C., W. R. Callen, and W. T. Rhodes (1978). *Introduction to Lasers and Their Applications*, Addison-Wesley, Reading, MA, section 4.2.
v. Goodman, J. W. (1985). *Statistical Optics*, Wiley-InterScience, New York, section 7.5.

CHAPTER 3

Waves and Phases

Introduction

Our discussion so far has described light waves as three-dimensional phenomena, especially their electric fields as functions of (x, y, z, t). But for what follows we want to concentrate on the behavior of waves as they cross specific two-dimensional planes, with their time dependence suppressed (because they will all come from the same laser, and have the same frequency). We need a way to describe the shape of the wave in mathematical terms. To do this, we will introduce the notion of the *phase* of the wavefront, as determined by its relative delay in arriving at various points on the measurement plane. The observation of the pulsing from a point source in space is delayed, relative to the source, by a time proportional to the straight-line path length between the source and the observation point. If the source is repetitive or cyclic, from a sine-wave source for example, then we can also express the delay as a fraction of the repetition time or period, T. This, in turn, is represented as a fraction of 360 degrees or 2π radians, the angle that a wheel turns in generating a full cycle of a sinusoid. As a rule, we are not interested in the number of whole cycles of delay, but in the fraction beyond the nearest whole cycle.

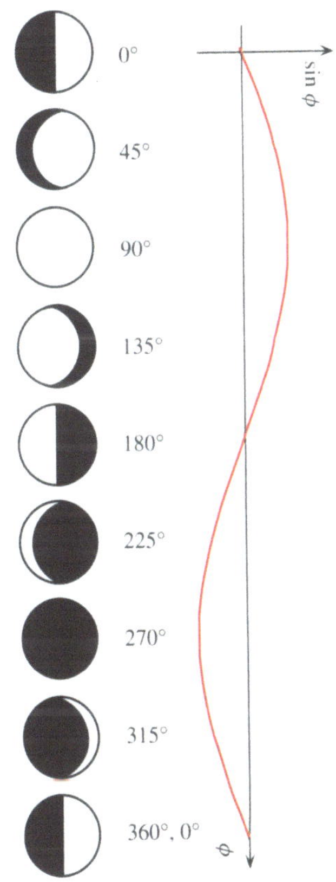

Wave Phase

Our most common notions of phase probably come from the "phases of the moon," and that is not a bad place to start! The moon goes through its cycles very reliably and repetitively, with a period of 28 days. We think of the phases as full, half (coming and going, or waxing and waning), and new moon (with gibbous in between somewhere, and ignoring eclipses). We can think of wave phases in much the same way, except that the fraction of a full cycle is measured in *degrees*, with 360° representing a full cycle, or the entire repetition time (or "period"). Describing the moon's cycle by a sinusoidal variation that is roughly the illuminated area (more nearly, $0.5 + 0.5\sin(2\pi t/T)$), a waxing half moon would be the zero degree mark, a full moon at the 90° mark, a waning half moon at 180°, a new moon at 270°, and the waxing new moon at 360° or 0° (they look the same in this modulo-360° math).

Formally, we would describe a wave as having an amplitude and a phase, with the frequency, ν (the Greek letter "nu"), usually being left implicit. Thus we would say for a spherical electromagnetic wave in 3-D space:

$$E(r,t) = \frac{E_0}{r}\sin\left(2\pi\frac{\left(t-\frac{r}{c}\right)}{T}\right) = \frac{E_0}{r}\sin\left(2\pi\left(\nu t-\frac{r}{\lambda}\right)\right) \qquad (1)$$

$$= E(r)\sin\left(2\pi\nu t-\phi(r)\right)$$

where the electric-field amplitude (as a function of distance from the source, r), $E(r)$, is

$$E(r) = \frac{E_0}{r} = \frac{E_0}{\sqrt{x^2 + y^2 + z^2}} \tag{2}$$

and the phase, denoted by ϕ (the Greek letter "phi") as a function of r, is

$$\phi(r) = \frac{2\pi}{\lambda} r = \frac{2\pi}{\lambda}\sqrt{x^2 + y^2 + z^2} \tag{3}$$

The fun begins when we look at specific analytical expressions for $\phi(r)$ and try to guess what kind of wave produced them, and where it is going! To do this, we will limit our attention to a single x-y plane in space, and try to identify the characteristic phase patterns, or "phase footprints" of some typical waves.

On-axis spherical wave

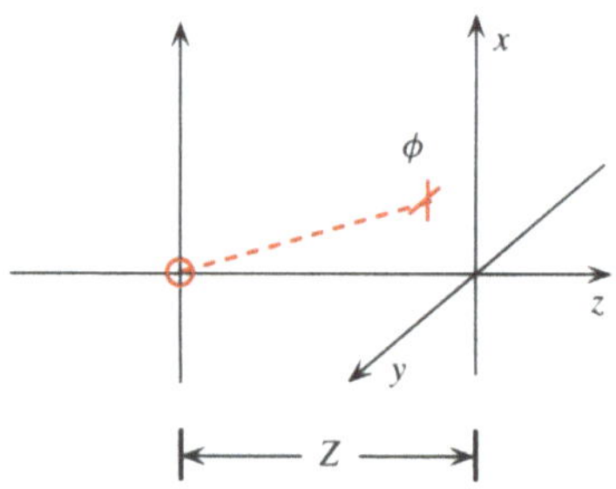

Imagine that a point source is located at (0, 0, 0), and our observation plane is located at $z = +Z$, as sketched on the left. What then is the form of the phase function, $\phi(x, y)$, in this plane? We have seen that the phase increases linearly with distance from the source, and so increases as we move away from the $(x, y) = (0, 0)$ point, on the z-axis. Further, the phase stays the same as we move in a circle around the z-axis, because the distance from the source is a constant as the line from the source to the observation point sweeps out a cone. Plugging in the expression from Eq. (2), which we will use over and over again, we find

$$\begin{aligned}\phi(r) &= \frac{2\pi}{\lambda}\sqrt{x^2 + y^2 + z^2} \\ &= \frac{2\pi}{\lambda}\sqrt{x^2 + y^2 + Z^2}\end{aligned} \tag{4}$$

To simplify the equations we must apply some approximations: namely, that the angles of the lines from the source to the observation plane are small, so that the x, y distances are much smaller than Z. In this case, we can use the binomial theorem to simplify the equations to:

$$\begin{aligned}\phi(x,y,\mathrm{Z}) &= \frac{2\pi}{\lambda}\sqrt{x^2 + y^2 + \mathrm{Z}^2} \\ &= \frac{2\pi}{\lambda}\mathrm{Z}\sqrt{1 + \left(\frac{x}{\mathrm{Z}}\right)^2 + \left(\frac{y}{\mathrm{Z}}\right)^2} \\ &\approx \frac{2\pi}{\lambda}\mathrm{Z} + \frac{\pi}{\lambda \mathrm{Z}}\left(x^2 + y^2\right)\end{aligned} \tag{5}$$

Now, we can identify Z as equal to the radius of curvature, R, of the spherical wave when it first reaches the observation plane. And, because this observation plane may be anywhere along the z-axis, we will consider the phase to be a function only of the coordinates within the plane, x and y. The phase pattern of a diverging spherical wave, what we will call its "phase footprint," is then:

$$\phi(x,y) = \frac{2\pi}{\lambda}R + \frac{\pi}{\lambda R}\left(x^2 + y^2\right) \tag{6}$$

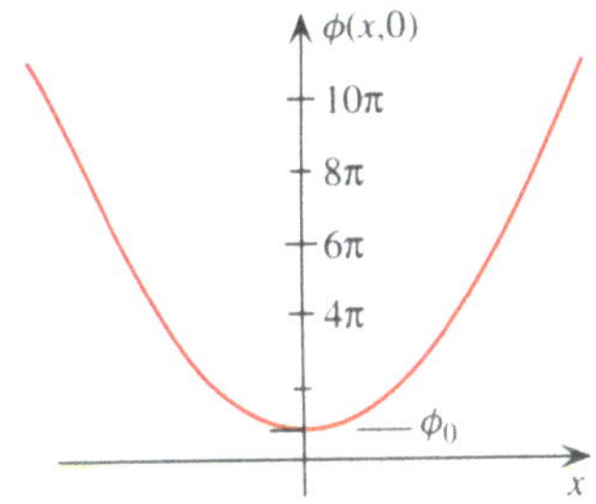

The first term represents a constant (over the plane) phase delay due to the time that it took the wave to get to the plane at all; we can call it ϕ_0. Because the phase of the source is itself unknown, we will usually ignore this constant phase, and emphasize the term that has variation with x and y, calling it the "footprint" pattern that will reveal the shape of the wave that caused it. In this case it is a "parabolic" term, with only second order terms, but in both x and y, and with the same coefficient for both directions.

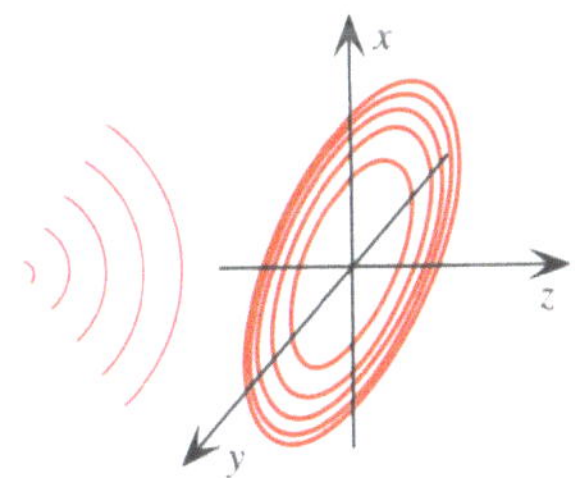

On-axis spherical wave phase footprint: This phase pattern is worth looking at in more graphical detail, as seen from the $+z$ direction and sketched here. The concentric circles represent the loci of points of equal phase, which we assign to be zero degrees (*i.e.* successive multiples of 360°) for simplicity. The lines are circles because the phase is constant wherever $x^2 + y^2$ is a constant. But the circles have radii that increase more and more slowly as the phase increases; if the n-th circle represents a phase of $n \cdot 360°$ greater than the center, then the radius of the n-th circle is proportional to the square root of n. That means that the area between successive circles is a constant, by the way!

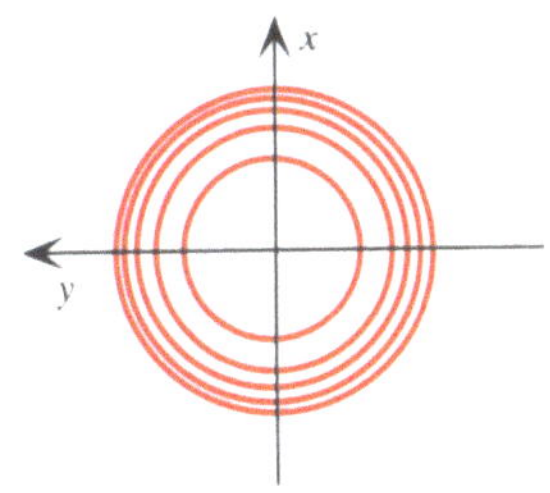

Off-axis plane wave

Let's now consider the case of a point source so far away that the wavefronts hitting the observation plane (at $z = 0$) are effectively flat. The source is at ($-X_0$, $-Z_0$), which are both very large compared to the size of the observation plane, but in a ratio that determines the inclination of the wave vector—which is perpendicular to the wavefront—to be θ_0 degrees to the z-axis, where $\theta_0 = \tan^{-1}(X_0/Z_0)$. The waves cross the x-y plane as seen in the illustration in the margin of the page.

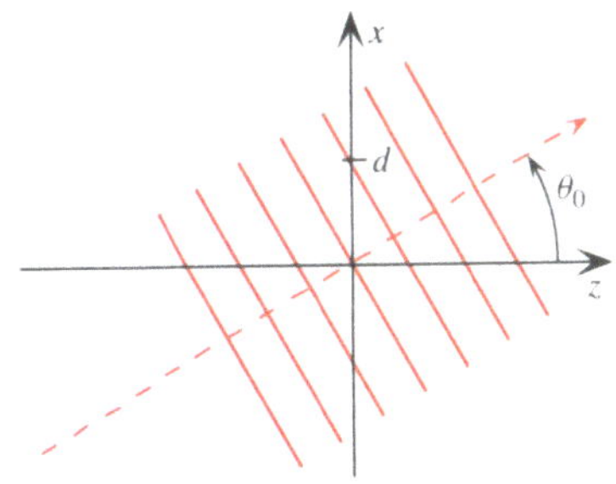

The higher up the x-axis we go, the farther away from the source we find ourselves, so that the wavefront phase increases with increasing x. If ϕ_0 is subtracted from the phase at $x = 0$ (as usual), the phase increases linearly with distance, x. Further, from the geometry (or by substitution into Eq. (3)), we can see that

$$\phi(x,y) = \phi_0 + \frac{2\pi}{\lambda}\sin\theta_0 \cdot x \tag{7}$$

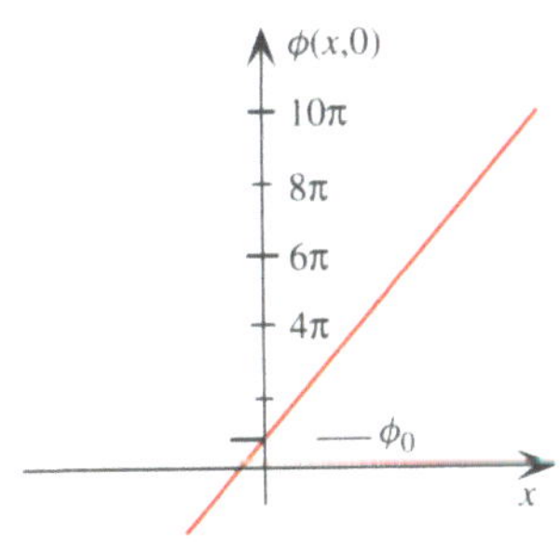

Off-axis plane wave phase footprint: The phase is constant along any $x = constant$ line, independent of y, because the plane wave is inclined purely vertically. As seen from the $+z$ direction, we can sketch the lines where the phase is equal to zero degrees (again), which are now straight horizontal lines. We find that the spacing of the lines, d, is inversely proportional to the (sine of the) angle of inclination, θ_0:

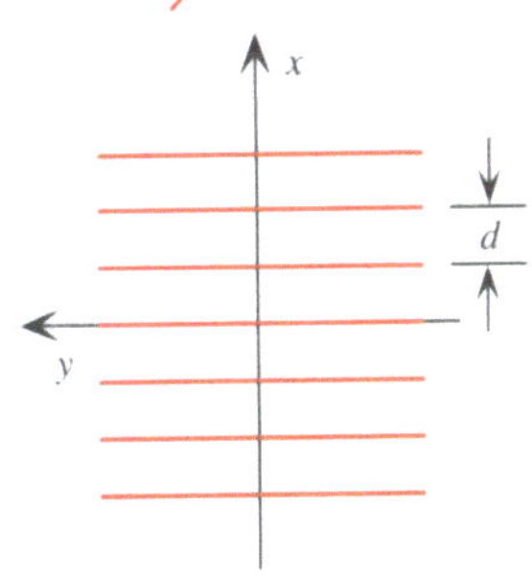

$$d = \frac{\lambda}{\sin\theta_0} \tag{8}$$

Note: If instead the source were off to the side, somewhere along the y-axis ($X_0 = 0$), the constant-ϕ lines would become vertical. The spacing of the constant-phase lines depends only on the angle of the wave vector to the z-axis; as the source revolves around the z-axis, the lines rotate around the $(x, y) = (0, 0)$ origin, staying perpendicular to the plane defined by the z-axis and the wave vector. We will only rarely consider waves not in the x-z plane, though.

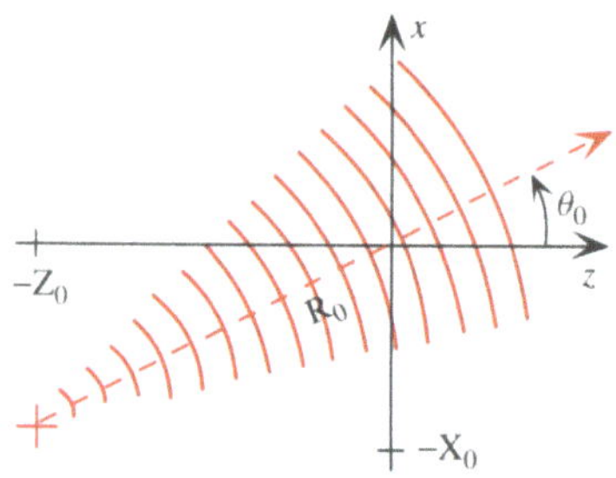

Off-axis spherical wave

The most general case of a spherical wave is the *off-axis* diverging wave, which forces us to grapple with a few strange new ideas. Now, we consider the source to be at $(-X_0, -Z_0)$, where these are not very large numbers (*i.e.*, we have to take wavefront curvature into account at last).

Converting to polar coordinates, we can express the radius of curvature and inclination of the wavefront at the origin as:

$$R_0 = \sqrt{X_0^2 + Z_0^2}$$
$$\theta_0 = \tan^{-1}\left(\frac{X_0}{Z_0}\right) \tag{9}$$

Now the problem is to find r (the distance from the source) as a function of x and y (the location in the $z = 0$ plane). Because we are interested in only a small area around the origin, we can express r as a power series expansion with acceptable accuracy:

$$r(x,y) = \left((x + X_0)^2 + y^2 + (z + Z_0)^2\right)^{\frac{1}{2}}$$
$$\approx R_0 + Ax + By + Cx^2 + Dy^2 + Exy + \ldots \tag{10}$$

The derivation of the coefficients is left as an exercise for the reader, but the answers are shown without too much difficulty to be:

$$A = \sin\theta_0$$
$$B = 0$$
$$C = \frac{\cos^2\theta_0}{2R_0} \tag{11}$$
$$D = \frac{1}{2R_0}$$
$$E = 0$$

Plugging these into the expression for the phase then gives

$$\phi(x,y) = \phi_0 + \frac{2\pi}{\lambda}\sin\theta_0 \cdot x + \frac{\pi}{\lambda R_0}\left(\cos^2\theta_0 \cdot x^2 + y^2\right) \tag{12}$$

which is what we will use in all our future work. Note that it seems like a logical combination of the on-axis spherical waves and off-axis plane waves we have seen so far, except for the peculiar $\cos^2\theta_0$ term; you should satisfy yourself that this term is correct, if not logical, before going any further! The wavefront itself has, by the definition of a spherical wave, equal physical curvatures in the x- and y-directions, namely R_0. However, the "mathematical curvatures," or coefficients of the second-order phase terms, are different, which is likely to be confusing.

Off-axis spherical wave phase footprint: Our sketch exaggerates the difference, but connotes how the "phase footprint" of this generalized spherical wave might look. You should satisfy yourself that it reduces to both the on-axis spherical wave case and the off-axis plane wave case under the proper conditions (namely, $\theta_0 = 0$ or $R_0 = \infty$).

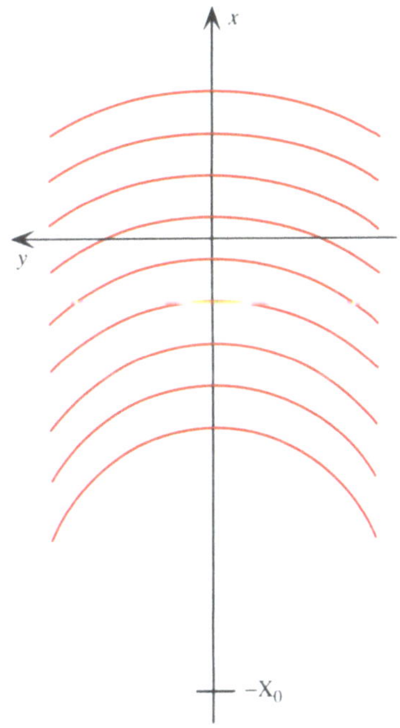

Radius of Curvature

Given a sketch of a wavefront, or better yet its analytical expression, we are now prepared to work backwards and determine its radius of curvature and inclination in a general spherical-wave case. We can summarize what we know about curvature of wavefronts fairly simply: if a spherical wavefront is coming from a point source (diverging) it has a positive radius of curvature, if it is focused onto a point (converging) we will say that its radius of curvature is negative, and if it the wavefront is planar (neither diverging nor converging) we will say that its radius of curvature is infinity. A planar wavefront doesn't actually have to come from a point source infinitely far away; it can be (and because we don't have infinitely large rooms in which to work, almost always is!) created from a diverging source by means of a lens we call a *collimator*.

So far we're talking only about spherical wavefronts, which can be described with a single radius of curvature. It may be that a wavefront has two radii of curvature, different ones in the x- and y-axis directions, but that is a story that has yet to come, and when it does it will be called *astigmatism*.

Local Inclination and Divergence of a Complex Wave

In analyzing holograms, we will often be dealing with complex waves that are reflected by complex three-dimensional objects. We will model such waves in two different ways:

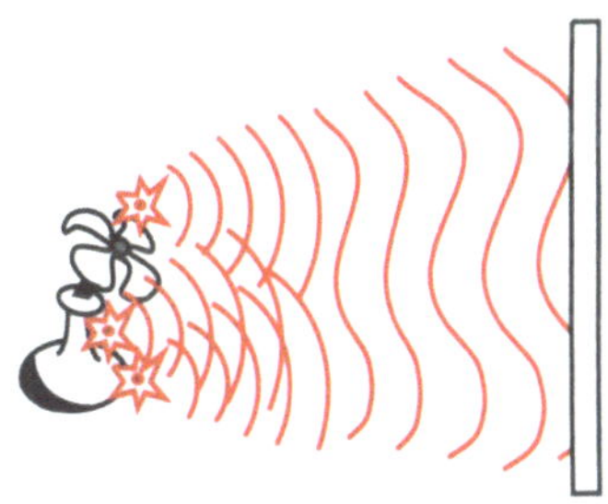

1. Most often, we will treat complex wavefronts as the composite of many spherical wavefronts emitted by point-like areas on the surface. When the recording and playback processes can be shown to be roughly linear, what is true for each of the spherical waves individually will be true for their combination, even for millions of them at a time. This approach is akin to the linear systems theory style of electrical engineering analysis, where complex waveforms are built up from a superposition of sinusoidal (Fourier) elemental components.
2. In other cases, we will consider even a complex wave to have a slowly varying amplitude and modest wavefront curvature at every point on the recording medium. Which is to say that the amplitude and curvatures are constant over areas that are several wavelengths on a side. This, in turn, requires that the complex objects subtend an angle that is much less than 180° in any direction (usually well under 30°). Then, we can model the wave at every small area as a spherical wavelet (or perhaps an astigmatic wavelet). We can then calculate the wavefront for that area at the output side of the hologram, stitch the small areas together, and predict what

the entire output wave will be like (or at least what its relationship to the input wave will be). This is the "patchwise-spherical" model, and will be especially useful if we have to deal with strong non-linearities in the recording/playback response.

Conclusions

There are many ways to describe waves, each of which highlights an aspect that is important to a certain kind of problems. Laser light is highly coherent, which means that it seems to come from a well-defined point source, and that it is of a single frequency, so we don't have to worry about dealing carefully with the frequency or wavelength of the light, and can concentrate instead on its spatial variations. Our concerns with interference and diffraction will make a careful account of the *phases* of wavefronts especially important. Phase will tell us "where the light goes." We will be a little less interested in the wave amplitudes, as we are a little less interested in "how much light gets there," for the time being anyway.

CHAPTER 4

Two-Beam Interference

Introduction

Because light waves are repetitive, with electric fields that swing alternately positive and negative (*i.e.*, they reverse direction sinusoidally), interesting things can happen when two (or more) of them arrive at the same place, but are delayed by differing amounts of time. When they arrive "in phase," so that crests meet crests, their effects "add up" or reinforce each other, and we get "constructive interference." However, if one wave arrives half a cycle behind the other (or 180° "out of phase"), so that crest meets trough, their effects cancel so that there is no net vibration, and we get "destructive interference." For intermediate phase shifts, intermediate results occur, so that the total vibration intensity can be negligible or enormous, depending on small time shifts between the waves.

We will begin by looking at a few examples of interference effects in everyday life. They are not extremely common, because the small time shifts involved are hard to control, and the effects average out if the phase shift varies during an observation time. Longer time delays can produce interference effects only for highly coherent or single frequency waves, which are also fairly rare (except in the holography lab!).

Soap films

Whenever kids blow soap bubbles, they enjoy the swirling play of colors on the soap film, which becomes increasingly intense until just before the film darkens and finally bursts. Similarly for oil films on water—the color of the reflected light depends on the thickness of the film. These effects are caused by the interference of waves reflected from the front with those from the back of the films. The same time delay can cause the interference to be additive or destructive, depending on the frequency or wavelength of the light, so that red light might be reinforced and blue light extinguished in one area, and vice versa in another with just a tiny difference in film thickness.

Colors observed through polarized glasses

Polarized sunglasses are wonderful for blocking reflected glare light, which tends to be horizontally polarized, but they cause strange color patterns to sometimes appear in car rear windows, stretched plastic sheets, and so forth. These color interference effects arise because glass and plastic sheets under mechanical stress will delay different polarizations of light by different amounts, and the sunglasses cause the two waves to combine and possibly interfere to form bands of color.

Radio fading

While driving around in hilly countryside, listening to the car radio, it is not uncommon for the signal to fade in and out almost randomly (this is more common with FM radios). This is caused not because the car is moving in and out of "radio shadows" caused by the hills, but because waves are reflected by the hills and combine to reinforce or cancel, depending on location (called "multi-path reception"). Similarly, a propeller plane flying over a TV antenna can cause a fluttering of the image due to the reception of multiple weak signals.

Audio beats

Familiar to musicians is the phenomenon of "acoustical beats," often used for tuning up stringed instruments such as guitars and pianos. Two strings are plucked, and one tuned until the "vibrato" or "beating" effect becomes slower and slower, and eventually freezes. When there is beating, the two strings are at nearly the same frequency, but slowly going in and out of phase. Their emissions thus add up and cancel alternately. When they are exactly the same frequency, and are "phase-synchronized" or "coherent," the sound can be weak or strong depending on whether they are "in" or "out" of phase, or somewhere in between. Usually, they don't stay tuned for long, though! Likewise, the amount of sound from a tuning fork varies markedly as it turns—the tines are out of phase at some angles.

Moiré fringes

You may not have thought of it this way, but the *moiré* patterns (the commonly heard "more-ay" is an English mispronunciation of French "mwah-ray," a watermarked taffeta fabric, whose name came from the English word "mohair" which, in turn, came from an Arabic word for a choice fabric) that are formed between two repetitive optical patterns is also a case of wave interference. You may have seen these between two pieces of window screening, chain link fences at a distance, muslin curtains, and so forth (printers worry about them in color half-tone printing, too). The mathematics of multiplying two repetitive or wave-like patterns is the same as the mathematics of adding and squaring them (*i.e.*, interference!), a fact that we will exploit on several occasions.

Wave interference is taught in many high school physics classes with the aid of ripple tanks. There, two bobbing corks launch shallow waves across a "pond" of constant depth, and the overall depth of the combined waves is a measure of the total "intensity." That is a wonderful way to observe this effect, and if you've never done so, we recommend that you find (or create out of a tray or pan or bathtub or something similar) a ripple tank to play with if at all possible. In the meantime, we will have to depend on a simpler optical demonstration of the same effects. Luckily, we can see most of the relevant phenomena almost as clearly with moiré fringe patterns. We can think of a pattern of concentric equally spaced circles as a "snapshot" of a slice through a spherical wave as it propagates outward from a central point source. A slow-motion movie would show the circles slowly expanding, and a new wave emerging from the center, until the pattern looked just the same as seen one oscillation period earlier.

When two spherical wave patterns are laid on top of each other, a distinctive pattern of dark and light bands, or "moiré fringes" appears (a "fringe" is a border, and the reason for calling the banded components of these patterns "fringes" probably has to do with a visual similarity to ornamental borders—as on a rug—*made of loose* parallel threads or strings). The dark fringes occur where the dark rings of one pattern overlay the light rings of the other, and the lighter fringes occur where the dark rings overlay dark rings so that some of the light ring area is visible (we say that the rings are "in phase"). If the slow-motion movie of the waves were played on from this point, although the rings/waves would move outward at exactly the same speed, the regions of dark and light moiré fringes would remain in the same places. In the example shown here, one set of ripples comes from a source that is 2.5 wavelengths above the source of the other. Everywhere straight up and down, the waves arrive 2.5 cycles "out of phase" and so produce a dark region (as though the waves were "canceling each other out"). But in areas exactly to the right and left, the waves arrive "in phase" and so those areas are brighter. As we move around the edge of the overlapped circles, we move from areas where the waves are "in phase," half a cycle out of phase, a whole cycle out of phase (and thus back in phase), and so on, until we reach the maximum phase difference of 2.5 cycles.

Quantitative Discussion of Interference Contrast

We can fairly easily describe the interference effects between two mutually coherent light sources in more quantitative and mathematical terms (more sources is much harder!). By "mutually coherent," we mean not only that each source is a point source of well-defined frequency, but also that the oscillation of each source is locked into phase with that of the other. Typically, this means that the two beams of light came from the same laser, via a system of beamsplitters and mirrors, but we can think of them as two separate sources, S_1 and S_2, that are somehow synchronized (by atomic clocks, etc.). We can imagine that the phase of the emission from one can be adjusted at will, and that the observation point, P, can be freely moved about in space (well, 2-D space in this diagram) to change the time delays between it and either or both of the two sources.

Each source, S_i, emits a spherical wave, which arrives at the observation point, P, after a time delay of $\tau = r_i/c$. This causes a phase delay of $\phi = (2\pi/\lambda)\ r_i$. The absolute phase of each wave is unobservable, because optical frequencies are so high, but the phase *difference* between the two waves will determine whether there is no vibration intensity observed at P, a little, or a lot. If the amplitudes of the two waves are equal at P, and they arrive "in phase," they will add together and the total intensity (defined as the average of the square of their sum) will be four times as great as the intensity of either of the waves separately. If they arrive "out of phase," one will be positive when the other is negative, and they will cancel out exactly, and the total intensity will be zero! This itself is odd enough. Consider that if there were one laser death ray headed straight at you, and another (coherent) beam coming in from the side, there supposedly would exist places where there is *no* total intensity, where it might be safe to stand—but if one of the beams is suddenly blocked, you get fried! The in-between cases, where the waves are not equal in amplitude and the phase difference is somewhere between 0 and π

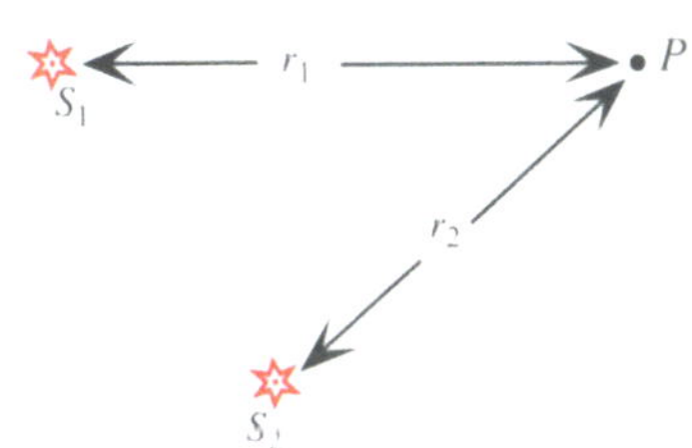

radians (0° and 180°) need more mathematics to be defined precisely.

Mathematical discussion
In this section we will grind through the derivation of the "interference equation" at a simple "shop math" level, so it will take about a page to finish. Depending on your own level of math background, you might be able to show the same results in only three lines using complex algebra and phasor notation; that approach will immediately follow this section so that you can see how we link the two together for possible future reference.

The expression for the wave amplitude (electric field) measured at point P from source S_1 is given by:

$$E_1(P,t) = \frac{U_1}{r_1}\sin\left(2\pi\nu t - \frac{2\pi}{\lambda}r_1\right) \tag{1}$$

Likewise, the amplitude for the second wave, from source S_2, is given by

$$E_2(P,t) = \frac{U_2}{r_2}\sin\left(2\pi\nu t - \frac{2\pi}{\lambda}r_2\right) \tag{2}$$

That is, the two waves have the same frequency, ν, and thus the same wavelength, λ. To make life a little more generalized, we will refer to these waves in the more general terms of their amplitudes and phases measured at P.

$$E_1(P,t) = a_1(P)\sin\left(2\pi\nu t - \phi_1(P)\right) \tag{3}$$

and similarly for the wave from S_2,

$$E_2(P,t) = a_2(P)\sin\left(2\pi\nu t - \phi_2(P)\right) \tag{4}$$

Intensity: The irradiance, or "intensity" as we will more commonly call it, is proportional to the average over time (a brief time, perhaps a few microseconds) of the square of the magnitude of the electric field vector of the total light field. The proportionality factor depends on the units of the discussion; we will use MKS for the time being, so that the factor becomes $\varepsilon_0 c$. It is the squaring and averaging that produces all of the interesting results, not the units.

The total electric wavefield is, summing the two waves:

$$E_{total}(P,t) = a_1(P)\sin\left(2\pi\nu t - \phi_1(P)\right) + a_2(P)\sin\left(2\pi\nu t - \phi_2(P)\right) \tag{5}$$

We are discussing the wave's electric field as a scalar quantity here, so the squared-magnitude is simply the arithmetic square (with P omitted on the right side to save space):

$$
\begin{aligned}
\left|E_{total}(P,t)\right|^2 &= \left[a_1 \sin(2\pi\nu t - \phi_1) + a_2 \sin(2\pi\nu t - \phi_2)\right] \\
&\quad \times \left[a_1 \sin(2\pi\nu t - \phi_1) + a_2 \sin(2\pi\nu t - \phi_2)\right] \\
&= \left[a_1^2 \sin^2(2\pi\nu t - \phi_1)\right] + \left[a_2^2 \sin^2(2\pi\nu t - \phi_2)\right] \\
&\quad + 2a_1 a_2 \sin(2\pi\nu t - \phi_1)\sin(2\pi\nu t - \phi_2) \\
&= \left[a_1^2 \sin^2(2\pi\nu t - \phi_1)\right] + \left[a_2^2 \sin^2(2\pi\nu t - \phi_2)\right] \\
&\quad + a_1 a_2 \cos(\phi_1 - \phi_2) - a_1 a_2 \cos(4\pi\nu t - \phi_1 - \phi_2)
\end{aligned} \tag{6}
$$

Note that the last step invokes some familiar trig identities. Recalling that the time average of sin t (and cos t) is 0.0, and the time average of $\sin^2 t$ is 0.5, we find that

$$
\begin{aligned}
&\left\langle \left|E_{total}(P,t)\right|^2 \right\rangle_{\text{time avg.}} \\
&= \frac{a_1^2(P)}{2} + \frac{a_2^2(P)}{2} + a_1(P)\, a_2(P) \cos(\phi_1(P) - \phi_2(P))
\end{aligned} \tag{7}
$$

The total intensity is then (in MKS units),

$$
\begin{aligned}
I_{\text{total}}(P) &\equiv \varepsilon_0 c \left\langle \left|E_{\text{total}}(P,t)\right|^2 \right\rangle_{\substack{\text{time}\\ \text{average}}} \quad (\text{W/m}^2) \\
&= \varepsilon_0 c \frac{a_1^2(P)}{2} + \varepsilon_0 c \frac{a_2^2(P)}{2} \\
&\quad + \varepsilon_0 c\, a_1(P) a_2(P) \cos(\phi_1(P) - \phi_2(P)) \\
&= I_1(P) + I_2(P) + 2\sqrt{I_1(P) \cdot I_2(P)} \cos(\phi_1(P) - \phi_2(P))
\end{aligned} \tag{8}
$$

$$I_{\text{total}} = I_1 + I_2 + 2\sqrt{I_1 \cdot I_2} \cos(\phi_1 - \phi_2)$$
The "Master Interference Equation"

It is the last form that is the most familiar in optics, in which the proportionality constants even out and the result is expressed in terms of the intensities of the waves by themselves and the cosine of the phase difference between them. We will be making repeated use of this interference equation later in this chapter, and in chapters to come as well.

The same proof can be compressed if we consider instead the complex amplitude, $u_i(P)$, of each of the waves. The complex amplitude of each wave, and its complex conjugate (denoted by an asterisk), are defined as (using Gaskill's (1978) notation here),[i]

$$
\begin{aligned}
u_i(P) &= a_i(P) e^{j\phi_i(P)} \\
u_i^*(P) &= a_i(P) e^{\;j\phi_i(P)}
\end{aligned} \tag{9}
$$

The real measurable field may be recovered as

$$E_i(P,t) = \mathrm{Im}\left\{u_i^*(P) e^{j2\pi\nu t}\right\} \tag{10}$$

Consistently with Eq. (8), we define the intensity of a single wave in terms of its complex amplitude as

$$I_i(P) \equiv \frac{\varepsilon_0 c}{2} u_i(P)\, u_i^*(P) = \frac{\varepsilon_0 c}{2} a_i^2 \tag{11}$$

Similarly, for a summation of many waves, the total intensity in terms of the total complex amplitude, $u_{\text{total}}(P) = \sum_i u_i(P)$, is

$$I_{\text{total}}(P) = \frac{\varepsilon_0 c}{2}\left|u_{\text{total}}(P)\right|^2 = \frac{\varepsilon_0 c}{2} u_{\text{total}}(P)\, u^*_{\text{total}}(P) \tag{12}$$

With these preliminaries in place, we deal entirely in terms of the complex amplitudes, and can readily show that, letting $i = 1$ and 2, in turn:

$$\begin{aligned}\left|u_{\text{total}}(P)\right|^2 &= u_{\text{total}}(P)\cdot u^*_{\text{total}}(P)\\ &= \left(a_1 e^{j\phi_1} + a_2 e^{j\phi_2}\right)\left(a_1 e^{-j\phi_1} + a_2 e^{-j\phi_2}\right)\\ &= a_1^2 + a_2^2 + 2a_1 a_2 \cos\left(\phi_1 - \phi_2\right)\end{aligned} \tag{13}$$

This is the desired result when plugged into the definition of the intensity.

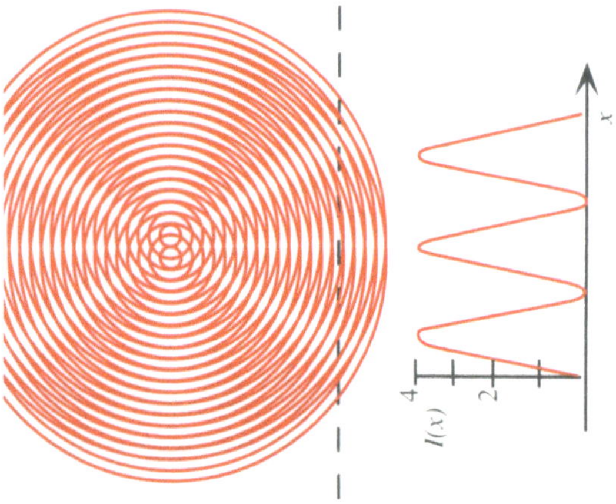

Equal beam case, conservation of energy

In the case seen above, we can imagine sketching the intensity observed along the right-hand edge of the interference pattern. Assuming that the intensities of the two beams are unity when measured separately, we see that when turned on together, we do not get a uniform reading of *two*, but rather that the energy "bunches up" to give *four* in some places, and *zero* in others. Simple interference patterns pose some of the mostly deeply reaching questions of modern physics. Here we see that the principle of *conservation of energy* does not always apply in the micro-scale, but only as an average over several cycles of the interference pattern.

Unequal beams; heterodyne gain

An interesting effect in interference patterns is that the variations of intensity are usually much greater than the intensity of the weaker of the two beams. That is, if a weak beam overlaps a strong beam the contrast of the fringe pattern, or its "visibility," is usually much greater than the visibility of the weak beam by itself; interference provides a kind of amplification, analogous to the "heterodyne gain" of radio electronics. Let the ratio of the beam intensities be given by $K = I_{\text{strong}}/I_{\text{weak}}$. The variation of intensity is then given by

$$I_{\max} - I_{\min} = 2\sqrt{I_{\text{weak}} I_{\text{strong}}} = 2\sqrt{K}\, I_{\text{weak}} \tag{14}$$

The "visibility" of a fringe pattern is defined as the ratio of the variation of the total intensity to its average intensity, times two:

$$V = \frac{I_{\max} - I_{\min}}{I_{\max} + I_{\min}} \tag{15}$$

A V of 0.01 is usually near the threshold of visibility of the human eye, depending on the fringe spacing, which means that a beam that is only one forty-thousandth (1/40,000) the intensity of the stronger beam (completely invisible as an incoherent addition) could produce an easily visible interference pattern! This causes lots of problems when we try to make holograms in the laboratory.

Geometry of Interference Fringes

We have learned about the magnitudes of interference effects, and their extreme sensitivity to weak beams, but we will generally be more interested in the geometry of these fringe effects. In particular, we will want to know *where* these moiré-like fringes are formed, and what their *spacings* and *orientations* are. These will eventually determine *where* light goes when it is diffracted by a hologram (as opposed to *how much* light goes there).

#1

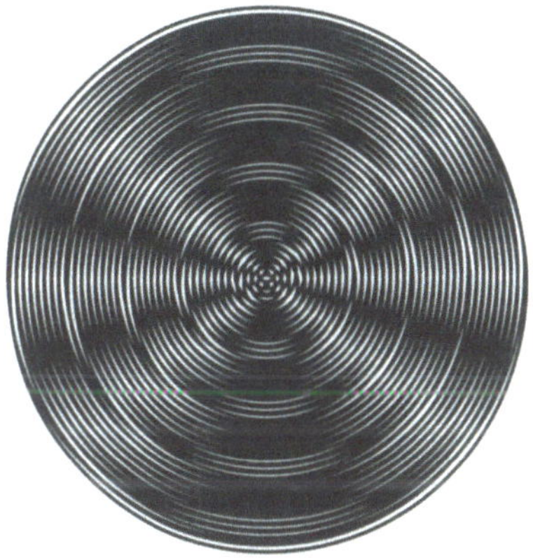

We have already been introduced to the moiré fringe analog of the ripple tank, and the two-point interference patterns that it produces. Now we will look at the same phenomena with a finer grating scale, in order to reduce the visibility of the circular rings and emphasize the fringe patterns that they create.

As the vertical separation of the two sources increases, the number of fringes around the perimeter of the circle increases. In the first sketch (#1), the sources are 1.5 wavelengths apart, so that there are two dark fringes between the 12 o'clock and 3 o'clock positions (centered at 12:00 and 2:20). In the second sketch (#2), the sources are 4.5 fringes apart, so there are five dark fringes in the same angular region (at 12, 1:15, 1:45, 2:15, and 2:45). As the sources separate further, more fringes emerge, and the angular spacing between them decreases. This kind of experimenting is best done with samples of such patterns right in your hands.

#2

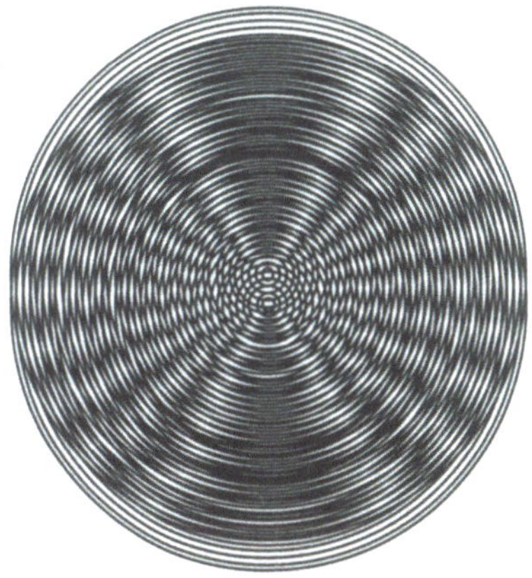

The fringe patterns are a little indistinct, especially as illustrated here, but we can draw center lines through them with the aid of a little mathematical insight.

Near the edges of the circles, the fringes seem to be straight lines, aimed between the two sources. In fact, they are mathematical hyperbolas, and arc around between the sources to emerge on the other side. The fringes are the loci of points of equal path difference between the two source points (the foci of the hyperbolas). If the sources really were point sources in 3-dimensional space, these fringes would be hyperbolas of revolution nested one within the other. Between the two sources, the fringes are equally spaced at half-wavelength intervals. As they move outward, they approach the straight-line asymptotes typical of hyperbolas.

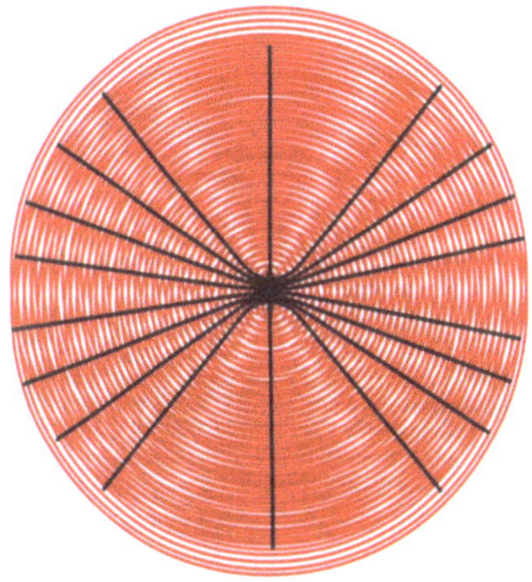

Spherical waves

As a rule, we will be dealing with sources that radiate light in only a fairly limited angle, perhaps 30° for light spread by a microscope lens. Thus we are interested at any one time in only a small region of the patterns we have been describing so far. Even so, we can use the overall pattern as a kind of "road map" of the various domains of holography, in which we will consider just one area at a time, as shown on the next sketch.

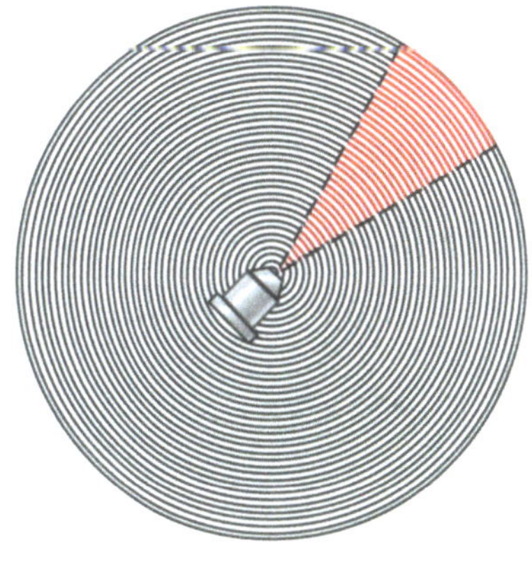

Here, the various types of holography are mapped out as domains with respect to the locations of the two sources used. If S1 and S2 are our sources, an observation of the patterns on a plane (or exposing a holographic plate) at "A" signifies "diffraction gratings," which we will study first. Then comes "B," the "holographic lenses," or "in-line" or "Gabor" holograms (where S1 becomes the prototype for the "object" and S2 for the "reference source"). Combined, their

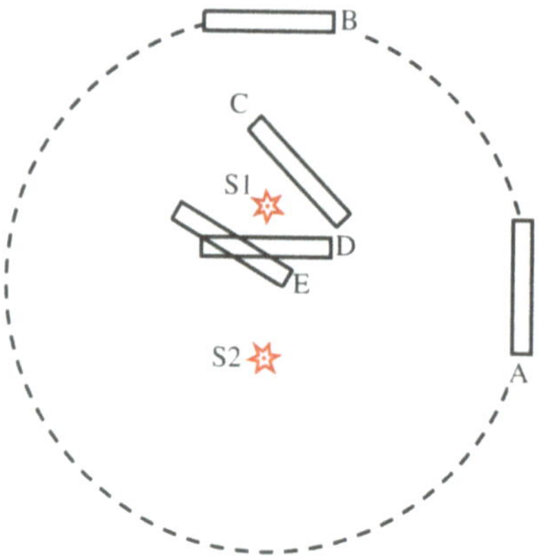

mathematics allows us to discuss "off-axis transmission" or "Leith-Upatnieks holograms" at location "C," which will extend to include "image plane" and "rainbow" holograms. Then we will move to reflection holograms, first the "single beam" or "Denisyuk" type, at location "D," and then the "off-axis" reflection hologram at location "E."

Side-by-side: linear fringes: When we are at location "A," the interference fringes are straight lines radiating from a point midway between the sources, and they intersect the recording plane at equally-spaced points, which become lines if we consider them in three-dimensional space (contentions that we will prove later on).

In-line: Fresnel zone plate: At location "B," the interference fringes are also straight lines radiating from a point midway between the sources, but they intersect the recording plane in circles, not lines, and the circles are not equally spaced—they become closer and closer as we move away from the line that passes through both sources. What happens at other positions where we're at some angle to the sources is a bit more complicated, but we'll deal with the mathematics there shortly.

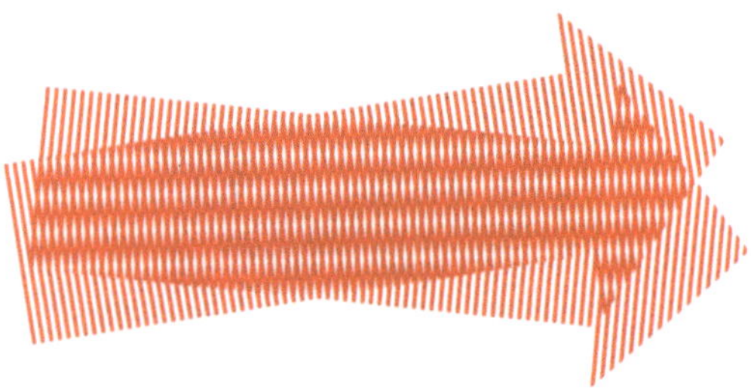

Plane waves (in x-z *plane)*

If we are far from a point source of radiation, and considering the waves only over a limited region, the waves can be approximated as flat or "plane" wavefronts. In this region, we say that the light is "collimated" or that the "rays" are all parallel. This is a common case for star light, for example, but in the laboratory it is often quite difficult to produce exact plane waves. We usually mean that waves are "plane" if their departure from exact planarity is small compared to a wavelength of light over the aperture we are interested in (a quarter of a wavelength tolerance is typical). We often sketch a portion of such a wave as a large arrow, pointed perpendicular to the wavefronts (which is the direction of propagation of the plane wave in most media), with the wavefronts more or less visible within it, and loosely refer to this as a light "ray" (a "ray bundle" might be more accurate).

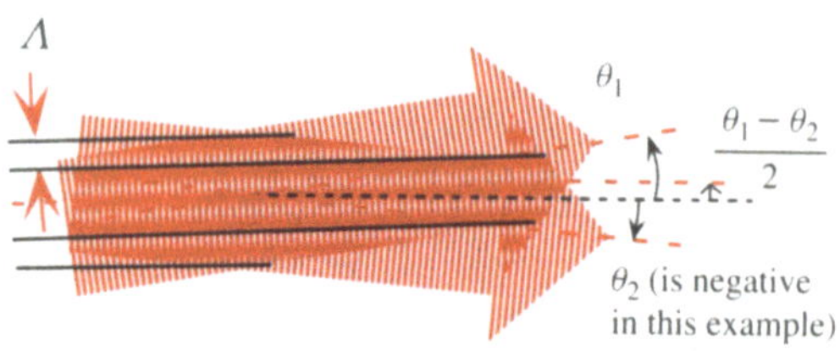

When two plane waves cross, the interference pattern between them takes on a fairly simple characteristic shape. The fringes are now strictly straight lines (the graphics here may make them wobble a bit) that are parallel and equally spaced. Their spacing decreases as the angle between the rays increases, and the line of the fringes bisects the angle between the two rays. These effects are really best explored by working with moiré patterns between pieces cut from parallel-line patterns on acetate (it helps the contrast if the ratio of dark/clear areas is around 1:1).

To get a little more quantitative about it, this is probably the time to state that the angle of the fringes is the average of the two ray angles,

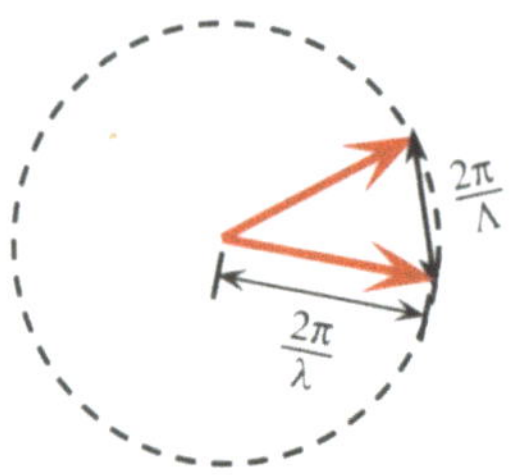

$$\theta_{\text{fringe}} = \frac{\theta_1 + \theta_2}{2} \tag{16}$$

and the spacing between the fringes, which we will call Λ, is determined by the angle between the rays and the wavelength of the light:

$$\frac{2\pi}{\Lambda} = \frac{2\pi}{\lambda} 2\sin\left(\frac{|\theta_1 - \theta_2|}{2}\right) \tag{17}$$

(The distance Λ is related to the grating spacing, d, that we will see later.)

It is sometimes easier to remember these in geometrical terms, with a vector representing the fringe pattern that is the difference between the vectors representing the two rays. These vectors all have lengths that are proportional to the reciprocal of the scale of the pattern they represent (here the wavelengths, λ and Λ), and a direction that is perpendicular to their wavefronts or fringes, and are generally known as **K**-vectors when the 2π is included. They are our introduction to *reciprocal space*!

Laser speckle

We have already briefly mentioned "laser speckle;" now we are in a position to understand it as an interference phenomenon. It is the gritty or sandy appearance of laser beams when played upon a diffusing or matte surface (like paper or paint). The microscopic roughness of the surface, which is what causes it to scatter light in all directions, creates many, many overlapping waves with randomized phases. When these waves cross again, such as when focused by the lens of your eye, they produce a randomized intensity pattern with high contrast. Try looking at a speckle pattern through a pinhole (made by pinching your fingertips together) and seeing how the size changes; watch how the speckles move as you move your head from side to side (repeat without your glasses, if you usually wear them). A rigorous discussion of laser speckle requires the mathematics of random process theory, but practicing holographers generally have some "rules of thumb" for minimizing it, and Prof. Gabor once referred to laser speckle as "holographic enemy number ONE!"

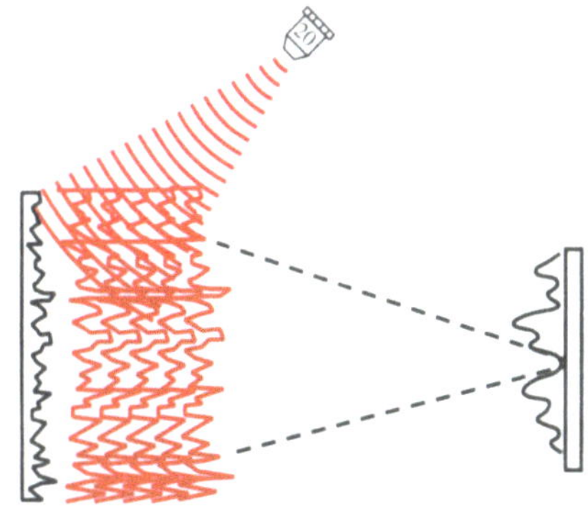

Simple Interference Patterns

With this background, we can now consider a few interference patterns produced by simple optical setups, using the expression of Eq. (8) in slightly different form to emphasize the usefulness of the "phase footprints" found in Chapter 3, so that the phases, and resulting total intensity, are expressed as functions of x and y in the observation plane, usually at $z = 0$.

$$I_{\text{total}}(x,y) = I_1(x,y) + I_2(x,y) + 2\sqrt{I_1(x,y)\cdot I_2(x,y)}\cos\big(\phi_1(x,y) - \phi_2(x,y)\big) \tag{18}$$

Overlapping plane waves

Consider two plane waves incident at angles θ_1 and θ_2, as shown in the sketch. Each has unit intensity, so $I_1 = I_2 = 1.0$, and their phase footprints are:

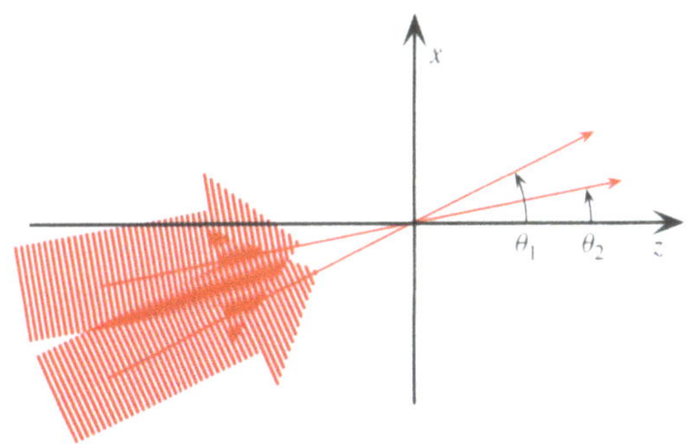

$$\phi_1(x,y) = \frac{2\pi}{\lambda} x \sin\theta_1$$
$$\phi_2(x,y) = \frac{2\pi}{\lambda} x \sin\theta_2 \tag{19}$$

Simply plugging this information into the expression above (again assuming that the intensity of each source at the hologram plane is unity) yields

$$\begin{aligned} I_{\text{total}}(x,y) &= 1+1+2\sqrt{1\cdot 1}\cos\left(\frac{2\pi}{\lambda}x\sin\theta_1 - \frac{2\pi}{\lambda}x\sin\theta_2\right) \\ &= 2+2\cos\left(\frac{2\pi}{\lambda}x\left(\sin\theta_1-\sin\theta_2\right)\right) \end{aligned} \tag{20}$$

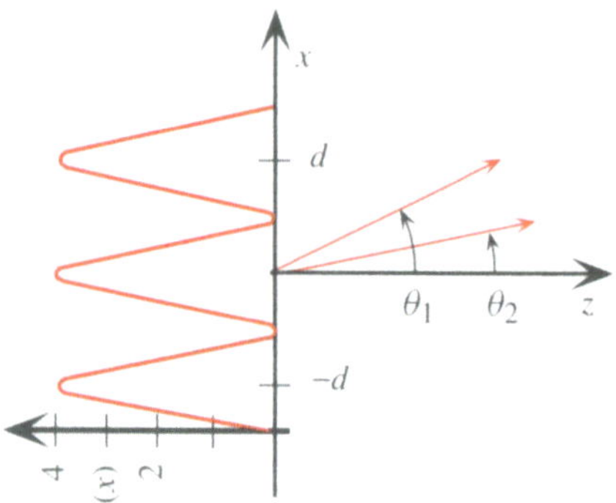

which is a sinusoidal variation of intensity as x increases, reaching a new peak at multiples of the distance d, given by

$$d = \frac{\lambda}{\sin\theta_1 - \sin\theta_2} \tag{21}$$

so that the spatial frequency of the pattern, f, is given by

$$f = \frac{\sin\theta_1 - \sin\theta_2}{\lambda} \tag{22}$$

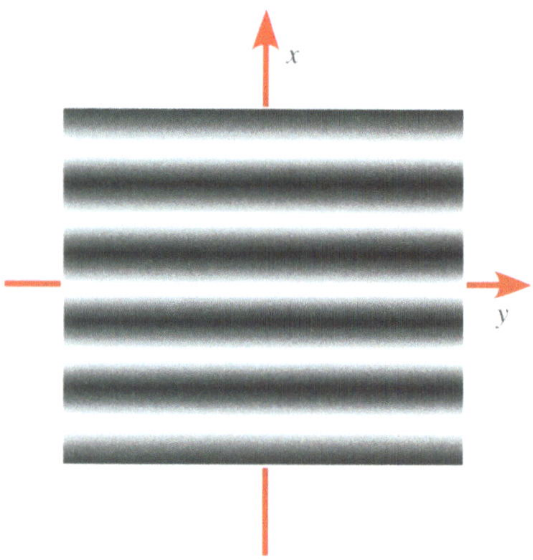

A comment about spatial frequency: Researchers in coherent optics often refer to patterns in terms of their "spatial frequency" (usually measured in cycles per millimeter), reflecting the grounding of the field in communication theory. As a two-dimensional (and occasionally three-dimensional) extension of temporal frequency concepts (cycles per second, referred to as Hertz or Hz), spatial frequency thinking makes the extension of signal analysis concepts fairly straightforward. Depending how we assign numbers to the beams, the results for d and f could well come out negative. By convention, we will always consider the spacing and spatial frequency to be *positive* numbers (negative frequencies are more common in linear systems theory), so there really should be "magnitude bars" around the right sides of Eqs. (21) and (22).

Note that there is no variation of either wave's phase in the y-direction, and thus no variation of I_{total} with y. The intensity pattern in the x, y plane will be a series of parallel bands of graded intensity.

Side-by-side point sources

Consider now the case where two coherent point sources of light, S_1 and S_2, are at the same distance from the hologram plane, at $z = -Z$, and at equal distances from the z-axis, at $x_1 = +s/2$ and $x_2 = s/2$. The intensity of each source at the hologram plane is unity, and their phase footprints are

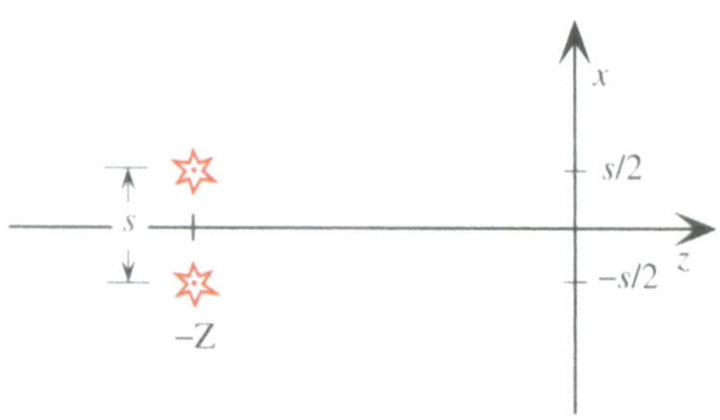

$$\begin{aligned} \phi_1(x,y) &= \frac{2\pi}{\lambda}Z + \frac{\pi}{\lambda Z}\left(\left(x-\frac{s}{2}\right)^2 + y^2\right) \\ \phi_2(x,y) &= \frac{2\pi}{\lambda}Z + \frac{\pi}{\lambda Z}\left(\left(x+\frac{s}{2}\right)^2 + y^2\right) \end{aligned} \tag{23}$$

Plugging these into the master interference equation (8) then gives

$$I_{\text{total}}(x,y) = 1+1+2\sqrt{1\cdot 1}\cos\left(\frac{\pi}{\lambda Z}\left(\left(x-\frac{s}{2}\right)^2+y^2\right)-\frac{\pi}{\lambda Z}\left(\left(x+\frac{s}{2}\right)^2+y^2\right)\right)$$
$$= 2+2\cos\left(\frac{2\pi}{\lambda}x\frac{s}{Z}\right) \tag{24}$$

This pattern has the same form as that shown above, and if we can arrange it so that (s/Z) is equal to the difference of the sines of the angles, the spatial frequency of the pattern will even be the same. Which is to say that the phase contributions due to the sphericity of the waves "cancel out" if the interfering sources have the same sphericity; *i.e.*, they are at the same distance. We caution that this is true only for small s/Z and for fringes near $(x, y) = (0, 0)$, which is often the case. The general principle that waves need not be exactly planar to make the plane wave approximation useful still stands, though.

In-line point sources

Here, the point sources are arranged one in front of the other, at $-Z_1$, and the second at $-Z_2$. The phase footprints are now

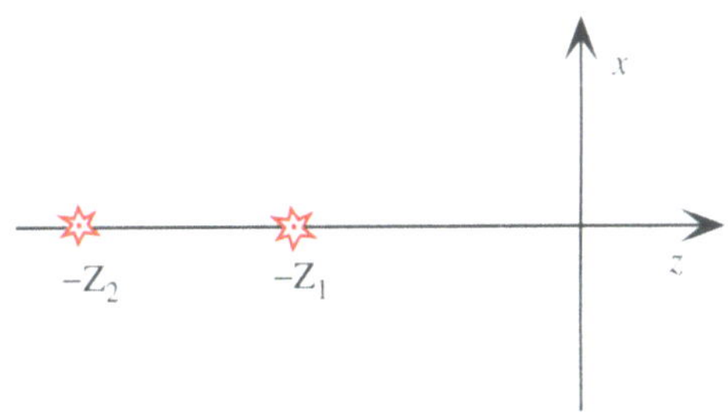

$$\phi_1(x,y) = \frac{2\pi}{\lambda}Z_1 + \frac{\pi}{\lambda Z_1}\left(x^2+y^2\right)$$
$$\phi_2(x,y) = \frac{2\pi}{\lambda}Z_2 + \frac{\pi}{\lambda Z_2}\left(x^2+y^2\right) \tag{25}$$

The leading terms in both are constant phases, and we will assume for the moment that they are both exact multiples of 2π, equivalent to zero, and can safely be ignored. Plugging the rest of the terms into the (8), the master interference equation (again assuming that the intensity of both waves at the hologram plane is unity) then gives a characteristic intensity pattern:

$$I_{\text{total}}(x,y) = 1+1+2\sqrt{1\cdot 1}\cos\left(\frac{\pi}{\lambda Z_1}\left(x^2+y^2\right)-\frac{\pi}{\lambda Z_2}\left(x^2+y^2\right)\right)$$
$$= 2+2\cos\left(\frac{\pi}{\lambda}\left(\frac{1}{Z_1}-\frac{1}{Z_2}\right)\left(x^2+y^2\right)\right) \tag{26}$$

Now we are dealing with something quite different! This pattern is a function of both x and y, and in a combination that makes it a function only of the distance, r, from the $(x, y) = (0, 0)$ point. That is, the pattern has rotational symmetry about the $(0, 0)$ point, and thus consists of some kind of pattern of concentric circles, however spaced. In fact, the spacing is also an important matter, so we will examine it in some detail.

We will consider first a general function of radius, r, described by

$$I(r) = 1+\cos\left(2\pi\frac{r^2}{a^2}\right) \tag{27}$$

This has a maximum at the origin, $r = 0$, and another maximum (or bright ring) at $r = a$. The third maximum is at $r = \sqrt{2}a$, and in gen-

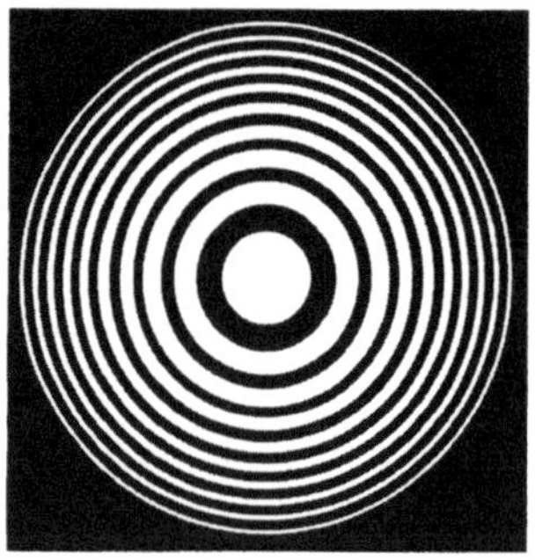

eral, the n-th maximum is at a radius $r = \sqrt{na}$. Which is to say that the bright rings are not equally spaced, but the spacing slowly shrinks as we move outward; in fact the area between successive bright rings is a constant. A pattern of this sort was first devised by the French mathematician Augustin Jean Fresnel (1788–1827), and generally bears the name "Fresnel zone plate" in his honor. Actually, Fresnel's zone plate is a binarized "on-or-off" version of this pattern, and we holographers tend to call this continuous-scale version a "Gabor zone plate." In the case of our interferometric exposure, the scale factor becomes

$$a = \sqrt{2\lambda \frac{z_1 z_2}{Z_2 - Z_1}} \tag{28}$$

When this pattern exposes a piece of film, the resulting transmittance pattern is found to have some interesting focusing patterns that we will soon explore in some detail.

Conclusions

The notion of "interference" defies some of our intuitive notions of "conservation of energy" on a small scale, but once it becomes a natural way of "seeing" things, it explains many interesting wave-optical phenomena. There are many, many categories of interference phenomena, as any book on physical optics will reveal. Here, we will limit our attention to the interference of waves from two spatially-separated coherent sources, as this is the simplest model for understanding holography. Later we will generalize from point-like sources to large-area diffuse sources, but the underlying concepts will stay the same.

With the help of the "phase footprints" of some common wavefronts, we can become quite quantitative about the intensities of some interference patterns of interest. But it is the *geometry* of the patterns—the directions, spacings, and shapes of the resulting fringes—that is of most interest to us for most of this course. That information follows directly from simply subtracting the "phase footprints," something we can do mathematically, or by looking at moiré fringes.

References

i. Gaskill, J. D. (1978). *Linear Systems, Fourier Transforms, and Optics,* John Wiley & Sons, New York, "Ch. 10: The Propagation and Diffraction of Optical Wave Fields." Contrast this with the classic Goodman, J. W. (1996), *Introduction to Fourier Optics,* McGraw-Hill Book Co., New York, "Ch. 3: Foundations of Scalar Diffraction Theory," which uses the opposite sign convention for spatial phase. It is sometimes said that the main differences between electrical engineers and physicists can be explained by their respective symbols for the square root of minus one having opposite signs.

CHAPTER 5

Diffraction

Introduction

Most of our intuition about light is based on *ray* or *geometrical* optical concepts. These are based, in turn, on three basic premises: 1) that light "particles" travel in straight lines (what we normally think of "rays") until they hit something, 2) that when they hit a reflector, the angle of reflection equals the angle of incidence, and 3) that when they hit a material of different refractive index, part of the light reflects (called the Fresnel reflection) and part continues at a different angle (measured to the perpendicular) determined by Snell's Law.[i] These three "laws" account for 99+% of what we see in everyday life, but they are only an approximation. When we deal with highly coherent light, such as from lasers, diffraction and interference effects become much more prominent than usual. These are usually described by *wave* or *physical* optical concepts, and are more complex and accurate than geometrical-optical concepts. Of course, in the limit of low coherence light the two approaches must both agree to within acceptable accuracy, and they do.

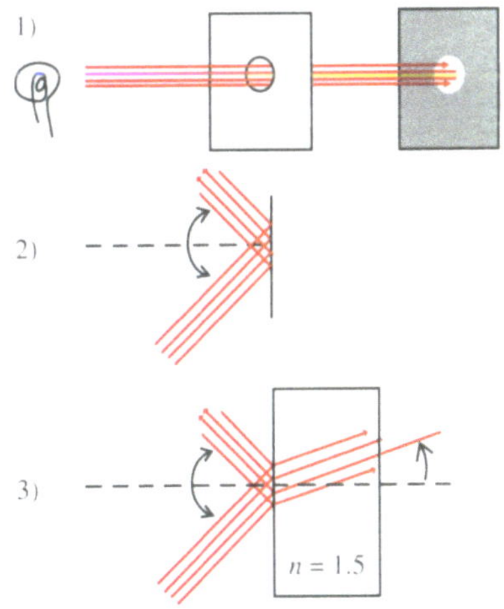

A simple experiment will show the limitations of the first premise, that light travels in completely straight lines. After this, you will believe that almost anything is possible! Consider an undiverged laser beam headed toward a white wall; it forms a single spot on the wall, perhaps 2 mm in diameter. Bring a razor blade slowly up into the beam, about a meter from the wall; mostly, you will see the spot being cut off, going through a half-round phase, and then being extinguished as the beam reflects off the solid blade. This is the "geometrical shadow" of the blade in the beam. But if you look closely while the blade edge is within the beam, you will see a streak of light above and below the geometrical shadow of the blade that carries a small percentage of the incident light. Now, you might think that the light *above* the shadow comes from a reflection from the razor blade's edge, as though it were a half-cylinder, but that doesn't explain the dark fringes in that light, and it certainly doesn't explain the light found *below* the shadow, which seems to veer around the edge as though deflected by some strange attractive force. If you put your eye in the streak (be careful to avoid the straight-through beam!), you will see that the light comes only from the edge of the razor blade.

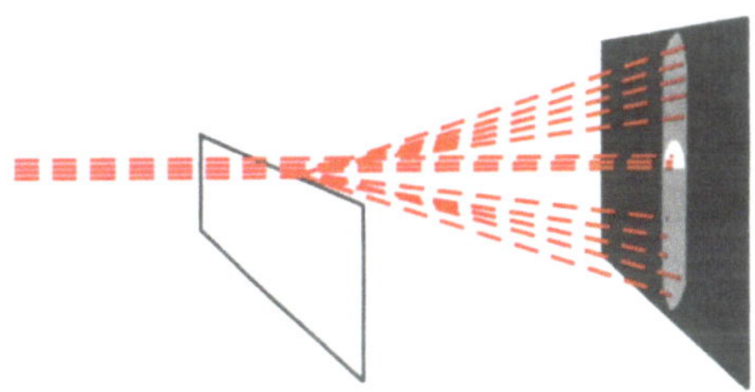

This "non-straight-line" behavior of light is a simple wave-optical phenomenon called *diffraction*. An explanation was first offered by Huyghens around 1678.[ii] He said that it was reasonable to consider air (and also vacuum) as a volume filled with imaginary spheres, like closely packed marbles, and that light was like a "nudge" from one of those spheres, which would nudge all the adjacent spheres, which would nudge all their neighbors, and so on and so forth (something like a pan full of marbles). If a sphere were equally nudged by neighbors to the left, above-left, and below-left, it would move to the right (a vector addition of the nudges), and nudge

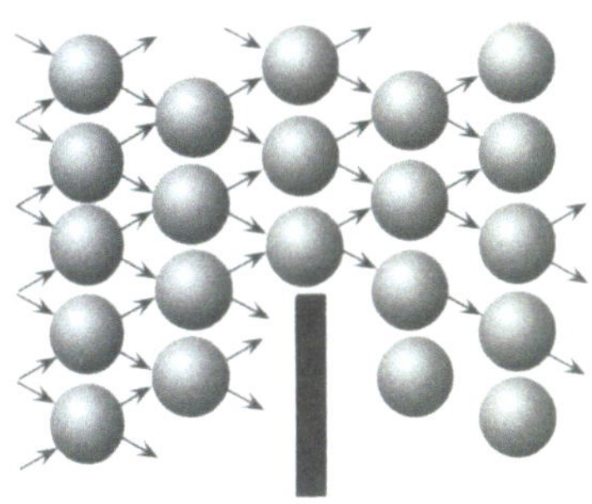

only that neighbor. Thus a wide nudge wave would propagate in a single direction. But, if a partial wall is stuck in among the spheres, a sphere just to the right of the top of the wall gets no nudges from below, and thus gets a net downward nudge from what is left of its neighbors. Thus a nudge starts propagating downward and forward into the "shadow" that should be cast by the wall. The story was fleshed out by Fresnel in the 1820s.[iii] He proposed that the nudges were periodic and even sinusoidal. Thomas Young had earlier anticipated some of the implications of periodicity, and argued that the "nudges" were actually side-to-side vibrations (lateral, not longitudinal excitations) of the medium.[iv] Maxwell then showed (in the 1870s) that these are lateral oscillations of coupled electrical and magnetic fields. It is all a fascinating story about the slow overcoming of a set of very strongly held beliefs in the particle theory of light established by Isaac Newton in the early 1700s, and we recommend browsing a book like *The Nature of Light,* by Ronchi (1970) for more of that history.[v]

Diffraction by Periodic Structures

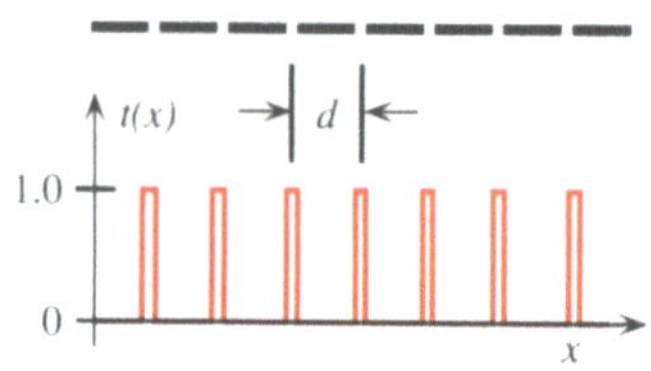

For discussing holography, we will concentrate on diffraction effects caused by repetitive or periodic structures, such as evenly-spaced slits in an opaque screen (like a picket fence). We usually describe these by a transmittance function, $t(x, y)$, and periodicity in the x-direction with spacing d means, in mathematical terms, that $t(x-md, y) = t(x, y)$, where m is any integer. Note that, because light will generally be moving from left to right across the page (for a while, at least), along the z-axis, the x-axis is drawn vertically here, as it will be in most of our sketches. For these examples, there will be no y-dependence of the transmittance pattern.

When an undiverged laser beam hits such a periodic structure, it breaks up into several laser beams deflected upwards and downwards by multiples of a certain angle determined by the spacing and the wavelength of the light. Actually, the trigonometric sines of the angles are multiples of the sine of a certain angle, up to the limit of ±1.0. The relationship is described by the simple form of the *diffraction equation* (which we will soon have to prove):

$$\sin\theta_{out,m} = m\sin\theta_{+1} = m\frac{\lambda}{d}, \qquad m = 0, \pm1, \pm2, \ldots \tag{1}$$

Single-Slit Diffraction

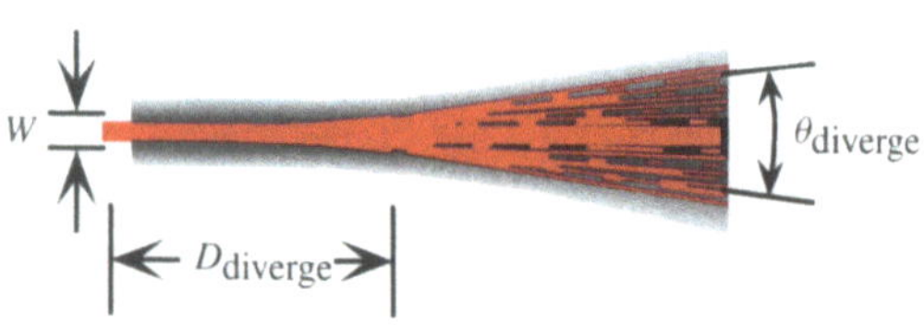

The situation for a single slit is actually more complex than it might seem at first glance. Even if the laser beam goes through an empty frame, it eventually starts to diverge, to expand at some constant diameter increase per distance, which is to say that it diverges with some constant angle, $\theta_{diverge}$. Even if the wavefronts are carefully collimated when they come out of the laser, by some distance downstream they will have become spherical, diverging from a point at the front of the laser. This is all due to the fact that the laser beam is like a uniform and infinite plane wave that has suddenly passed through a circular aperture or window in order to get out of the laser. That constriction at one end causes the beam to spread out at its other end, and the smaller the constriction width, W, the larger the angle of divergence, $\theta_{diverge}$. This effect is called *single-slit diffraction*, although in this case the "slit" is a smoothly tapering circular aperture, producing a beam with a *Gaussian* intensity profile (uniquely, the beam remains Gaussian as it diverges!).

If the beam from the laser passes through a nearby diffraction grating, all of the downstream orders eventually start to expand, and all with the same divergence angle. Thus the downstream spot pattern evolves at a large enough distance into a pattern that no longer changes shape, but only expands uniformly with distance. That pattern is called the *Fraunhofer*, or *far-field* diffraction pattern.[vi] Within some distance (the "far-field distance," naturally), the diffraction is characterized as *Fresnel* or *near-field* diffraction. That pattern changes mysteriously as a function of distance, generally requiring mathematical techniques that go beyond the scope of this book. As we approach the grating itself, we expect to see the geometrical shadow of the grating emerge. But not far downstream from there, we also find a negative image of the grating pattern!

Use of Lenses

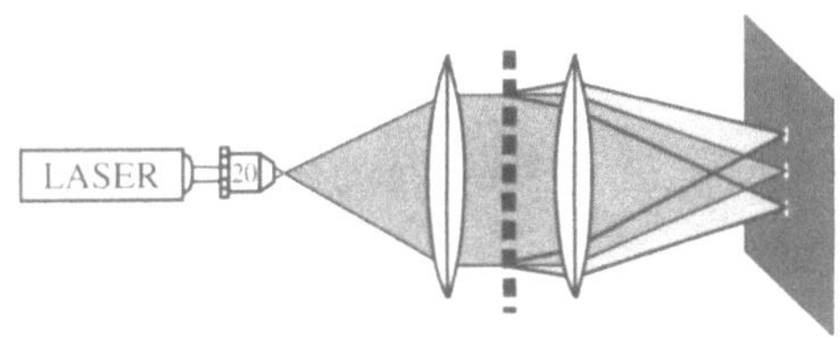

It is inconvenient to traipse far down a hallway to look at far-field diffraction patterns, so we often use lenses to bring them into focus at much closer distances. First, imagine that a microscope objective has been attached to the laser to produce a point source of diverging spherical waves. A first lens is then placed its own focal distance away from the point source, so as to produce collimated or plane wavefronts. Then comes the grating, which breaks up the incident plane wave into a series of plane waves at various angles. Then comes a second lens, which causes each of those plane waves to curve inward toward a focus one focal length behind the lens. If we put a white cardboard screen there, we will observe the same pattern that would appear in the far field of the grating, except scaled down in the ratio of the focal length to the distance to the far-field pattern.

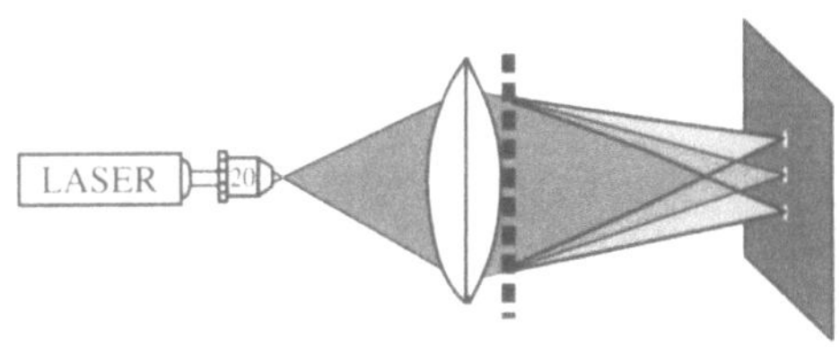

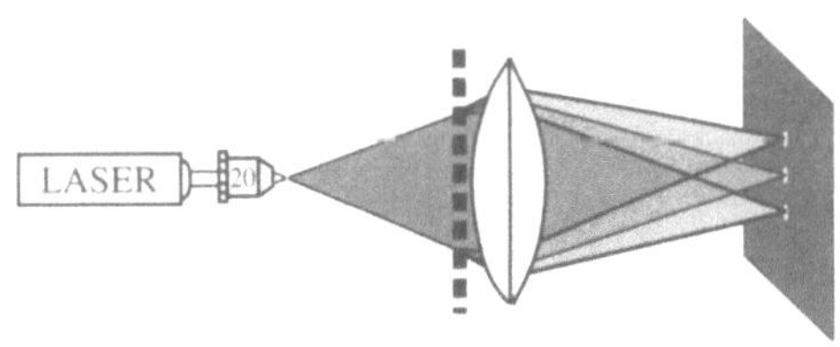

From a wave-optical point of view, a lens simply multiplies the amplitude of the wavefront by a factor that varies as the square of the distance from the center of the lens (it is a complex phase-only multiplication, which adds a phase that varies as r^2). The diffracting pattern also produces a multiplication of the wavefront, often by an amplitude-only function of x and y, and the second lens is again a phase-only multiplication. Now, because the results of these multiplications are invariant under interchange (or commutation) of the operations, it doesn't matter which happens first, or second, or third. Also, the effect of two lenses being exactly in tandem is the same as for a single lens of twice the thickness variation. So, all three of the sketched optical setups produce the same intensity pattern in the back focal plane of the lens!

Viewing Diffraction Patterns with the Eye

An important outcome of this discussion is the realization that the naked human eye can also be used to view diffraction patterns. If a point source of light is viewed in sharp focus, then a grating placed just in front of the eye will produce a "far field" diffraction pattern on the retina, which will appear as an array of spots in the same plane as the point light source, surrounding it like a halo. You can try this easily: shine a laser pointer on a wall across the room and look at the wall through a grating. Note how the spacing of the spots in the pattern changes as the distance from you to the wall changes. You could use this effect to measure the distance to the wall without getting out of your chair![vii]

Styles of Diffraction Analysis

"Every problem in optics becomes easy if you look at it the right way," the old maxim goes, and there are dozens of ways of looking at diffraction and trying to understand its effects. We will take a passing glance at two very different approaches, and then simply accept the mathematical rules that result without many further questions. If you want more detailed explanations, there are plenty of other books to refer to for other approaches, so you can probably find one that explains things in a way that makes you comfortable.[viii] They are all simplifications in one sense or another, as even the most basic problem in diffraction (by an opaque single edge) is not yet completely solved.

Graphical analysis

Here, we will look at the addition of contributions to the far-field intensity pattern as more and more equally spaced narrow slits are opened, showing that the pattern converges to a series of distinct spots in the limit of many, many slits. First we consider the contribution of a single isolated slit. As that slit narrows to an idealized line source, the transmitted wavefront becomes an idealized cylindrical wave with an amplitude that is equal in all directions (there may be a gradual cosine-theta falloff, which we will ignore). Now the question is: what happens if we open an identical slit parallel to the first and separated by a distance "d"? We consider, for simplicity's sake, the geometry shown here, with the slits one focal length of the lens (called "F") in front of the lens, and the observation plane one focal length behind the lens, so that parallel "rays" can be considered wherever possible.

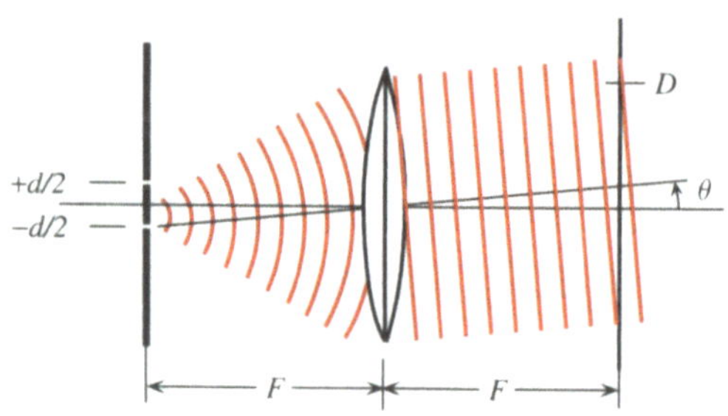

Two-slit diffraction, one at a time

Here we show the wave from the lower slit only; the wave from the upper slit will be symmetrical. Let the two slits be at equal distances from the z-axis, at $+d/2$ and $-d/2$, so that their equal contributions will arrive in phase at $x = 0$ in the back focal plane of the lens. First we open the lower slit by itself. We want to know the location, $x = D$, at which the wave from the lower slit will be exactly one-half cycle out of phase with (lagging behind) its value at $x = 0$:

$$D\sin\theta = \frac{\lambda}{2}$$

$$D\sin\left(\tan^{-1}\frac{d/2}{F}\right) = \frac{\lambda}{2} \qquad (2)$$

$$D \approx \lambda\frac{F}{d}$$

That is, the lower slit emits a cylindrical wave that the lens transforms into a tilted plane wave with an angle $\theta = \tan^{-1}(d/2F) \approx \sin^{-1}(d/2F)$. This creates a phase increase of $\phi = 2\pi(x/\lambda)\,\sin\theta$ in any plane behind the lens, including the back focal plane at F. The distance D then follows from $\pi = 2\pi\,(D/\lambda)\,\sin\theta$, giving our result, $D \approx \lambda F/d$.

Two-slit diffraction, two at a time

Now we open the upper slit too. The two plane waves overlap, and are in-phase along the z-axis (at $x = 0$) because of symmetry. Because the waves are equally but oppositely tilted, they become increasingly out of phase as x increases, until they are so far *out* of phase that they are back *in* phase again (the wavefronts cross peak over peak). The height at which the phase difference between the waves from the two slits is 2π is given by $D = \lambda/\sin\theta$. Note that the phase difference (and thus the interference pattern) is independent of the distance along the z-axis, but we will stay at the $z = F$ plane for this discussion.

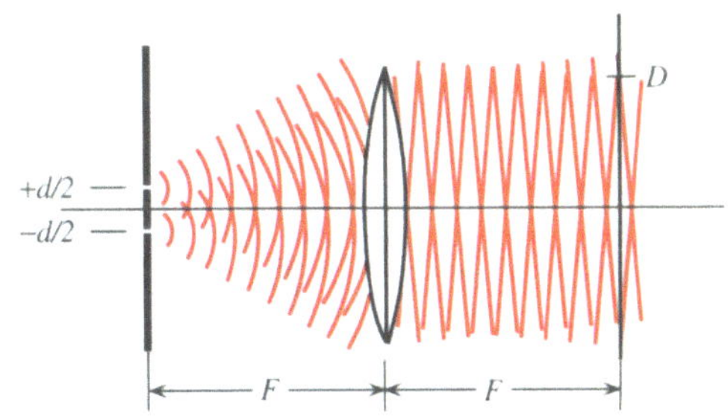

As the observation location moves between 0 and D, the waves from the two slits arrive increasingly out of phase, with the phase difference, $(\phi_1 - \phi_2)$, passing π radians or 180° at $D/2$, and continuing on the way to being 2π radians or 360° out of phase, which is back in phase, at $x = D$. The phase difference is a linear function of the height in the back focal plane. Thus the intensity of the interference pattern formed by the two equal-intensity waves varies according to our familiar interference equation:

$$I_{\text{total}}(x) = 2\,I_0\left[1 + \cos\left(2\pi\frac{x}{D}\right)\right] \qquad (3)$$

As the observation location moves onward from D to $2D$, the phase difference further increases from 2π to 4π, and the cosinusoidal fringe pattern continues through equally-spaced maxima and minima until the angles become so large that our paraxial approximations break down.

If we look below the axis in the back focal plane, the same phase variation happens, but with the opposite sign. Thus the cosinusoidal pattern extends for many cycles above and below the axis, producing a series of parallel bright "fringes" in the x-y plane that look a lot like furrows in a plowed field.

The pattern we have been talking about is usually called "Young's double-slit fringes," honoring their first explanation by Thomas Young in 1801. Young based his arguments for the wave theory of light on those patterns, leading to the work of Fresnel and Maxwell, and finally overcoming hundreds of years of domination by Newton's particle theory. For us, they are also the building blocks of a theory, this time of holographic imaging! Fortunately, we don't have to contradict any giants in the field; Gabor, Leith, and Denisyuk have all agreed with these ideas about waves and light, especially where lasers are concerned.

Multiple-slit diffraction, N *at a time*

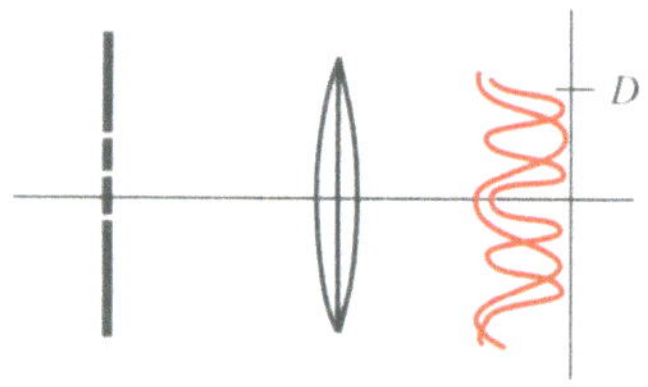

The really interesting part begins when we introduce a third slit, spaced a distance d above the first slit. There are now *three* interference patterns formed, one for every possible pair of slits, and one of those patterns (between the first and third slits) has *two* intensity maxima between zero and D on the x-axis. And, one of those maxima is centered at $x = 0$, right on top of the maximum formed by the first two slits (and on top of the maximum formed by the second and third slits). That is, every pair of adjacent slits produces a pattern that has a maximum at 0, D, $2D$, and so forth, but the patterns that are formed by slits further apart have other maxima in between. The sum of all the patterns has principal peaks that are much narrower than for the two-slit pattern, with a weaker "secondary maximum" in between.

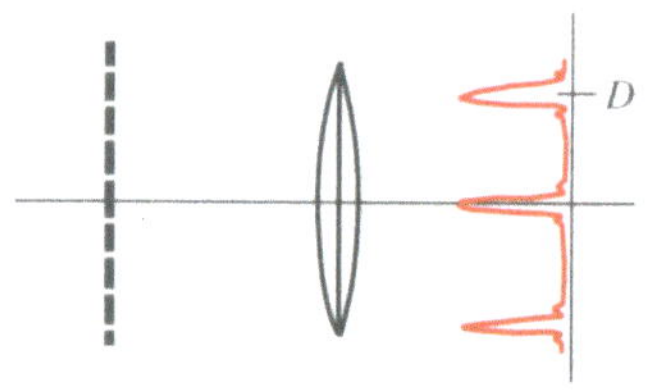

As even more equally-spaced slits are opened above and below the first two slits, further components of the intensity pattern are introduced. We state without proof that the interference pattern formed by even more slits is equal to the interference patterns formed by all possible pairs of the slits, minus a constant equal to the sum of the intensities of the individual slits (readers may want to try working this out for themselves). The pattern from the furthest-apart slits (let's say that they are $N{\cdot}d$ apart) has the finest fringes, with $N-2$ maxima between 0 and D. As more and more slits are opened, and further fringe patterns are added to the overall pattern, the overall intensity pattern converges to a characteristic shape, with principal maxima that are N^2 as high and $1/N$ as wide as for the $N = 2$ case, and with $N-2$ smaller maxima in between. As N becomes a few hundred, the light concentrates almost entirely in the peaks, one right on axis and others spaced equally up and down on the output plane, separated by the distance D.

Mathematical proofs of the properties of diffraction gratings are certainly in order here, but we will save our effort for a different approach. Various analyses can be found in optics textbooks that emphasize one or another point of view. For the time being, let's explore the general properties we have described so far.

The grating can be considered as an optical component that breaks an incoming plane wave into a set of outgoing plane waves at roughly-equally-spaced angles, a "fan" of beams or rays. The *amount* of energy in each member of the fan of beams depends on the details of the shape of the grating slits, whether they are hard-edged or soft-edged, or just slow down the wave a bit. But the *angle* of deflection of each component of the fan depends only on the spacing of the slits, and that is the aspect that we will explore in this chapter.

Grating Equation

The previous discussion showed that an incoming plane perpendicular to the grating results in one plane wave angled up at an angle "theta" (θ) given by $\sin(\theta) = \lambda/d$, and another wave angled downward by the same angle, at $-\theta$. And above and below those are more waves deflected by larger angles, given by $\sin(\theta) = \pm 2\lambda/d$. If the slits have a suitable shape, much larger deflection angles (up to 90°) can be produced. We describe these beams as various "orders" of deflected or diffracted waves, with the "first order" beams being those closest to the straight-through or "zero order" beam. The next set, if those beams exist at all, are the "second order" beams, and so forth, through the third, fourth, fifth, and higher orders.

To begin, let's look at the on-axis grating equation, which describes what happens to a plane wave that comes in perpendicular to the grating. A "fan" of plane waves emerges, consisting of pairs of waves deflected or diffracted through equal but opposite angles. These angles are given by the first or "on-axis" form of the *grating equation:*

$$\sin\theta_{\text{out},m} = m\frac{\lambda}{d} \tag{4}$$

where m is the "order number" of a particular beam of interest. The relationship breaks down when the sine of the diffracted angle goes beyond unity (for sufficiently large $|m|$, for example), and there is no wave that propagates past the grating corresponding to that order. Instead, a so-called "evanescent" wave travels slightly beyond the grating, turns around, and re-enters the grating to contribute to the specularly reflected light.

Wavelength variation (red rotates radically)

One of the most important observations from the grating equation is that long-wavelength or "red" light is deflected through a larger angle than mid-wavelength or "green" light, which is in turn deflected through a larger angle than short-wavelength or "blue" light. This qualitative fact is important to remember in more complicated situations too, and one mnemonic for it is to remember "the three **R**s": "Red Rotates Radically!" Landscape paintings occasionally show upside-down rainbows, and holographers' sketches sometimes show upside-down spectra! So it is important to have an easy way to remember which way is "up" in diffraction.

Red
Rotates
Radically!

Spatial Frequency

We have been describing diffraction gratings so far in terms of their repeat distance, d, which gives a kind of concreteness to the discussion. However, from now on we will almost always describe them instead in terms of their "spatial frequency" in cycles per millimeter. This allows the analogies with temporal frequency in electrical engineering to become more obvious, and we describe the spatial frequency by the variable $f = 1/d$. We can also include the orientation of the grating by letting f become a two-dimensional vector, $\mathbf{f}$, oriented perpendicularly to the grating's grooves, with magnitude $f = |\mathbf{f}| = 1/d$.

However, the orientation of the grating will always be clear in our discussions (in the x-direction, unless otherwise noted), and we will try to stick with the scalar f wherever possible, in the spirit of "shop math" calculations.

Grating Example

As an example, let's consider a grating of spatial frequency $f = 450$ cycles/mm, or $d = 2.2\ \mu$m. A He-Ne laser beam ($\lambda = 633$ nm) incident at $\theta = 0°$ produces seven output beams: $\theta_0 = 0°$, $\theta_{\pm1} = \pm16.5°$, $\theta_{\pm2} = \pm34.7°$, $\theta_{\pm3} = \pm58.7°$, $\theta_{\pm4} =$ evanescent.

Off-Axis Grating Equation

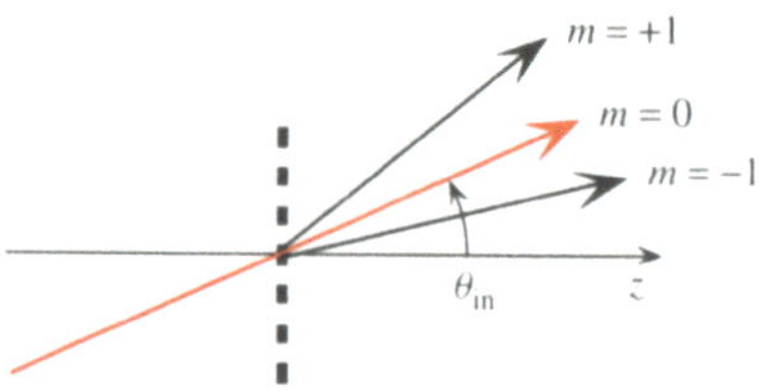

$$\sin\theta_{out,m} = m\frac{\lambda}{d} + \sin\theta_{in}$$

The Off-Axis Grating Equation

When the incident beam comes in at an angle to the perpendicular, the output fan of beam *roughly* follows it around, staying centered about the continuing beam—but upon closer look the angles between some of the beams increase and others decrease, sometimes significantly. Also, some beams may try to come out at angles beyond 90°, becoming evanescent in the process, and others may "emerge" from the grating on the other side of the beam fan. The details are described by the "off-axis grating equation" below, in which θ_{in} is the angle of the incident beam normal to the grating perpendicular, as shown in the diagram:

$$\sin\theta_{out,m} = m\frac{\lambda}{d} + \sin\theta_{in} \tag{5}$$

Or restated in spatial frequency terms,

$$\sin\theta_{out,m} = m\,\lambda\,f + \sin\theta_{in} \tag{6}$$

We will "prove" this relationship in combination with an interesting result in diffraction theory in the next section.

Diffraction by a Sinusoidal Grating

Our arguments for physical reasonableness have been built on a model of the grating as a series of narrow slits, with assurances that the angles depend only on the spatial frequency of the slits and not on their width or other properties. Now we will examine this premise for a special kind of slit, one that attenuates the wavefront according to a smoothly varying (specifically, sinusoidal) function. Such a pattern might be produced by exposing a piece of photo film to a two-beam interference pattern, for example.

Transmittance of a grating

We describe the sinusoidal transmittance pattern of the grating/film as

$$t_{amp}(x,y) = a + b\cos(2\pi fx) \tag{7}$$

where t_{amp} is the ratio of the amplitude of the electric field of the light wave just after and just before the grating. That is, at every point,

$$E_{out}(x,y) = t_{amp}(x,y)\cdot E_{in}(x,y) \tag{8}$$

Note that in photography we usually consider the transmittance of the *intensity* of the light wave (or its negative logarithm, the *opti-*

cal density), which would be the square of the amplitude transmittance we are considering here (a point to consider when attempting to measure transmittances directly). If we assume that the amplitude transmittance must always be between zero and unity (a non-amplifying medium, you might say), there are limitations on the size of a compared to b; for example:

$$\begin{aligned} 0 \le a + b \le 1 \\ 0 \le a - b \le 1 \end{aligned} \tag{9}$$

In a more elaborate view of amplitude transmittance, the phase of the output wavefront can also be manipulated, leading to the description of the amplitude transmittance by complex numbers. For example, if the wave is delayed by one-half of a cycle, the amplitude transmittance is effectively minus one. Further discussion will be deferred to the chapter concerning the diffraction efficiency of such gratings; the physical principles can be illustrated by positive-real amplitude transmittances.

Effect of illumination

The grating is illuminated by a plane wave inclined at an angle, θ_{in}, as shown in the sketch. The mathematical description of that wave is, as shown in Chapter 2, Eqs. (2) and (11), where now $z = 0$ at the observation plane:

$$E_{in}(x,y,t) = \sin\left(2\pi\nu t - \frac{2\pi}{\lambda}x\cdot\sin\theta_{in}\right) \tag{10}$$

The output wave is then simply given by

$$\begin{aligned} E_{out}(x,y,t) &= E_{in}(x,y,t)\cdot t_{amp}(x,y) \\ &= \sin\left(2\pi\nu t - \frac{2\pi}{\lambda}x\cdot\sin\theta_{in}\right)\cdot\left(a + b\cos(2\pi f x)\right) \end{aligned} \tag{11}$$

Application of a familiar trig identity,

$$\sin\alpha\cdot\cos\beta = \frac{1}{2}\left[\sin(\alpha+\beta) + \sin(\alpha-\beta)\right] \tag{12}$$

to the product provides the output wave as the sum of three components:

$$\begin{aligned} E_{out}(x,y,t) = a\cdot&\sin\left(2\pi\nu t - \frac{2\pi}{\lambda}x\cdot\sin\theta_{in}\right) \\ &+ \frac{b}{2}\sin\left[2\pi\nu t - \left(\frac{2\pi}{\lambda}x\cdot\sin\theta_{in} + 2\pi fx\right)\right] \\ &+ \frac{b}{2}\sin\left[2\pi\nu t - \left(\frac{2\pi}{\lambda}x\cdot\sin\theta_{in} - 2\pi fx\right)\right] \end{aligned} \tag{13}$$

We identify these (by phase-footprint inspection) as three plane waves, which we distinguish by order number m equal to zero, plus one, and minus one:

$$\begin{aligned} E_{\text{out}}(x,y,t) &= E_0 \sin\left(2\pi\nu t - \frac{2\pi}{\lambda} x \cdot \sin\theta_0\right) \\ &+ E_{+1} \sin\left[2\pi\nu t - \left(\frac{2\pi}{\lambda} x \cdot \sin\theta_{+1}\right)\right] \\ &+ E_{-1} \sin\left[2\pi\nu t - \left(\frac{2\pi}{\lambda} x \cdot \sin\theta_{-1}\right)\right] \end{aligned} \tag{14}$$

where the output angles are given by

$$\begin{aligned} \sin\theta_0 &= \sin\theta_{\text{in}} \\ \theta_0 &= \theta_{\text{in}} \\ \sin\theta_{+1} &= \lambda f + \sin\theta_{\text{in}} \\ \sin\theta_{-1} &= -\lambda f + \sin\theta_{\text{in}} \end{aligned} \tag{15}$$

This motivates/justifies the generalization of the *grating equation* to the off-axis case:

$$\sin\theta_{\text{out},m} = m\lambda f + \sin\theta_{\text{in}} \tag{16}$$

However, a more far-reaching observation is that the amplitudes of the two first orders of diffracted waves are $b^2/4$, and those of all the higher order waves are *zero*! That is, a purely sinusoidal transmittance grating diffracts light *only* into the first orders (and the zero or "straight-through" order). More complex gratings (such as slit-like gratings) will diffract light into many orders, but these gratings can be thought of as combinations of mathematically-simpler sinusoidal gratings, each having a frequency that is a multiple of the first-order frequency, and each giving rise to a single pair of diffracted orders. This is equivalent to analyzing the complex transmittance pattern as a Fourier series, which has led to a whole new field of optical theories based on communication theory.

Conclusions

What happens when light hits a picket fence structure (that is, a "grating") is truly amazing, and raises all kinds of doubts about our real-world physical intuition. Suppose we let the light trickle through one photon at a time? Which direction does an individual photon take? Such questions are the meat of quantum physics, and we are using a classical optics approach, so shelve them for now and try to think about the way light behaves as described by the equations we have developed.

What is it exactly that's happening? The beam breaks up into a number of distinct beams that go in very distinct and well-defined directions given by the grating equation. That equation involves the spatial frequency of the grating (in cycles per millimeter, analogous to the cycles/second of radio and TV signals) and the trigonometric sines of all the angles involved. Luckily, the sine is nearly a linear function of the angle, for small enough angles, and we can get a fairly simple general idea before relegating the calculations to a computer program.

The amount of energy in the various beams is another interesting story that will occupy an entire chapter of its own. Suffice to say that the simplest type of grating to analyze is a sinusoidal variation of transmittance between zero and unity, and it sends all its light into the plus and minus first orders, plus the straight-through zero order (and light absorbed in the grating). None of the possible higher orders are stimulated! Enter Fourier...

Soon it will be time to combine the stories of interference and diffraction to learn about "holography," after a digression about *how much* light goes into these various beams.

References

i. Willebrod Snell van Roijen (1591–1626) (of Leyden, Holland) succeeded in giving an exact form to the law of refraction, as did René Descartes (1596-1630) shortly afterward.

ii. Christiaan Huyghens or Huygens (1629–1695), Dutch mathematician, physicist, and astronomer who discovered Saturn's rings (1655), pioneered the use of the pendulum in clocks (1657), and formulated Huyghens' principle (ca. 1678).

iii. Jean Pierre Augstin Fresnel (1788–1827), French physicist and government civil engineer. First memoir on diffraction submitted on October 15, 1815 (at age 27).

iv. Thomas Young (1773–1829), English physicist, physician, and Egyptologist. He was professor of natural philosophy (1801-1803) at the Royal Institution of Great Britain, where he presented the modern physical concept of energy, and was elected (1811) a staff member of St. George's Hospital, London. In 1807 he stated a theory of color vision known as the Young-Helmholtz theory (the 3-primary-color theory) and described the vision defect called *astigmatism*. Young conducted experiments in diffraction and interference (1801) that could only be explained by the wave theory of light, finally overturning Newton's corpuscular theory. He also established a coefficient of elasticity (Young's modulus) and helped to decipher the Rosetta Stone. He was hounded out of physics by a hostile journal editor, and spent most of his life as a medical doctor.

v. Ronchi, V. (1970). *The Nature of Light,* Harvard Univ. Press, Cambridge, MA.

vi. Joseph von Fraunhofer (1787–1826), German physicist.

vii. An acquaintance of ours has a patent on a system for doing that: T. DeWitt, *Range Finding by Diffraction,* US Patent 4,678,324.

viii. Two different approaches can be found in the well-illustrated Hecht, E. and A. Zajac (1997). *Optics,* Addison-Wesley, Reading, MA, and the math-heavy classic Born, M. and E. Wolf (1980). *Principles of Optics,* Pergamon Press, Oxford, UK.

CHAPTER 6

Diffraction Efficiency of Gratings

Introduction

For most of our discussions in this book, we will be worrying about *where* the light goes. But every once in a while, we have to worry about *how much* light gets there too. After all, no hologram means much if it is so dim that no one can see it! This brings us to the subject of *diffraction efficiency*, what determines it, and how it might be maximized.

We begin by considering the diffraction efficiency of a few simple gratings, for the cases of absorbing and phase-retarding gratings (unbleached and bleached holograms), gradually developing some general rules. This chapter can offer only a very preliminary pass at understanding the theoretical underpinnings of the brightness and contrast of holographic images, but it will be a good start!

Thin gratings

At the outset, we have to add that all of this chapter's remarks will be limited to "thin" gratings and holograms. That is to say that the thickness of the diffracting structures is small, compared to the grating spacing, *d*, and that angular selectivity effects (the Bragg angle effects)[i] are not significant. This situation is often described by the "thickness parameter," Q (which is also a function of the incident illumination angle, θ_0, and the wavelength, λ) being much less than unity (strong modulations of absorption or refractive index can also increase the apparent thickness of the grating):

$$Q = \frac{2\pi \cdot \lambda \cdot \text{thickness}}{n \cdot d^2 \cdot \cos\theta_0} << 1 \qquad (1)$$

$$Q = \frac{2\pi \cdot \lambda \cdot \text{thickness}}{n \cdot d^2 \cdot \cos\theta_0} << 1$$

The Thickness Parameter

If the hologram thickens, the diffraction efficiency generally increases when it is properly angled, but the theories describing these conditions become very complicated. We will visit this special domain when we talk about reflection holograms, later in the book.

Definition of Diffraction Efficiency

Usually, we mean by diffraction efficiency (designated as DE_{+1}, or η_{+1} "eta-plus-one") the ratio of the intensities of the desired (generally the plus-first order) diffracted beam and the illuminating beam, measured when both beams are large enough to overfill the area of the detector being used.[ii] We ignore any power losses due to reflections at the surfaces of the grating or hologram; practical *DE* measurements also take these surface reflections (approximately 4% per surface, for uncoated glass) into account. This chapter's simple mathematical models overlook polarization effects, large-diffraction-angle effects, and many other subtleties of rigorous electromagnetic theory, but their conclusions can generally be extended into those domains by more detailed math, so these results serve as useful guides nonetheless.

The local intensity of a beam of light is proportional to the time-average of the square of the magnitude of its electric field. Thus the ratio of the intensity of the output and input beams is equal to the square of the magnitude of the electric field transmittance of the hologram. Because the output field consists of several beams that eventually separate, we are interested in accounting for them one by one. That means breaking the transmittance pattern down into components that correspond to each of the beams. Finding the amplitude of those transmittance components is the principal concern of the rest of this chapter.

Transmission Patterns

We describe a grating or hologram by its two-dimensional *transmittance* pattern. By transmittance we mean the ratio of the electrical wave fields just after and just before the grating at the same (x, y) location:

$$t_{\text{amp}}(x,y) = \frac{E_{\text{out}}(x,y)}{E_{\text{in}}(x,y)} \tag{2}$$

In the simplest case, the wave is simply attenuated, so that its electric field amplitude diminishes. This is called an *amplitude* transmittance grating.

Note that this is *not* the transmittance that one might usually think of measuring with a photographic light meter, because a light meter responds to the *intensity* of a beam, which is proportional to the *square* of the wave amplitude.

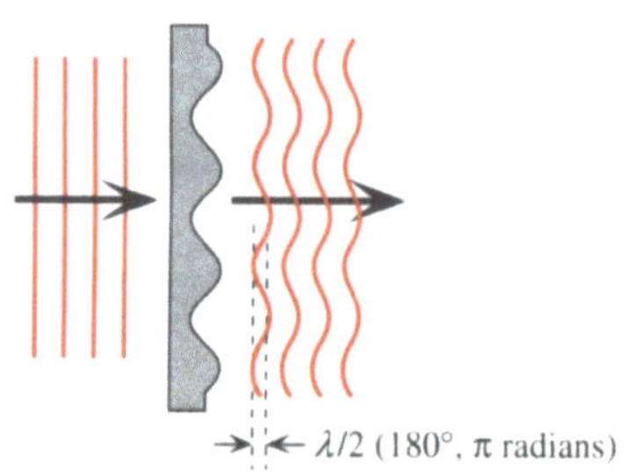

In so-called coherent optical systems (where the illumination is monochromatic and from a point-like source, generally from a laser), the *delays* that a wave encounters in passing through a grating are also very important. For example, if they are great enough to retard the wave by half a cycle in some places, then those waves would cancel out waves from other parts! Retarding effects are described as variations of *phase* transmittance, with the amount of delay being measured in degrees or radians, or sometimes in terms of wavelengths (or fractions thereof). Phase delays can be caused by variations in either the local microscopic thickness of a grating (with the phase delay increasing with thickness) or the local refractive index, n, or both.

Usually, amplitude and phase transmittance variations, or modulations, are linked together in practical cases, but it is useful to first think of them as separate cases, with unbleached holograms being amplitude-only gratings, and bleached holograms being phase-only gratings.

For this discussion, the modulation patterns will vary with x only; that is, the patterns will be horizontal, with the "furrows" extending out of the page. The attenuation or phase delay variations of the pattern will be described as a graph and/or as an analytical expression, such as shown here. A few simple cases can help us find guidelines that will predict the behavior of a wider variety of holograms. However, true speckley-object-beam holograms (such as of 3-D objects) have a random transmittance pattern that requires a more complex analysis. The patterns we are talking about now are for "gratings" that have no randomness—images of single points, in effect.

Sinusoidal transmittance grating

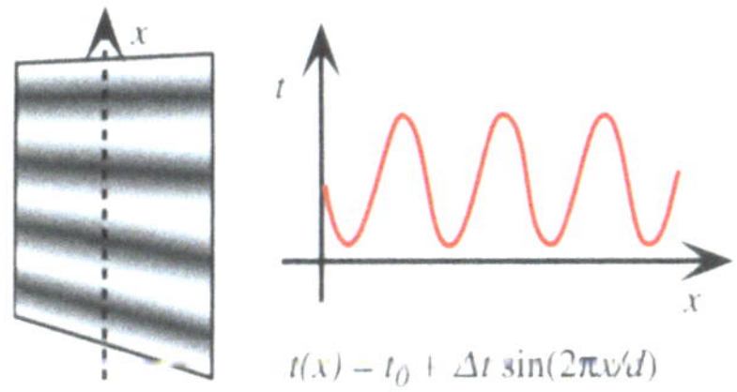

$t(x) = t_0 + \Delta t \sin(2\pi x/d)$

Here, the amplitude-only transmittance is a perfectly-smoothly-varying sinusoidal function of position, an ideal simplest case. Such a pattern could be produced by a low-contrast interference pattern exposure, for example. The most striking geometrical property of diffraction by such a grating is that there are only two output beams, the $m = +1$ and $m = -1$ orders, on either side of the straight-through $m = 0$ beam. In the far-field pattern, we see only one spot of light on each side of the zero-order beam for each sinusoidal component of the grating's amplitude transmittance; it acts as a kind of Fourier transformer![iii] This property can readily be proven by matching the amplitudes and phases of sets of waves on both sides of the grating (boundary condition matching).

The intensity of each of the $m = \pm 1$ beams varies as the *square* of the amount of swing of the sinusoidal modulation:

$$DE_{m=0} = t_0^2$$
$$DE_{m=\pm 1} = \left(\frac{\Delta t}{2}\right)^2 = 6.25\%_{\max}$$
$$DE_{|m|>1} = 0 \tag{3}$$

Because the transmittance has to stay between zero and one, the maximum value of Δt is 0.5 (only possible if $t_0 = 0.5$ also), and the maximum value of the diffraction efficiency is then 6.25%. Four times that amount emerges in the straight-through beam, and the rest is absorbed in the grating, gently warming it. This low maximum *DE* is not very encouraging for the brightness of display holograms! Unbleached holograms can be bright enough to be impressive under controlled lighting conditions, but it is usually quite difficult to consistently produce the maximum possible diffraction efficiency.

Square-wave transmittance grating

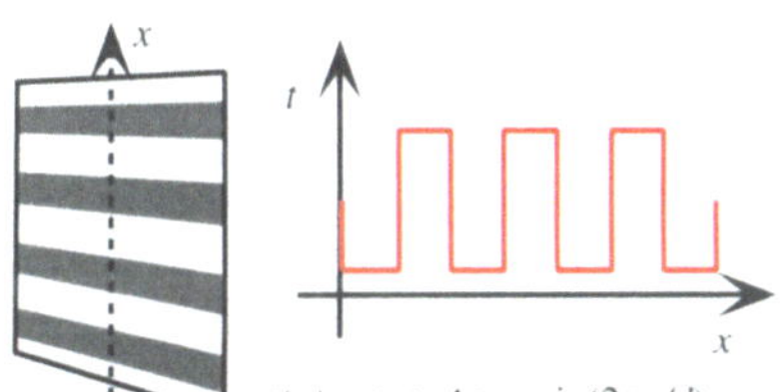

$t(x) = t_0 + \Delta t$ sq-sin$(2\pi x/d)$

Often, there are non-linearities in the exposure response of photographic materials that distort the purely sinusoidal nature of a transmittance pattern, in much the way that "fuzz boxes" can distort electrical guitar sounds. An extreme is a "hard-clipped" sine wave, which we will refer to here as a square wave or "squared-up sine wave" (*i.e.*, it is "high" 50% of the time, and "low" the other 50%), sometimes denoted as "sq-sinθ" (unpronounceable).

Such a grating can be considered as a summation of many ideal sinusoidal gratings, one with the same period as the square wave, and then gratings with integer fractions of that period (or multiples of that spatial frequency—the "higher harmonics," one might say). Each sinusoid diffracts two beams of light, so that many points of light now appear in a straight line alongside the straight-through beam. But in spite of the energy going into the extra beams, the first-order beams are brighter than before! This is because the "fundamental sinusoidal component" of a square wave has a magnitude that is larger than the magnitude of the square wave itself by a factor of

$4/\pi$. So, we get transmittance values greater than unity and less than zero for that particular grating component, a physical paradox.

The application of Fourier theory produces these predictions of the diffraction efficiency: more light in the first order image, by 62% (when $\Delta t = 0.5$) giving over ten percent diffraction efficiency, plus some higher orders. Note that there are no *even* orders though; this depends on the grating being exactly 50/50 open/closed. Where does the extra total energy come from? Only one-half of the grating is dark, in the highest DE (black/clear) case, and therefore only 50% of the total energy gets absorbed, versus 62.5% in the sinusoidal case:

$$\begin{aligned} DE_{m=0} &= t_0^2 \\ DE_{m=\pm 1} &= \left(\frac{2\Delta t}{\pi}\right)^2 = 10.1\%_{\max} \\ DE_{m=\text{even}} &= 0 \\ DE_{m=\text{odd}} &= \frac{1}{m^2} DE_{+1} \\ \sum_{m\neq 0} DE_m &= \Delta t^2 = 24\%_{\max} \end{aligned} \tag{4}$$

So, it looks as though non-linearities can work to our advantage! Unfortunately, in more complex images non-linearities produce noise images that strongly degrade the desired image.

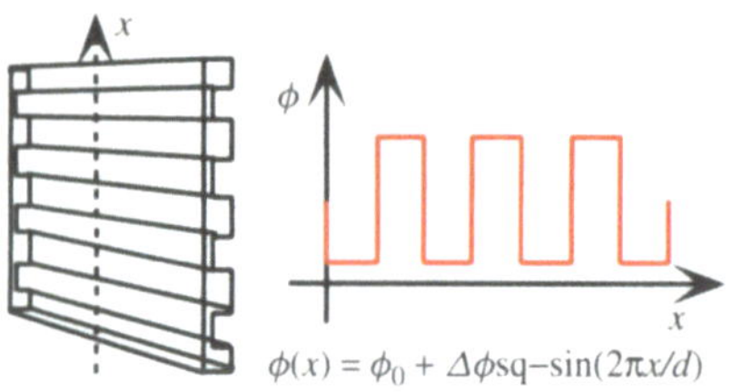

Square-wave phase grating

One reason for dwelling on the square-wave grating is that it offers a good introduction to simple *phase-only* gratings. Such gratings work by retarding the wavefronts as a function of position, and the results are hard to analyze for most modulation shapes. But if the grating comprises only two phase levels, such as 0 and π, the results follow from the same analysis used for square-wave amplitude-only gratings.

Phase-only gratings absorb no light energy, so that the total amount of diffracted light can reach 100% when summed over all the orders:

$$\begin{aligned} DE_{m=0} &= \cos^2 \Delta\phi \\ DE_{m=\pm 1} &= \left(\frac{2}{\pi}\sin\Delta\phi\right)^2 = 40.5\%_{\max} \\ DE_{m=\text{even}} &= 0 \\ DE_{m=\text{odd}} &= \frac{1}{m^2} DE_{+1} \\ \sum_{m\neq 0} DE_m &= \sin^2 \Delta\phi = 100\%_{\max} \end{aligned} \tag{5}$$

The modulation possible for the fundamental transmittance component becomes twice what it was in the amplitude-only transmission case, ranging from +1 to −1 in effect (when $\Delta\phi = \pi/2$), so that the maximum diffraction efficiency can quadruple to over forty percent!

Sine-wave phase grating

Now we will come almost full circle, from sinusoid to square and back: the depth of the phase-retarding structure varies smoothly, exactly as a sinusoid (the refractive index might vary sinusoidally instead, which is more common in bleached holograms). This turns out to be one of the few cases where the diffraction efficiency can be calculated analytically without much trouble, even though the link between phase and complex transmittance becomes highly nonlinear for only moderate modulations. For small phase modulations, the results should resemble those for sinusoidal amplitude gratings, although the phases of the first-order diffracted waves are different by 90° from the unbleached case (which hardly ever matters).

The diffraction efficiencies are expressed in terms of zero- and first-order Bessel functions of the first kind, which are a lot like cosine and sine functions except that they damp down for large $\Delta\phi$, are not strictly periodic, and the maxima of J_1 do not lie at the minima of J_0. Nevertheless, the general behavior is as expected: as the modulation increases the zero-order beam weakens and the first-order beams strengthen to a maximum *DE* of 33.8% (when $\Delta\phi = 0.59\pi$):

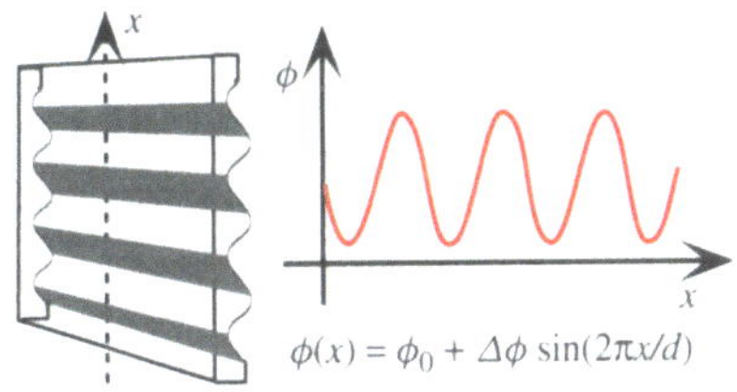

$$DE_{m=0} = J_0^2(\Delta\phi)$$

$$DE_{m=\pm 1} = J_1^2(\Delta\phi) = 33.8\%_{\max}$$

$$\sum_{m \neq 0} DE_m = 1 - J_0^2(\Delta\phi) = 100\%_{\max} \qquad (6)$$

Because a sinusoidal-phase grating is a distorted sinusoid in amplitude transmittance terms, higher order beams begin to appear too, each described by a higher-order Bessel function, $J_m^2(\Delta\phi)$. For a more detailed look at this question, see for example Collier, Burckhardt and Lin (1971).[iv]

Generalized gratings

If the transmittance variation is neither smoothly sinusoidal nor stepwise constant, the diffraction efficiency can be difficult to compute even within the limited accuracy of this simple thin-hologram approach. However, there are a *few* things we can say in general that help tie the just-previous results together, and extend them in interesting ways toward image holograms. These ideas are simple to comprehend for amplitude gratings, a little harder for phase-only gratings, and the general mixed case requires a lot of dabbling in the unit circle of complex variable mathematics.

The fraction of the optical power transmitted at each point of the grating is given by the magnitude-squared of the amplitude transmittance at that point, or $|t(x, y)|^2$. That power finds its way into the variously diffracted beams, so that the first number we can find is the sum over all the orders, including the zero order, of the diffraction efficiencies.

The amplitude of the zero-order beam by itself is given by the average of the transmittance over the entire hologram area, which we call $\bar{t}$ (this was t_0 in the previous amplitude-only examples; in general, it is a complex number, $t_0 e^{j\phi_0}$). The power in the zero-order beam is the magnitude-squared of that average, or t_0^2 in this case.

The difference between the total diffracted power and the power in the zero-order beam must be the total power in all the diffracted beams! If the transmittance is a constant over the hologram area, then the average of the square of the magnitude of the transmittance will be equal to the square of the magnitude of the average transmittance, and the diffracted power will be zero, as expected. If the transmittance fluctuates as a function of position, diffraction begins. The difference between the average of the magnitude-squared and the magnitude-squared of the average is termed the *variance* of the random fluctuations, or the square of their *standard deviation*. This is equal to the sum of the diffraction efficiencies in all the orders.

$$\begin{aligned} DE_{m=0} &= \left|t_{\text{avg}}\right|^2 \\ \sum_{m\neq 0} DE_m &= \overline{\left|t(x)\right|^2} - DE_0 = \overline{\left|\delta t(x)\right|^2} \\ &= \operatorname{var} t = \sigma_t^2 \end{aligned} \tag{7}$$

Telling how much power goes into any specific order, and into the $m = +1$ order in particular, is trickier. If the fluctuating transmittance pattern can be decomposed into various spatial frequency components, then the variance can be interpreted as a sum over a power spectrum, where each component of the power spectrum corresponds to the power in one diffracted order (an application of the Wiener[v]-Khinchine theorem of communication theory). That precise decomposition requires finding the Fourier transform of the transmittance fluctuations, which is beyond the scope of this discussion.

There are some interesting special cases, though. For example, can you think of a transmittance pattern that diffracts all of the light into one of the first-order beams?

Thick Gratings

All of the above discussion assumes that the modulation of the grating has been crammed into a layer that is infinitely thin. In real holograms, the emulsion is several wavelengths of light thick (silver-halide emulsion thickness of 5–7 μm is typical), and the modulation is spread over fringe surfaces that are wide enough to act something like mirrors—that is, they may diffract light more into the +1 order than the −1 order, or vice versa. This angular selectivity is usually called the "Bragg effect," and its analysis would take us far beyond the mathematical scope of this book.[vi,vii,viii] In general, the trend is to increase the diffraction efficiency of one beam at the expense of the others, and to make the hologram quite sensitive to its angle to the illuminating beam. Volume reflection holograms, described in later chapters, are at the opposite extreme, where the emulsion layer is considered as being extremely thick in the simplest case.

References

i. The Bragg angle is the angle (for each wavelength of light) at which selection effects due to the thickness of a hologram maximize its diffraction efficiency.

Named after the Braggs, father and son: Bragg, Sir William Henry (1862–1942) and Sir William Lawrence Bragg (1890–1971), who shared a 1915 Nobel Prize for the analysis of x-ray spectra and the structure of crystals.

ii. Another ratio that can be calculated is beam power diffraction efficiency: the ratio of the powers in the desired (usually plus-first order) diffracted beam and the illuminating beam, measured when both beams are small enough to fit into the area of the detector being used (undiverged laser beams are generally used). However, the total power in the beam is equal to the product of its intensity and its cross-sectional area.

iii. The Fourier transform, to oversimplify a bit, recognizes that an arbitrary signal can be made up of varying amplitudes and phases of sinusoidal components, and is a way of figuring out how much of each frequency is in the signal. Analogously here, each sinusoid in the grating pattern makes its presence known by making a spot of diffracted light.

iv. Collier, R. J., C. B. Burkhardt, and L. H. Lin (1971). *Optical Holography,* Academic Press, San Diego, Section 8.5.

v. Norbert Wiener (1894-1964), known for founding the theory of *cybernetics* and for his many contributions to the development of computers, Wiener also did research in probability theory and the foundations of mathematics. He was one of the few child prodigies whose later lives fulfilled their early promise.

vi. Kogelnik, H., (1969). "Coupled Wave Theory for Thick Hologram Gratings," *Bell System Technical Journal,* **48**, pp. 2909-2947.

vii. Collier, R. J., *et al., op. cit.,* Chapter 9.

viii. Hariharan, P. (1996). *Optical Holography: Principles, Techniques, and Applications,* Cambridge University Press, Cambridge, UK, Chapter 4.

CHAPTER 7

"Platonic" Holography

Introduction

With simple concepts of interference and diffraction, we are ready to "prove" the validity of holography in a fairly simple and interesting way, based only on mathematics. The generality of the proof may come in handy later on, but the lack of practicality of the argument deprives it of much practical value in solving problems. If Gabor had lived in a cave, as Plato proposed to do, his proof might have looked something like this (actually, it does anyway!). But luckily for us he also spent plenty of time in the laboratory, and showed us how to produce pictures that would convince the doubters that this revolutionary approach to imaging could actually work.

Object Beam

We represent the optical wave scattered by a generalized diffusely-reflecting object as a wave having an amplitude and phase that are random variables of x and y, and call it the *object beam*.[i] The object beam is usually incident roughly perpendicularly to the recording plate. We will assume that the diffuse object reflection preserves the polarization of the beam (*e.g.*, that the object is aluminum spray-painted). The object wave has wavelength λ_1 (the recording wavelength), and the corresponding temporal frequency, ν_1; the expression for the object beam is

$$E_{\text{obj}}(x,y,t) = a_{\text{obj}}(x,y)\sqrt{\frac{2}{\varepsilon_0 c}}\,\sin\!\left(2\pi\nu_1 t - \phi_{\text{obj}}(x,y)\right) \qquad (1)$$

The average of the square of the amplitude, a_{obj}, is unity, so that the average *intensity* of the object beam is unity (which is why we included the $\varepsilon_0 c$ term in the expression; see Eq. (14) of Chapter 2). Ordinarily, a_{obj} will have an exponential probability distribution function, and the variation of its autocorrelation function with distance will be closely related to the distribution of intensity in the object as measured by the angle it subtends at the plate, which determines the size of the "speckles" or intensity non-uniformities in the object beam. The object beam's phase is also a random variable, uniformly distributed over $[0, 2\pi]$, so that it has a meaningless average. Note that although a_{obj} and ϕ_{obj} are random functions of x and y, they do not change with time—that is, the exposure system is stable during the exposing time.

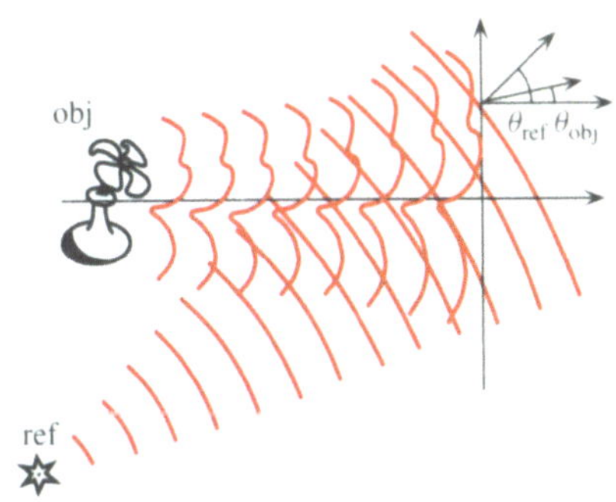

Reference Beam

By contrast, the reference beam is constant in intensity over the plate, but can have any phase variation with x and y. For simplicity, we will assume that it is a plane wave incident at an angle θ_{ref} (perhaps 30°). The reference beam intensity has to be greater than that of the object beam by some factor, K, which we call the "beam ratio."

This typically varies from 5 to 50. We can express this reference wave in the form

$$E_{\text{ref}}(x,y,z,t) = \sqrt{K}\sqrt{\frac{2}{\varepsilon_0 c}}\ \sin\left(2\pi \nu_1 t - \frac{2\pi}{\lambda_1} x \sin\theta_{\text{ref}}\right) \tag{2}$$

Interference Pattern

Where the object and reference beams overlap, an interference pattern is formed between them. This can be considered as a simple two-beam interference pattern, although now the amplitude and phase of one of the beams is gradually varying with x and y. Thus, continuing from Eq. (7) of Chapter 4, we find the total intensity pattern to be given by

$$I_{\text{total}}(x,y) = K + a_{\text{obj}}^2(x,y) + 2\sqrt{K}\,a_{\text{obj}}(x,y)\cos\left(\phi_{\text{obj}}(x,y) - \frac{2\pi}{\lambda_1} x \sin\theta_{\text{ref}}\right) \tag{3}$$

Here, the first two terms are the intensities that would be found if the reference or object beams were turned on separately from each other. The third term is the *holographic* term, the fringe pattern that arises from interference between the two beams. It is the recording of this pattern that provides the necessary information for the reconstruction of an accurate three-dimensional image.

Holographic Recording Material

The link between the exposure pattern and the reconstructed image is the recording material and its processing. The exposure is a positive real variable, a scalar, as no known material responds to anything but the "heat" of the exposure, the integration of its local intensity or irradiance over the exposure time. Ordinarily, the intensity is a constant over the duration of the exposure, which is gated by a shutter somewhere in front of the laser or by turning a semiconductor laser on and off. The effect of the exposure is to bring about some chemical or physical change in the material, which produces a change in the optical properties of the material (usually after some further steps called "processing"). The properties we are concerned with most are the amplitude and phase transmittances of the material, as discussed in Chapter 6.

Silver halide materials

Later chapters of this book will touch upon some other materials that can be used to record holograms, but the most commonly-used medium is silver-halide photographic film/plates, a suspension on a plastic or glass substrate of very small (approximately 35 nm diameter) silver bromide (though for non-holographic applications they may also be silver chloride or iodide) micro-crystals in gelatin, plus sensitizers and other odds and ends. The absorption of photons creates tiny clusters of silver atoms on the grain surfaces. At this stage the emulsion contains a "latent" image, which is turned into an actual image by conversion (chemical reduction) of the entire microcrystal into a "grain" of metallic silver during "development" (which is essentially an amplification step)—the exposed grain then becomes "black" or light absorbing. Since an entire crystal will be exposed if only a small part of is hit by photons, it follows that bigger crystals (with more area to catch photons) make for a more light-sensitive emulsion. Emulsions suitable for holography have very small crystals in order to be able to resolve fine fringe patterns and thus aren't very sensitive.

Processing

Processing of exposed silver-halide media for holography should be familiar to those who have done black-and-white film processing. A *developer* solution turns the (colorless) latent image into the (opaque) developed image. Because the developer is a mild base, its action is stopped after an appropriate time by moving the plate to a mildly acidic *stop bath*. *Fixer* then removes unexposed silver halide, leaving only silver. The plates are then washed in water (or in several baths that gradually increase the proportion of methanol, which will evaporate from the emulsion faster than water and permit viewing the plate more quickly). More information on the chemistry of processing for holography can be found in various references such as Saxby (2004).[ii]

As we saw in Chapter 6, phase gratings can have very high diffraction efficiencies (which makes a nice, bright hologram), so if we want to make a phase hologram instead of an amplitude hologram we can perform a further processing step called *bleaching*. This works because silver halide has a different index of refraction from gelatin, so we can get a transparent phase grating by one of several methods: washing away the silver from an unfixed plate and leaving the silver halide behind (a *reversal* bleach, potassium dichromate), converting the silver on an unfixed plate to silver halide and redepositing it where there is already unexposed silver halide (a *physical transfer* bleach, ferric sodium EDTA), or turning the silver on a fixed plate back to silver halide (a *rehalogenating* bleach, the nasty and dangerous bromine water). Removing material from the emulsion can cause the emulsion to shrink, which can change the color of a reflection hologram to be "greener" or "bluer." If we don't want the emulsion to shrink, we can use a physical transfer bleach which rearranges the silver halide rather than washing it away or we can do another trick: the substance triethanolamine (often abbreviated TEA) applied to the emulsion before exposure will swell it so that when it shrinks during processing the result is the desired thickness and thus color. As a nice side effect, TEA also increases the sensitivity of the plate by a factor of 3 or more.

Response

We will talk here mostly about the *amplitude transmittance*, the ratio of the electric field amplitudes just after and just before the film layer, denoted as t_{amp}.[iii] It is important that the resolution of the material be high enough to allow the film to "follow" very fine-scale local variations of exposure. For very low exposures the transmittance is nearly unity, and as the exposure increases the transmittance drops monotonically to less than 0.1.

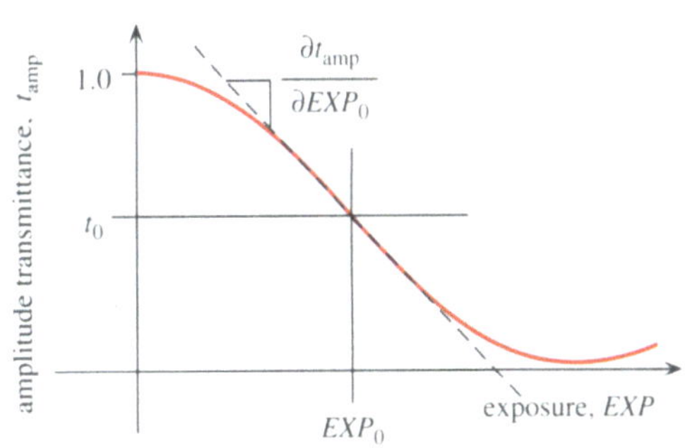

The response of a recording material can be expressed graphically as a relationship between the amplitude transmittance, t_{amp}, and the exposure, *EXP*, which is the product of the exposing intensity and the exposure time. That relationship is generally non-linear, and perhaps not even monotonic, as sketched in the margin. However, over some limited range of exposures the transmittance varies nearly linearly with exposure, and can be approximated by a straight line; this is the range of exposures in which photographers try to place their pictures, and where we will try to make holograms. Thus a *linearized* model of a recording material expresses this mathematically as

$$t_{\text{amp}}(x,y) - t_0 = \left.\frac{\partial t_{\text{amp}}}{\partial EXP}\right|_{E=E_0} \left(EXP(x,y) - EXP_0\right) \tag{4}$$

where EXP_0 is the so-called "bias" exposure around which the response is reasonably linear, and t_0 is the transmittance produced by a uniform exposure at that bias level.

The holographic recording material will be exposed to this pattern for some period of time, T_0, so as to bring the spatially-averaged exposure to the required level, EXP_0. The exposure at any point is given by $EXP(x, y) = I_{\text{total}}(x, y) \cdot T_0$, so the needed exposure time, T_0, is given by

$$T_0 = \frac{EXP_0}{\left\langle I_{\text{total}}(x,y)\right\rangle_{\substack{\text{spatial}\\ \text{average}}}} = \frac{EXP_0}{K+1} \tag{5}$$

The amplitude transmittance as a function of intensity then becomes

$$\begin{aligned} t_{\text{amp}}(x,y) &= \frac{1}{K+1}\left(EXP_0 \left.\frac{\partial t_{\text{amp}}}{\partial E}\right|_{EXP=EXP_0}\right) I(x,y) \\ &\quad + \left(t_0 - \left.\frac{\partial t_{\text{amp}}}{\partial E}\right|_{EXP=EXP_0} EXP_0 \right) \\ &= \frac{1}{K+1}\left(\left.\frac{\partial t_{\text{amp}}}{\partial \ln EXP}\right|_{EXP=EXP_0}\right) I(x,y) + \left(t_0 - \left.\frac{\partial t_{\text{amp}}}{\partial \ln EXP}\right|_{EXP=EXP_0}\right) \end{aligned} \tag{6}$$

Holographic Transmittance Pattern

Inserting the expression for the holographic intensity gives us the resulting amplitude transmittance pattern. The relevant characteristic of the recording material, the slope of the curve of its amplitude transmittance versus the natural logarithm of its exposure, is usually referred to as the "beta" of the material, β. It is sometimes multiplied by the "modulation transfer function" or MTF of the material at the resolution scale of the hologram (clearly the contrast of the image is going to decline when the pattern approaches the size of the grain of the recording material; the MTF describes the percentage response to a sine-wave intensity exposure as a function of spatial frequency). Thus, substituting Eq. (3) into Eq. (6), we obtain

$$\begin{aligned} t_{\text{amp}}(x,y) &= \frac{1}{K+1}\beta\, I(x,y) + (t_0 - \beta) \\ &= \beta\frac{K}{K+1} + (t_0 - \beta) \\ &\quad + \beta\frac{1}{K+1}\, a_{\text{obj}}^2(x,y) \\ &\quad + 2\beta\frac{\sqrt{K}}{K+1}\, a_{\text{obj}}(x,y)\cos\left(\phi_{\text{obj}}(x,y) - \frac{2\pi}{\lambda_1}x\sin\theta_{\text{ref}}\right) \end{aligned} \tag{7}$$

This is the "hologram!" Within its transmittance pattern is embedded a precise description of the object beam, along with several other terms, awaiting only illumination by a suitable beam to release its information.

Illuminating Beam

The illumination beam, like the reference beam, may be any uniform-intensity beam (with an arbitrary phase distribution), but we will limit our discussion to a unit-amplitude plane wave inclined at angle θ_{ill}. It has wavelength λ_2, the reconstruction wavelength.

$$E_{\text{ill}}(x,y,t) = \sin\left(2\pi\nu_2 t - \frac{2\pi}{\lambda_2}x\sin\theta_{\text{ill}}\right) \tag{8}$$

The diffracted output from the hologram is then given by the product of the hologram amplitude transmittance and the illumination amplitude,

$$\begin{aligned} E_{\text{out}}(x,y,t) &= t_{\text{amp}}(x,y)\cdot E_{\text{ill}}(x,y,t) \\ &= \left(\beta\frac{K}{K+1} + (t_0 - \beta)\right)\sin\left(2\pi\nu_2 t - \frac{2\pi}{\lambda_2}x\sin\theta_{\text{ill}}\right) \\ &\quad + \beta\frac{1}{K+1}\, a_{\text{obj}}^2(x,y)\sin\left(2\pi\nu_2 t - \frac{2\pi}{\lambda_2}x\sin\theta_{\text{ill}}\right) \\ &\quad + 2\beta\frac{\sqrt{K}}{K+1}\, a_{\text{obj}}(x,y)\cos\left(\phi_{\text{obj}}(x,y) - \frac{2\pi}{\lambda_1}x\sin\theta_{\text{ref}}\right) \\ &\quad \times\sin\left(2\pi\nu_2 t - \frac{2\pi}{\lambda_2}x\sin\theta_{ill}\right) \end{aligned} \tag{9}$$

It is the *last* of these terms that is of special interest to us, and to explore it we need to apply the same trigonometric identity used previously, $\sin\alpha\cdot\cos\beta = (1/2)\left[\sin(\alpha+\beta) + \sin(\alpha-\beta)\right]$:

$$\begin{aligned} E_{\text{last}}(x,y,t) &= 2\beta\frac{\sqrt{K}}{K+1}\,a_{\text{obj}}(x,y)\cos\left(\phi_{\text{obj}}(x,y)-\frac{2\pi}{\lambda_1}x\sin\theta_{\text{ref}}\right) \\ &\quad\times\sin\left(2\pi\nu_2 t-\frac{2\pi}{\lambda_2}x\sin\theta_{\text{ill}}\right) \\ &= \beta\frac{\sqrt{K}}{K+1}\,a_{\text{obj}}(x,y)\sin\begin{pmatrix}2\pi\nu_2 t+\phi_{\text{obj}}(x,y) \\ -\frac{2\pi}{\lambda_1}x\sin\theta_{\text{ref}}-\frac{2\pi}{\lambda_2}x\sin\theta_{\text{ill}}\end{pmatrix} \\ &\quad+\beta\frac{\sqrt{K}}{K+1}\,a_{\text{obj}}(x,y)\sin\begin{pmatrix}2\pi\nu_2 t-\phi_{\text{obj}}(x,y) \\ +\frac{2\pi}{\lambda_1}x\sin\theta_{\text{ref}}-\frac{2\pi}{\lambda_2}x\sin\theta_{\text{ill}}\end{pmatrix} \end{aligned} \tag{10}$$

We will represent these components as a sum over a variable, m, the "order number," so that

$$E_{\text{out}}(x,y,t)=\sum_{m=-\infty}^{+\infty}E_m(x,y,t) \tag{11}$$

where

$$\begin{aligned} E_0(x,y,t) &= \left(\beta\frac{K}{K+1}+(t_0-\beta)\right)\sin\left(2\pi\nu_2 t-\frac{2\pi}{\lambda_2}x\sin\theta_{\text{ill}}\right) \\ &\quad+\beta\frac{1}{K+1}\,a^2_{\text{obj}}(x,y)\sin\left(2\pi\nu_2 t-\frac{2\pi}{\lambda_2}x\sin\theta_{\text{ill}}\right) \\ E_{+1}(x,y,t) &= \beta\frac{\sqrt{K}}{K+1}\,a_{\text{obj}}(x,y)\sin\begin{pmatrix}2\pi\nu_2 t-\phi_{\text{obj}}(x,y) \\ +\frac{2\pi}{\lambda_1}x\sin\theta_{\text{ref}}-\frac{2\pi}{\lambda_2}x\sin\theta_{\text{ill}}\end{pmatrix} \\ E_{-1}(x,y,t) &= \beta\frac{\sqrt{K}}{K+1}\,a_{\text{obj}}(x,y)\sin\begin{pmatrix}2\pi\nu_2 t+\phi_{\text{obj}}(x,y) \\ -\frac{2\pi}{\lambda_1}x\sin\theta_{\text{ref}}-\frac{2\pi}{\lambda_2}x\sin\theta_{\text{ill}}\end{pmatrix} \end{aligned} \tag{12}$$

In general, there will be several higher-order components. It is only our assumption of linearity of the response of the recording material that has limited us to finding only the 0, +1, and –1 terms here. Also, either or both of the first orders may not actually exist, as they may turn out to be *evanescent* upon further analysis.

A Proof of Holography

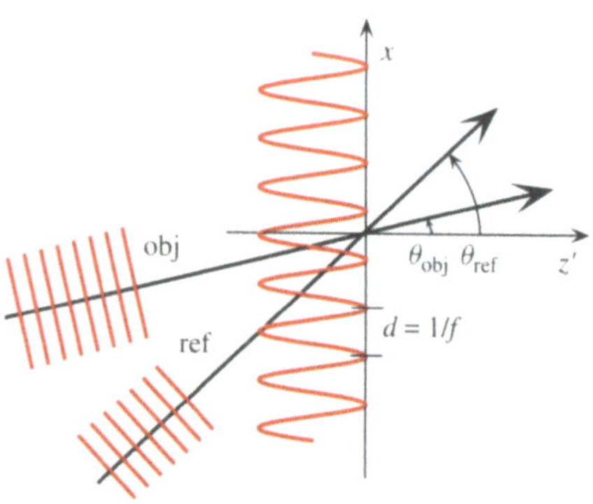

It is the $m = +1$ diffracted wave that is the potential reconstruction of the object wave. If the angle and wavelength of the illumination beam are made equal to the angle and wavelength of the reference beam, the last two terms in the parentheses cancel out, leaving only the amplitude and phase terms identical to those of the object wave. These are the conditions that we refer to as "perfect reconstruction." That is,

$$\begin{aligned} &\text{if}: \lambda_2=\lambda_1 \text{ and } \theta_{\text{ill}}=\theta_{\text{ref}}, \text{then} \\ &\mathrm{E}_{+1}(x,y,t)=(\text{constant})\,a_{\text{obj}}(x,y)\sin\left(2\pi\nu_1 t-\phi_{\text{obj}}(x,y)\right) \end{aligned} \tag{13}$$

This represents a general statement of the central property of holography, that it can reproduce an exact replica of the amplitude and phase of the object wave under very general circumstances. The constant term reflects the diffraction efficiency of the hologram, or the brightness of the image it produces. If the object wavefront was produced by a three-dimensional scene, the reconstructed wavefront will be focused by the eyes to produce a three-dimensional perception of that scene. There is no "illusion" involved, the eyes are not being tricked—they are enjoying the same information that the scene itself would have provided, were it still there.

Note that part of the originally-inclined illumination wave has been deflected to travel along the z-axis, in the direction of the object beam's light. This change of direction is caused by diffraction by an overall grating pattern caused by interference between the object and reference beams, and is sometimes referred to as a "spatial carrier wave" to make an analogy to the radio carrier wave used in AM and FM modulation. Its spatial frequency is determined mainly by the angle of the reference beam; that is, $f_{\text{carrier}} \approx \sin\theta_{\text{ref}} / \lambda_1$. A reference beam angle of 30° thus creates a grating of 790 cy/mm, or a grating spacing d of 1.27 μm (using a 633 nm He-Ne laser). This tiny spacing presents most of the practical challenges of high-quality holography.

There do seem to be some physical paradoxes involved, of course. A purely two-dimensional recording is reconstructing information about a three-dimensional volume, for example! But this is a consequence of Huyghens' principle (or the ellipticity of the wave equation, if you prefer) that a specification of the amplitude and phase boundary conditions specifies the wave throughout the enclosed volume. And, there is the paradox of reconstructing the amplitude and phase of a quantity from a purely scalar (intensity) recording. This is resolved by noting that we are also reconstructing some other terms that can be regarded as the "extra baggage" required to resolve this paradox.

Other Reconstructed Components

The most interesting of the "extra baggage" terms is the $m = -1$ component, which is termed the "conjugate" or "twin" image. Note that under "perfect reconstruction" its terms are

$$E_{-1}(x,y,t) =$$

$$(\text{constant})\, a_{\text{obj}}(x,y) \sin\left(2\pi \nu_1 t - \left(-\phi_{\text{obj}}(x,y)\right) - \frac{4\pi}{\lambda_1} x \sin\theta_{\text{ref}}\right) \quad (14)$$

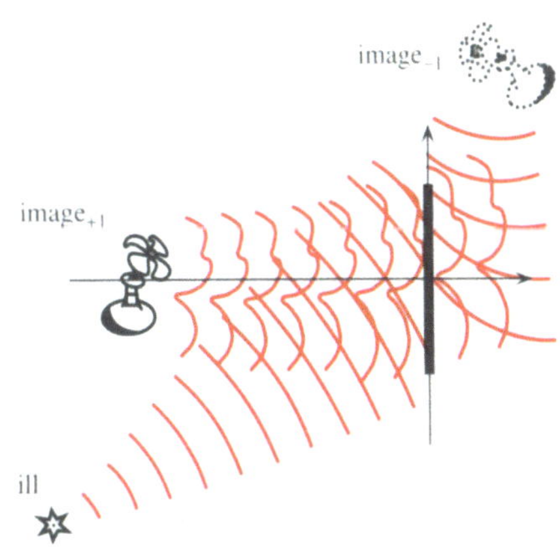

Which is to say that although the amplitude is the same as for the object beam, the phase has the *opposite* sign. That is, a diverging object-beam wavefront will produce a converging wavefront in its conjugate image, focusing toward a point on the viewer's sideof the hologram. This focus represents a *real* image, focused in space and visible on a white card if it is held in the right place. In the early history of Gabor-style in-line holography, this real image caused considerable corruption of the desired, true, or *virtual* image, the one corresponding to the $m = +1$ term. The introduction of off-axis reference and illumination beams by Leith and Upatnieks caused the output angle of the conjugate wave to be significantly different from that of the desired wave. If the object wave were an on-axis plane wave, $\phi(x, y) = 0$, the output angle of that term would be

$$\theta_{\text{out},-1} = \sin^{-1}\left(2\sin\theta_{\text{ref}}\right) \tag{15}$$

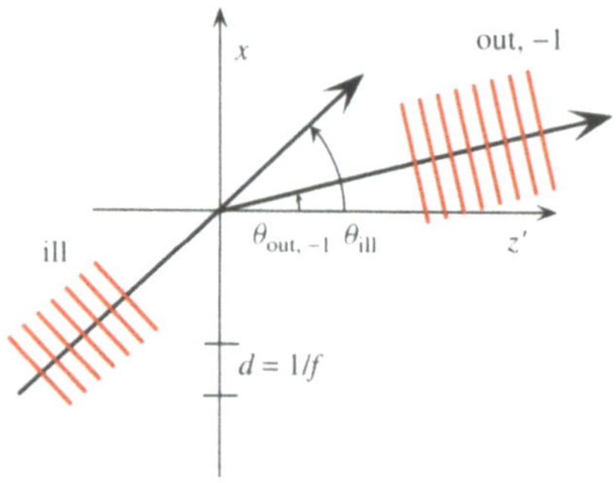

Note that for reference beam angles of 30° and above, this term is evanescent, and doesn't propagate at all.

The other "extra baggage" is the zero-order component, which has two terms. The first is simply an undiffracted, attenuated version of the illumination beam, headed in the same direction that the illumination was headed before the hologram was placed in the beam. Any energy left in this beam is not available for the desired reconstruction beam, so some effort usually goes into minimizing the zero-order beam to make bright holographic images.

The other zero-order term is more subtle, and deserves a description of its own. We usually call it the "halo" component. This beam is diffracted by the $a_{\text{obj}}^2(x,y)$ term, which is the same transmittance term that would be produced by exposing the hologram to the object alone, without the reference beam. That speckled exposure pattern contains grating patterns caused by interference between all possible pairs of points in the object, and the finest pattern (highest spatial frequency) will be produced by those object points that are the farthest apart. Let's say that these points subtend an angle ω as seen from the hologram plane. Assuming that ω is fairly small, that grating will have a spatial frequency of $\rho = \sin\omega/\lambda$. Including that grating in the hologram means that this modest spatial frequency will diffract the illumination beam over modest angles, roughly equal to ω on either side of the central direction of the beam. Even if the reference beam angle is large enough to allow the illumination beam to clear the desired image beam without overlapping it, the "halo" terms can scatter image-degrading light into that beam. Thus we will have to pay some attention to this component!

The analogies between diffraction by a hologram and diffraction by a simple grating should be becoming clearer to you. Interference of the light from the object with the reference beam creates a grating that is a generalized diffraction grating—that is, it has some variation or modulation of the contrast and location of its fringes (corresponding to amplitude and phase modulation of radio waves). When an illuminating plane wave is scattered by such a grating, it breaks up into the three components we normally see from simple gratings, except that each now has some trace of the object information impressed upon it. The $m = +1$ and $m = -1$ waves correspond to the same orders we observe with diffraction gratings, and most of our analysis will build on these similarities. The third component includes the zero-order and halo terms.

Arbitrary Wavefronts

This analysis can readily be extended to include reference and illumination beams of any wavefront shape—we only require that their amplitudes be reasonably uniform across the area of the hologram (if they are not, then the amplitude of the output wave will be modulated by the product of the two amplitude variations, which will generally degrade its image). The phase of the various reconstructed components can then be shown to be given by

$$\phi_{\text{out},m}(x,y) = m\left(\phi_{\text{obj}}(x,y) - \phi_{\text{ref}}(x,y)\right) + \phi_{\text{ill}}(x,y) \qquad (16)$$

$\phi_{\text{out},m}(x,y) =$
$m\left(\phi_{\text{obj}}(x,y) - \phi_{\text{ref}}(x,y)\right) + \phi_{\text{ill}}(x,y)$
General formulation of the principle of holography

Thus, whenever the wavefront of the illumination is identical to that of the reference beam, the phase-footprint of the object beam will be reconstructed. If the wavelength of the reconstruction is the same as that of the recording, then the physical properties of the image corresponding to that phase-footprint will be the same as those of the recorded object. This is perhaps the most general formulation of the holographic principle, one that we will use occasionally for fairly high-level proofs; some people have even called it the "Heisenberg's Equation of Holography."

Diffraction Efficiency

Although we won't worry about just how bright our holograms are (or ought to be) for a while, we can already come to some conclusions about the diffraction efficiency of the Platonic holograms we have just described. Note that the ratio of the intensity of the $m = +1$ output beam to the intensity of the illumination beam is given by the ratio of their averaged-squared amplitudes. We define this ratio to be the *diffraction efficiency*, and note that for large K:

$$DE_{+1} = \eta_{+1} = \left(\beta\frac{\sqrt{K}}{K+1}\right)^2 \left(\left\langle a_{\text{obj}}^2(x,y)\right\rangle_{\substack{\text{spatial}\\\text{average}}} = 1\right) \qquad (17)$$

$$\Rightarrow \frac{1}{K}\beta^2 \quad \text{for large K}$$

Thus the fraction of the illumination energy that finds its way into the desired image beam decreases as the beam ratio increases, and depends critically on the slope of the t_{amp}-ln(EXP) curve, which we have dubbed the β of the material, and which is similar to its "contrast."

Reconstruction ratio

Another way of thinking about the diffraction efficiency for diffuse objects, and a handy way of gauging it in practice, is to illuminate a processed hologram with the reference beam that originally exposed it (or a replica of it); that is, a beam that has a uniform intensity of K. The diffracted intensity is then divided by the intensity of the original object beam (unity in our case) to yield the ratio of the *luminance* of the image (roughly its brightness) to the luminance of the object, which we call the "reconstruction ratio," denoted by RR. Substituting into the above equation (17) gives

$$RR = \beta^2 \qquad (18)$$

All of the absolute and relative beam intensities cancel out and we can aspire, with good reason, to make holographic images that are actually brighter than the objects that created them! It is only a matter of properly chemically processing the material to give a $|\beta| > 1$, and making sure that the holographic setup is tied down tightly enough so that the recorded fringes are as contrasty as they are supposed to be.

Conclusions

A generalized analysis can be very satisfying, and reassuring that we haven't just stumbled across some special case or circumstance. But idealized analyses are often useless for solving practical problems. For instance, Eq. (12) tells us nothing about what happens to the $m = +1$ or "true" image if the illumination beam is misaligned a little, or the wavelength isn't quite right, or its radius of curvature is not correct. Those answers are implicit in that equation, of course, but we need a more directly physically-based approach to build up the sense of physical reasonability that will allow us to understand our experimental results, and to predict the likely outcome of proposed new experiments. Thus, we will abandon this domain of modest theoretical luxury, and descend into the dark and greasy pit of slippery approximations and hasty assumptions, with these more precise results safe in our pocket lest we should lose our way.

References

i. There is some debate as to whether these should be called the *subject* and the *subject beam* instead; we will adopt the more common Leith and Upatnieks convention of *object* and *object beam*.

ii. Saxby, G. (2004). *Practical Holography,* Institute of Physics Publishing, Bristol UK.

iii. Most applications of photographic film concentrate instead on the *intensity transmittance* of the layer, the ratio of the irradiances just after and just in front of the film, or the "photographic density," which is the negative base-10 logarithm of the intensity transmittance (typically varying between zero and three).

CHAPTER 8

Ray-Tracing Analysis of Holography

Introduction

Rather than tackling a generalized and global proof of the wavefront reconstruction properties of holograms, we can instead look at the recording and reconstruction of wavefronts at every small area of a hologram, using just the simple ideas of two-beam interference and diffraction by periodic structures, with the coordinate system continually re-centered on the small region of momentary interest. In this approach we say either that the object beam is locally planar—because its radius of curvature is so much larger than the diameter of the region of interest—or that the object beam can be considered as the sum of a number of plane waves from point sources far from the hologram plane, and we consider them one at a time. Either way, we use a single plane wave as a localized "stand-in" for a more complex 3-D image-bearing object wave. Likewise, we will consider the reference wave to be locally planar; it is usually a long-radius spherical wave, so this is a very good approximation.

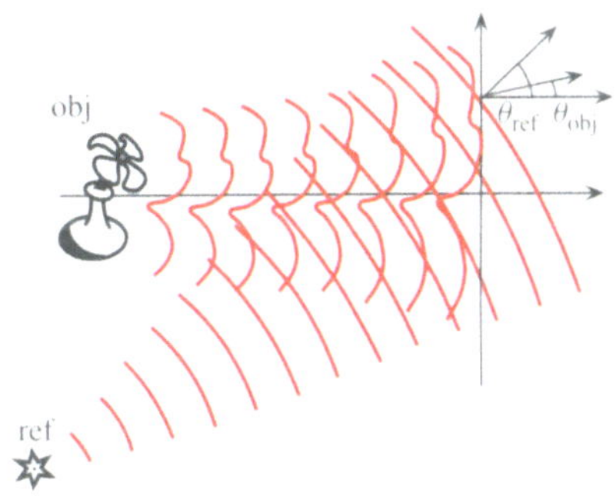

In our sketches, we will indicate the diverging spherical wave by a cluster of arrows perpendicular to the surface of the wavefront. If we examine each arrow carefully, we might find that it has wavefronts within it that are too small to see. Thus these arrows are not the **k**-vectors, or the "rays," or even the "ray bundles" you sometimes see mentioned in optics texts, although they resemble them all and are parallel to them ("rays," for example, are the line trajectories of imaginary light particles). The arrows are instead "mini-wave-beams" of a new sort, which let us draw accurate wave-optical pictures in familiar ray-optical ways. What they leave out is that the wavefronts within the arrows have a particular and fixed phase relationship; this doesn't usually matter for imaging calculations. Whatever we ought to call them, though, we will probably be careless and call them "rays" anyway, and it might even be useful to think of them as an extended variety of a generalized ray. Enough semantics!

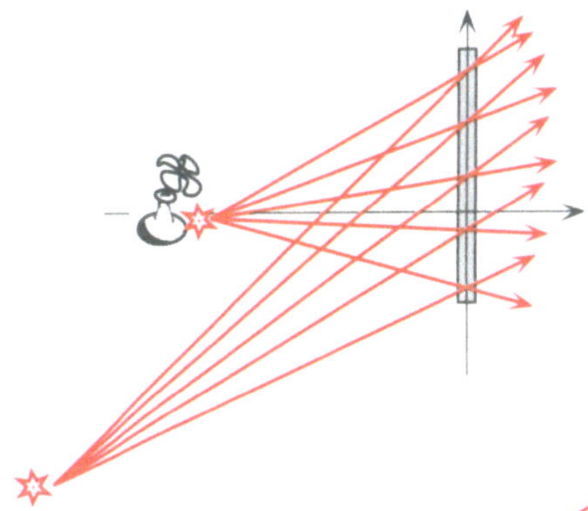

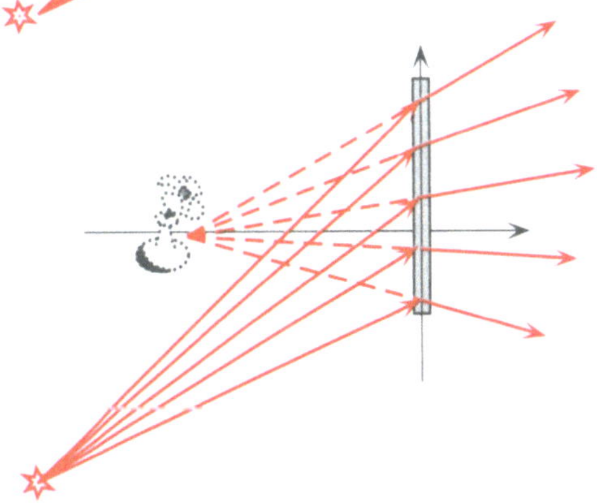

Suffice to say that we consider each point of the source to be emitting a diverging fan of rays, as does the reference point source. Where the object and reference rays cross, we compute the spatial frequency of the interference fringe pattern within their (not insignificant) width according to the grating equation, as they are both plane waves (locally). That pattern exposes the hologram plate, which is then processed to produce a modulating structure (or grating) with the same spatial frequency. A ray from the point source of illumination then strikes that grating, and becomes diffracted into several orders. The rays from any one order, such as the $m = +1$ order, can be traced back from several different locations on the plate, and their intersection will define the apparent location of a single source that produces them all, the "virtual" image of the point. If all goes well, holographically speaking, that location will be at the location in 3-D space of the original object point. We can do the same for all points on the object (arguing by linearity that the hologram can hold all the little gratings without their affecting one another), and trace out the virtual image in 3-D space point by point. There are some subtleties here though: how does the eye know where, along the possible line of locations for the part of the ray that it receives, the intersection is? A mystery of visual perception perhaps, but at least consistent with simple triangulation. Note that there is much more to spatial perception than triangulation, however!.

Mathematical Ray-Tracing

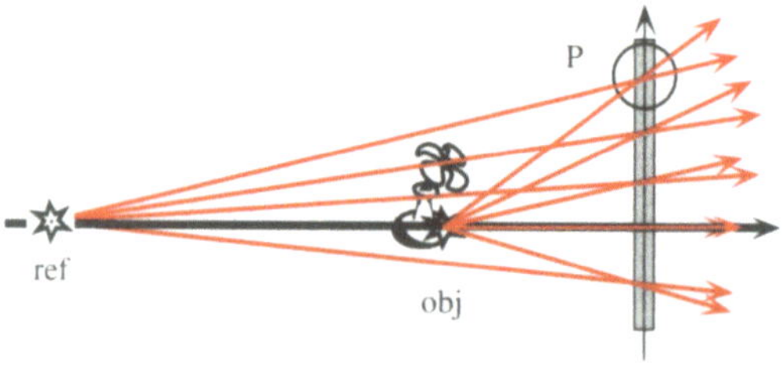

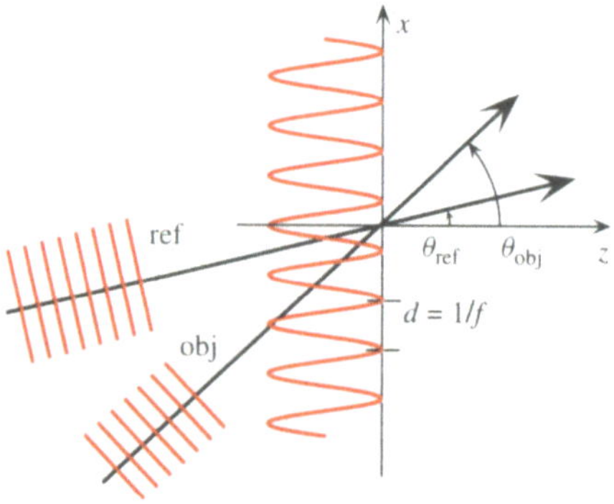

The problem has now been reduced to keeping track of the fates of a few plane waves during interference and diffraction, something we are now well set up to handle. Consider this sketch of an "in-line" or "Gabor" hologram (this is the kind of setup that Dennis Gabor was experimenting with in 1947 when he invented holography), the first type we will analyze in detail. We examine the beams or rays crossing a point, P, above the z-axis, where we construct a local coordinate system with axes x' and z'. An object point is on the z-axis at some distance, and the reference beam source is farther away, still on the z-axis. We examine the area around P, some distance up the x-axis, where the beams take on their local angle values, θ_{obj} and θ_{ref}. Where they overlap, a pattern of spatial frequency f is generated, where

$$f = \frac{\sin\theta_{obj} - \sin\theta_{ref}}{\lambda_1} \tag{1}$$

This becomes the spatial frequency of the grating created at P by exposure and processing of the holographic plate (that is, the plate doesn't expand or contract).

The plate is then illuminated by a point source at some other distance, producing the local illumination angle θ_{ill}. The output angles are therefore given by

$$\sin\theta_{out,m} = m\,\lambda_2\, f + \sin\theta_{ill} \tag{2}$$

Combining the relevant equations yields

$$\sin\theta_{out,m} = m\frac{\lambda_2}{\lambda_1}\left(\sin\theta_{obj} - \sin\theta_{ref}\right) + \sin\theta_{ill} \tag{3}$$

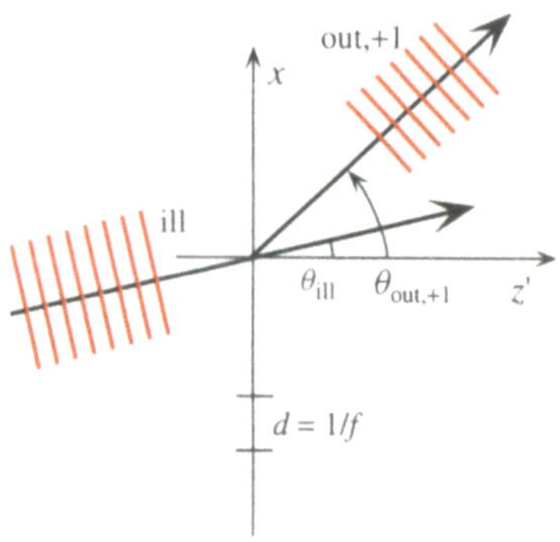

This is our general ray-tracing equation, applied at every $(x, y, 0)$ location on the hologram plate.

Now, considering the $m = +1$ term, it is clear that if $\lambda_2 = \lambda_1$ and $\theta_{ill} = \theta_{ref}$ (as it would be if the illumination source location were the same as the reference source location), then $\theta_{out,+1} = \theta_{obj}$. If this is true at every (x, y) point on the hologram surface, then the angle of the wavefront will have been reproduced everywhere on the hologram surface, and so will the wavefront itself (give or take an overall constant). If the reconstruction conditions are changed from "perfect," then the output angle follows the general relationship,

$$\sin\theta_{out,+1} = \frac{\lambda_2}{\lambda_1}\left(\sin\theta_{obj} - \sin\theta_{ref}\right) + \sin\theta_{ill} \tag{4}$$

and the location of the image has to be determined by more careful numerical triangulation (as we do in the next section). In general, the rays of any order might not all diverge from (or converge to) an exact or single common point; in such a case the focus is said to be "aberrated" and its location is not well defined. For now, we will assume that the point *is* well defined, and all wavefronts will be spherical.

Equation (3) represents the ray-tracing equivalent of a general statement of holography. Indeed, it can be considered as a reduced version of the general phase equation, namely the relationship between the first-order x-derivatives of the wavefront phases. Unfortunately, it is limited to points located on the $y = 0$ plane (the x-z plane), and thus is not a fully three-dimensional statement. We will discuss fully-three-dimensional ray-tracing near the end of this chapter, and find that it takes us well beyond "shop math!" Fortunately, most of the relationships that we care about in this course are limited to the x-z plane, or very close to it, and we can prove them using the simpler 2-D form, which is Eq. (3). As a last resort, we can appeal to elaborations on the phase equation, which at least involve only scalar variables.

Numerical Example

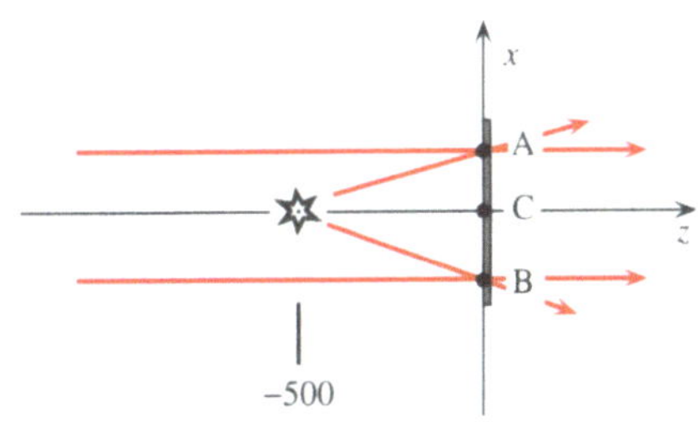

As an example, let's consider a specific case of recording a single point in the "in-line" or Gabor hologram geometry: the object is a point on the x-axis at $z_{obj} = -500$ mm—that is, at location $(x, y, z) = (0, 0, -500)$, as sketched in the margin. The reference beam is a point source an infinite distance away, so that it produces a plane wave at the hologram surface, and all its rays are horizontal, parallel to the z-axis. We can assume that the intensities of the two beams are equal at the plate (although the reference beam would typically be 5 to 30 times stronger when recording an actual image hologram).

Getting even more concrete, let's try to trace at least two rays through the hologram in order to do some image location by triangulation. If we do one ray, we will get another "for free" in this case, and there is a third that is also available nearly "for free," so we will soon have even more rays than we really need. Let's consider first the ray-tracing location at point A, which is 50 mm above the z-axis. There, the object beam's angle is

$$\theta_{obj} = \tan^{-1}\left(\frac{50}{500}\right) = 5.71° \qquad (6)$$

and the reference beam's angle is 0°. Assuming that we are using a He-Ne laser, the spatial frequency of the grating formed at point A is therefore

$$f_A = \frac{\sin\theta_{obj} - \sin\theta_{ref}}{\lambda_1} = \frac{0.1 - 0}{633 \times 10^{-6}} = 157 \text{ cy/mm.} \qquad (7)$$

By symmetry, the spatial frequency at point B, which is 10 mm below the z-axis, is also 157 cy/mm (note that there is a sign reversal, and that we are taking the *magnitude* of the result to be the spatial frequency).

The third "landmark" point of known spatial frequency is found by casting a line between the reference point source and the object point, and extending it to the hologram plane. In this example that location is at $(x, y, z) = (0, 0, 0)$, the point we are calling C. As seen from this location, the two waves are "in-line," and the angle between them is zero; thus the spatial frequency here is also zero! We will often refer to this as the "zero-frequency-point" or "ZFP" in our analyses. For reasons that will become clearer in a moment, it is also sometimes called the "hinge point" of the hologram.

Now, imagining that the hologram has been properly exposed and processed so that it has high contrast gratings everywhere, let's consider what happens when we illuminate the hologram again with the reference beam. Consider first the location A: the output angles are given by

$$\sin\theta_{\text{out},m} = m\,\lambda_2\,f + \sin\theta_{ill} = m\cdot 633\times10^{-6}\cdot\,157+0 \tag{8}$$

And the values work out to be, for m between –2 and +2, at A:

m:	–2	–1	0	+1	+2
θ_{out}:	–11.48°	–5.71°	0°	+5.71°	+11.48°

To calculate the angles at B, we have to be a little careful about which we call the $m = +1$ order. Although the spatial frequency is the same at B as at A, straightforward application of Eq. (1) would give a negative frequency. We have to apply the magnitude bars to get the same positive f. But in ray tracing we have to *avoid the magnitude bars*, and accept a negative frequency if we are to find the rays corresponding to the same m crossing at the same point. *Note:* This is not a problem for electrical engineers, who often deal with negative frequencies! The calculation at location B then produces the same results as at A, but with reversed signs (this follows from symmetry, without calculation in this case):

m:	–2	–1	0	+1	+2
θ_{out}:	+11.48°	+5.71°	0°	–5.71°	–11.48°

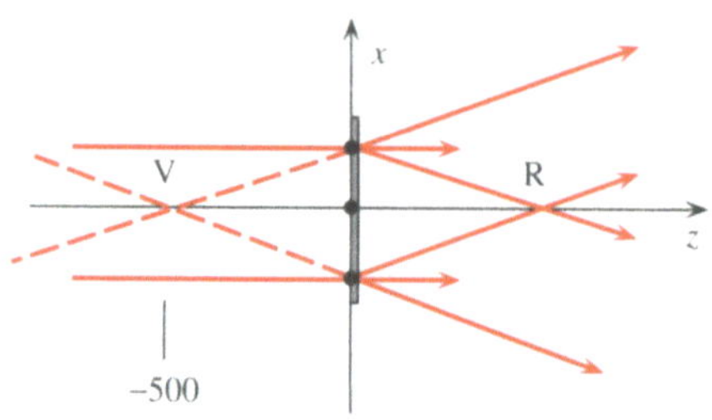

Point image locations—real and virtual

To find the locations of the focus points that represent the "images" produced by this hologram, let's look first at the easiest term, the $m = +1$ term. This produces rays propagating in the same direction as the original rays from the object at 5.71° away from the z-axis. Their "back-casted rays" cross each other, and the z-axis (recall that the ZFP ray will travel straight along the z-axis!) at $z = -500$ mm, so that a "virtual' image of a point will appear there, in the same location that the object point occupied. In this "perfect reconstruction" case, that location will be obtained no matter where we choose the ray-tracing locations A and B.

The $m = -1$ rays, on the other hand, are headed *toward* the z-axis at the same angle, 5.71°. Without much effort, we can predict that they will intersect at $z = +500$ mm, producing an on-axis focus in front of the hologram, a "real" image a distance in front of the hologram equal to the distance of the virtual image behind. It can be focused onto a card or ground glass as a bright point or (with much care) viewed directly with the eyes.

The angles of the two second-order rays are approximately twice as large as those of the two first-order rays, and produce virtual and real images that are roughly half the distance from the hologram.

These follow from the definition of the trigonometric tangent, where h is the height of the ray-tracing location (plus and minus 50 mm in this example),

$$\frac{h}{-z_m} = \tan\theta_m \tag{9}$$

so that from the A and B locations we get these z-values:

m:	−2	−1	0	+1	+2
z_m:	+246 mm	+500 mm	∞	−500 mm	−246 mm

Approximations
A full expression of the image distance is given by cascading these relationships to yield

$$\frac{1}{z_m} = \frac{1}{h}\tan\left(\sin^{-1}\left(m\sin\left(\tan^{-1}\frac{h}{z_{obj}}\right)\right)\right) \tag{10}$$

We note that an expansion of the trigonometric terms of this expression yields an interesting approximation for $1/z_m$:

$$\frac{1}{z_m} \approx \frac{m}{z_{obj}} + \frac{h^2}{z_{obj}^3}\left(\frac{m\left(m^2-1\right)}{2}\right) \tag{11}$$

If we let h decrease by a factor of 5, to 10 mm, then the second term decreases by a factor of 25, becoming negligible. We can describe this as a *paraxial* case, which involves only rays making small angles to the axis, so then $\tan\theta = \sin\theta = \theta$ (in radians), and which stay near enough to the axis so that h is small compared to the object and image distances. In this case the approximation reduces to

$$z_m = \frac{z_{obj}}{m} \tag{12}$$

However, in the general case we will have to do a more careful job of ray-tracing, yielding the deviations from the simple prediction that are shown in the table above.

Illumination wavelength effects
If the wavelength of the illumination light, λ_2, is changed, then the angles of diffraction will change, and so will the image locations. Assuming that the new wavelength is 550 nm, for example, we find by calculation that the angles at A and the corresponding image locations become:

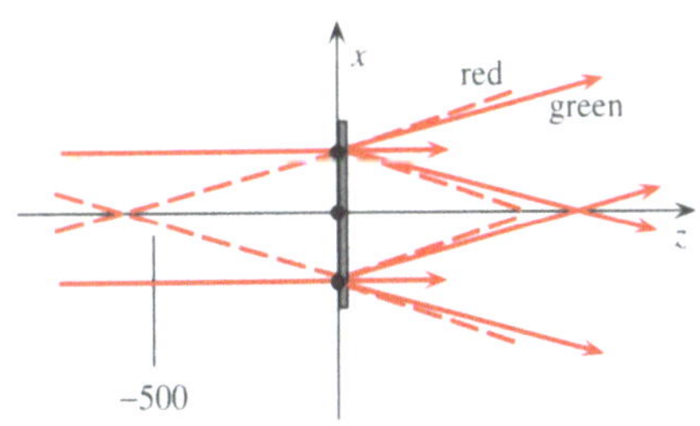

m:	−2	−1	0	+1	+2
θ_{out}:	−9.96°	−4.96°	0°	+4.96°	+9.96°
z_m:	+284 mm	+576 mm	∞	−576 mm	−284 mm

Note that, because the green light is deflected "less radically" than the red, the images are formed farther out on the positive and negative z-axis. The higher-order images are still formed closer in than the first-order images, but are also farther out than the red images were.

The distance-calculation equation is now given by

$$\frac{1}{z_m}=\frac{1}{h}\tan\left(\sin^{-1}\left(m\frac{\lambda_2}{\lambda_1}\sin\left(\tan^{-1}\frac{h}{z_{\text{obj}}}\right)\right)\right) \tag{13}$$

where λ_1 and λ_2 are the recording and reconstruction wavelengths, respectively. The expansion then becomes

$$\frac{1}{z_m}\approx\frac{m\dfrac{\lambda_2}{\lambda_1}}{z_{\text{obj}}}+\frac{h^2}{z_{\text{obj}}^3}\left(\frac{m\dfrac{\lambda_2}{\lambda_1}\left(\left(m\dfrac{\lambda_2}{\lambda_1}\right)^2-1\right)}{2}\right) \tag{14}$$

and the paraxial approximation form becomes

$$z_m=\frac{z_{\text{obj}}}{m}\frac{\lambda_1}{\lambda_2} \tag{15}$$

Source distance effects

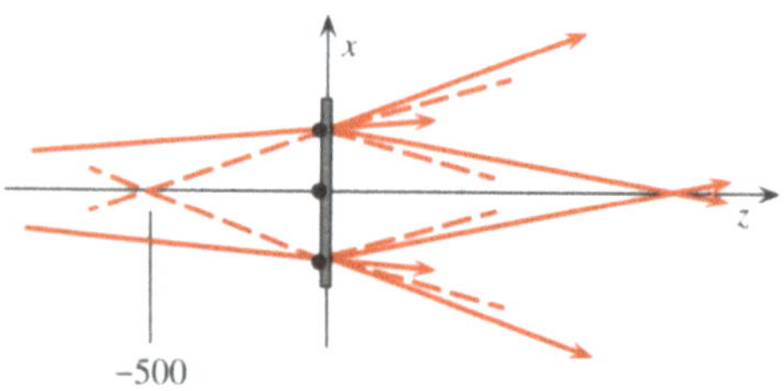

Now let's consider the effects of moving the illumination source closer to the hologram. Leaving the wavelength at 550 nm (green), let's put the illumination at five meters from the plate, at $(x, z) = (0, -5000)$. Now the illumination angle at A is 0.57°, which will rotate all of the diffracted beams by roughly that amount (greatly exaggerated in the sketch). As a consequence, the $m = +1$ rays will be traveling at a larger angle to the z-axis, and the resulting virtual image will move in toward the hologram, while the $m = -1$ rays will travel at a smaller angle to the z-axis, and cross it farther out from the hologram. Plugging values into the equations again, we find the relevant numbers to be:

m:	−2	−1	0	+1	+2
θ_{out}:	−9.38°	−4.38°	0.57°	+5.54°	+10.54°
z_m:	+303 mm	+652 mm	∞	−516 mm	−269 mm

Note that the first-order real image moves outward by almost 75 mm, and the first-order virtual image moves inward by only 60 mm.

The relevant cascaded mathematical expression is now

$$\frac{1}{z_m}=\frac{1}{h}\tan\left(\sin^{-1}\left(m\frac{\lambda_2}{\lambda_1}\left(\sin\left(\tan^{-1}\frac{h}{z_{\text{obj}}}\right)-\sin\left(\tan^{-1}\frac{h}{z_{\text{ref}}}\right)\right)+\sin\left(\tan^{-1}\left(\frac{h}{z_{\text{ill}}}\right)\right)\right)\right) \tag{16}$$

and the corresponding expansion becomes

$$\frac{1}{z_m} = m\frac{\lambda_2}{\lambda_1}\left(\frac{1}{z_{obj}} - \frac{1}{z_{ref}}\right) + \frac{1}{z_{ill}}$$
$$+ \frac{h^2}{2}\left(\left(m\frac{\lambda_2}{\lambda_1}\left(\frac{1}{z_{obj}} - \frac{1}{z_{ref}}\right) + \frac{1}{z_{ill}}\right)^3 - \left(m\frac{\lambda_2}{\lambda_1}\left(\frac{1}{z_{obj}^3} - \frac{1}{z_{ref}^3}\right) + \frac{1}{z_{ill}^3}\right)\right)$$
$$+ O[h^4] \tag{17}$$

of which the first of the three lines, which is the paraxial approximation, suffices for the first-order images and matches the calculations very well if they are made for a ray height of only 10 mm or so.

Aberrations of holograms—spherical aberration
We have been pursuing this example to demonstrate exact ray tracing, and to develop some approximations to its results. But a few secondary issues have also come to light. We note that if we calculate z_m for a variety of ray heights, h, that rays from the edges of the hologram do not always cross the axis at the same z_m as do rays from near the axis, the paraxial rays. The resulting degradation of the focus of the image is called an *aberration* of the wavefront, describing its departure from perfectly spherical behavior. This particular type of aberration is directly analogous to *spherical aberration* in simple glass lenses (a consequence of the spherical shape of their surfaces), and so has been given the same name in spite of there being no spheres involved.

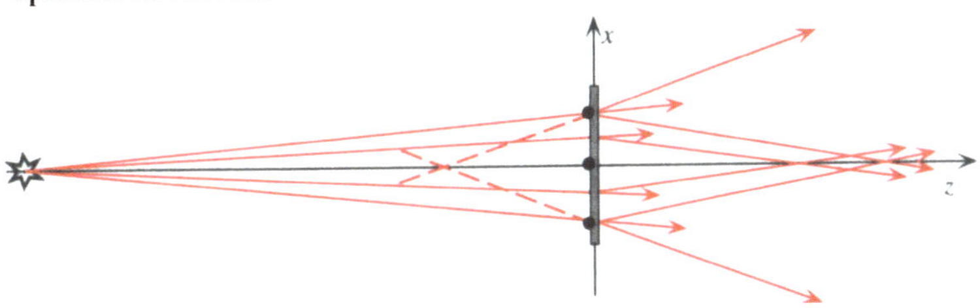

Aberrations of holograms—chromatic aberration
While we are at it, we might as well point out that if the hologram is illuminated with light of several different wavelengths, each wavelength will be focused to an image at a different distance, and the overall focus will be degraded. Glass lenses produce a similar effect, due to the prism-like nature of their edges, and the result is called *chromatic aberration.* It is a much stronger effect in holograms than in lenses, and not so readily correctable (as we shall see!).

Source angle effects
Consider the effect of moving the illuminating point source to one side of the z-axis, say upward by 50 mm. Now, the rays illuminating points A, B, and C all change angle by about the same amount, and with the same sign. The fan of diffracted output rays also rotates by roughly that angle at each location, and their intersections necessarily rotate also, so that the images are moved away from the z-axis.

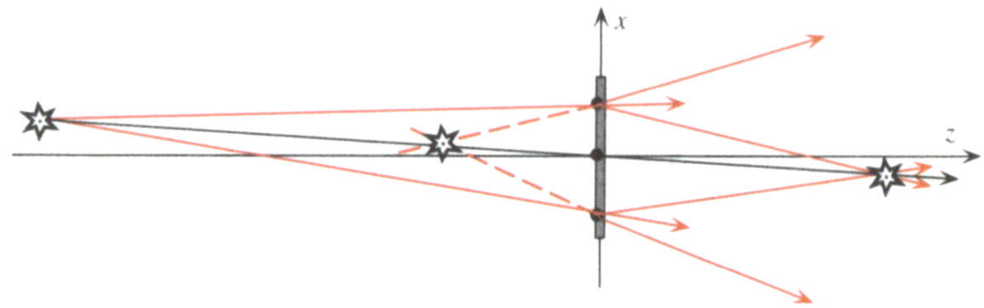

Source size effects

We can now imagine that if the illumination point moved continuously from the z-axis to a point 50 mm above, that the various images would move continuously from their original locations to the locations described in the above section. And, if an array of *several* sources were placed on a line connecting those two locations (each incoherent with all the others), then an array of several images would appear on a line between the two extreme images. Thus we can already begin to see how a spatially incoherent source can produce blur in an image.

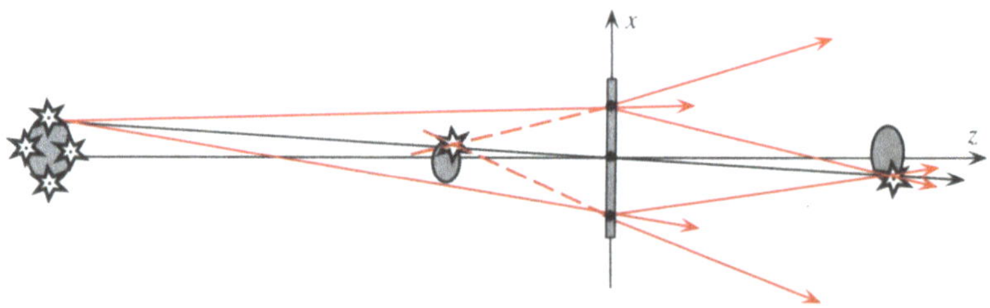

Comparison of Paraxial Hologram and Lens Optics

The expansion of the expression for the numerically ray-traced image location, Eq. (17), has as its first line of terms the *paraxial approximation* of holographic ray-tracing. This relationship is

$$\frac{1}{z_m} = m\frac{\lambda_2}{\lambda_1}\left(\frac{1}{z_{\text{obj}}} - \frac{1}{z_{\text{ref}}}\right) + \frac{1}{z_{\text{ill}}} \tag{18}$$

In the cases we have been describing, all of the z_{abc} are negative quantities, representing locations to the left of the x-axis, except for the z_m for negative m in most cases. The point of this section is that this formula is identical in form to the equation that describes focusing by a refracting or glass lens if the focal length is suitably described. The analogy between the elements of a hologram and conventional lenses is a very powerful one to those who are familiar with optical components. Assuming that this might not be the case here, we will demonstrate a few of the simpler principles along the way.

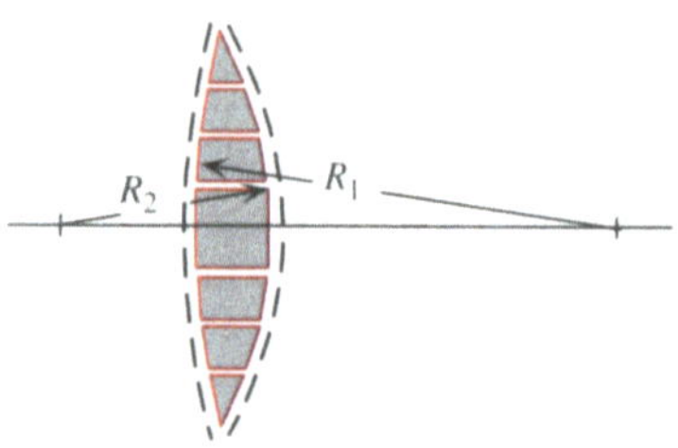

Definition of a glass lens

A normal glass lens (the same ideas will apply to plastic and liquid lenses) is defined by two spherical surfaces that cause the lens to be either thicker at the center than the edges (for a so-called *positive* lens), or thinner at the center (for a so-called *negative* lens). We denote the radii of curvature of the surfaces by R_1 and R_2, which are positive if they are convex to the right (or concave to the left). Thus, in the sketches here, R_1 is negative and R_2 is positive. Depending on which comes first, the lens might be positive or negative. The thickness at the edge or center doesn't affect the lens focusing in the thin-lens approximation, only the curvatures do, or rather the change in slopes of the surfaces as a function of height above the z-axis.

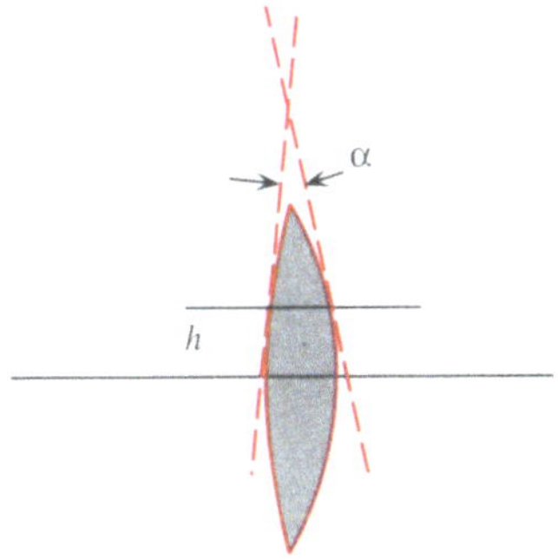

Such a lens can be approximated by a pile of prisms, as shown here. The central parallel-sided block of the prism doesn't deflect rays, only the refraction at the tilted surfaces at the edges does—which leads to a Fresnel-lens-like representation.

For a real lens, the surface slopes change continuously. The thickness at any height, $t(x, y)$, is given by (using the same approximations for spherical surfaces as for spherical wavefronts)

$$t(x,y) = t_0 + \frac{1}{2}\left(\frac{1}{R_1} - \frac{1}{R_2}\right)\left(x^2 + y^2\right) \tag{19}$$

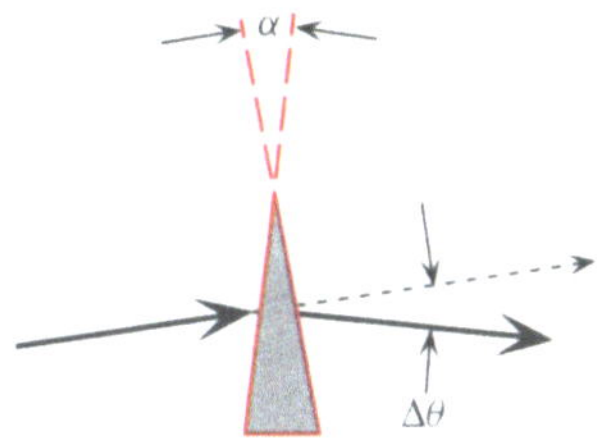

where t_0 is the center thickness. The downward-pointing angle, α, between the surfaces at height h (that is, at $(x, y) = (h, 0)$) is given by

$$\begin{aligned}\alpha(h) &= -\left.\frac{\partial t(x,y)}{\partial x}\right|_{x=h,y=0} \\ &= -\left(\frac{1}{R_1} - \frac{1}{R_2}\right)h\end{aligned} \tag{20}$$

which increases linearly with height for a positive lens.

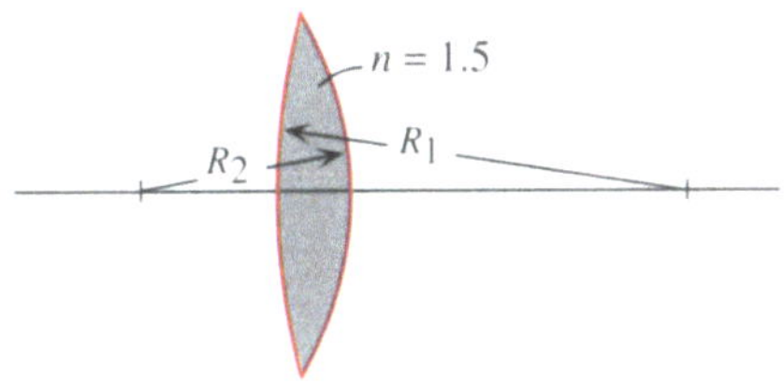

Ray Deflection by a Lens

Snell's Law[i] describes what happens to the direction of a plane wave when it passes from one material with index of refraction n_1 into another material with index n_2: its angle changes from the incident angle θ_1 to another angle θ_2 according to the following relationship:

$$n_1 \sin\theta_1 = n_2 \sin\theta_2 \tag{21}$$

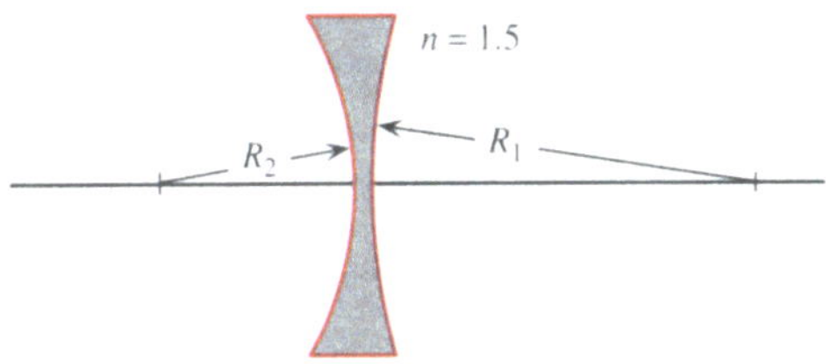

Ray deflection by a prism has to be determined fairly carefully, due to the non-linearities of the sine functions in Eq. (21). However, for rays roughly perpendicular to the surfaces, where $\sin\theta = \theta$, and further where we assume we're passing from air to glass and back again, a simple rule for the angle of deflection, $\Delta\theta$, as a function of the apex angle, α, and the index of refraction of the prism, n, can be used:

$$\Delta\theta = (n-1)\alpha \tag{22}$$

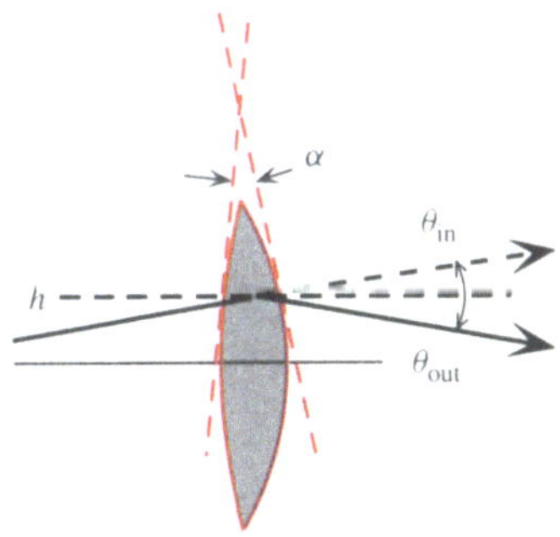

Image distances: the focus law

Now, if we adopt the same illumination source convention as above, a source ray striking the lens at height h will be incident at an angle θ_{ill} given by

$$\theta_{ill} = \tan^{-1}\left(\frac{h}{R_{ill}}\right) \tag{23}$$

The output angle, θ_{out}, will be given by $\theta_{ill} - \alpha$, and will appear to be coming from a location z_{image} given by, within the paraxial approximation,

$$\frac{1}{z_{\text{image}}} = -\frac{\tan\theta_{out}}{h} \approx -\frac{\theta_{\text{ill}} - \Delta\theta}{h} = -\frac{\dfrac{-h}{z_{\text{ill}}} + (n-1)\left(\dfrac{1}{R_1} - \dfrac{1}{R_2}\right)h}{h} = \frac{1}{z_{\text{ill}}} - (n-1)\left(\frac{1}{R_1} - \frac{1}{R_2}\right) \tag{24}$$

which is completely independent of h!

Now, if we let $z_{\text{ill}} \to \infty$, the image will be formed at a positive z_{image} (recall that R_1 is negative), at a distance that we will call the "focal length" of the lens, or FL. This is the distance at which an image of the sun will be formed by a burning glass, for example. Note that the focal length of a "positive lens" is a positive number, so that the focus is formed to the right of the lens—a real image of the illumination source. The focal length of the lens is given by the so-called "lens-maker's formula" as

$$\frac{1}{FL} = (n-1)\left(\frac{1}{-R_1} + \frac{1}{R_2}\right) \tag{25}$$

Note that a variety of combinations of curvatures can produce the same focal length lens; they differ only in their higher order optics—aberrations and so forth—so that dish-shaped or "meniscus" lenses are usually used for reading glasses, and flat-on-one-side or plano-convex lenses are used for collimators. If you read about this topic in other optics books, beware of differences in definitions of the signs of curvatures that might change some signs in the corresponding results.

z_{ill}	z_{image}
−∞	FL
−10FL	1.1FL
−2FL	2FL
−1.1FL	10FL
−FL	+∞
−FL/2	−FL

Substituting the focal length into Eq. (24) then gives the focusing equation well known in all of optics:

$$\frac{1}{z_{\text{image}}} = \frac{1}{FL} + \frac{1}{z_{\text{ill}}} \tag{26}$$

Some combinations of illumination and image distances are shown for reference in the margin.

Comparison to holographic focusing

Now lets re-examine the focusing law for the paraxial ray-trace of holograms and dwell on a few similarities:

$$\frac{1}{z_m} = m\frac{\lambda_2}{\lambda_1}\left(\frac{1}{z_{\text{obj}}} - \frac{1}{z_{\text{ref}}}\right) + \frac{1}{z_{\text{ill}}} = \frac{1}{FL_{\text{holo}}} + \frac{1}{z_{\text{ill}}} \tag{27}$$

where we define the focal length of the hologram, FL_{holo}, as

$$\frac{1}{FL_{\text{holo}}} = m\frac{\lambda_2}{\lambda_1}\left(\frac{1}{z_{\text{obj}}} - \frac{1}{z_{\text{ref}}}\right) \tag{28}$$

That is, each order of diffraction by the hologram corresponds to focusing by a different glass lens, where the focal lengths of the lenses are both positive and negative and the focal lengths of higher orders are integer fractions of the first-order focal length. Further, the focal length is inversely proportional to the wavelength of the light used for reconstruction, so that red light is focused closer to the hologram than blue light in each order.

The plus and minus first-order holographic lenses always have the same diffraction efficiency, and always appear together, occupying the same location and providing the effects of positive and negative lenses of equal and opposite focal length. It is as though two differently-shaped pieces of glass occupied the same physical space! This lens-pair model of a simple holographic lens will arise again and again in our discussions to come.

Depending on your own insight into conventional or refractive optics, the use of the lens analogy may or may not be useful, but it seems comforting to know that the results of diffraction by these simple holograms have at least a small resemblance to centuries-old optical principles!

Three-Dimensional Ray-Tracing

The extension to angles out of the x-z plane has been shown by Welford, (1975)[ii] among others, to be (adapted to our notation)

$$\mathbf{n}\times\left(\mathbf{r}_{\text{out},m}-\mathbf{r}_{\text{ill}}\right)=m\frac{\lambda_2}{\lambda_1}\mathbf{n}\times\left(\mathbf{r}_{\text{obj}}-\mathbf{r}_{\text{ref}}\right) \tag{29}$$

where $\mathbf{n}$ is a unit vector perpendicular to the hologram surface at the ray-tracing location, and the $\mathbf{r}_{\text{abc}}$ are four unit vectors in the directions of the corresponding obj, ref, ill, and out,m rays. The $\times$ denotes the vector cross product. Clearly this takes our discussion into vector algebra proofs for which "shop math" hasn't yet prepared us. However, the equation can be broken down into components that do resemble the equations we have been working with.

Let the individual ray unit vectors be represented by their components in the x-, y-, and z-directions, which are the cosines of the angles of the ray $\mathbf{r}_{\text{abc}}$ with the x-, y-, and z-axes, respectively, which we denote as ℓ_{abc}, m_{abc}, and n_{abc} (the reader will have to keep the distinction between "m, the order number," and "m_{abc}, the direction cosine" clearly in mind here). The ray unit vector could then be given by ($\ell_{\text{abc}}\mathbf{x}$, $m_{\text{abc}}\mathbf{y}$, and $n_{\text{abc}}\mathbf{z}$), where $\mathbf{x}$, $\mathbf{y}$, and $\mathbf{z}$ are unit vectors in the corresponding directions. However, we shall analyze Eq. (29) for the x- and y-components separately to give

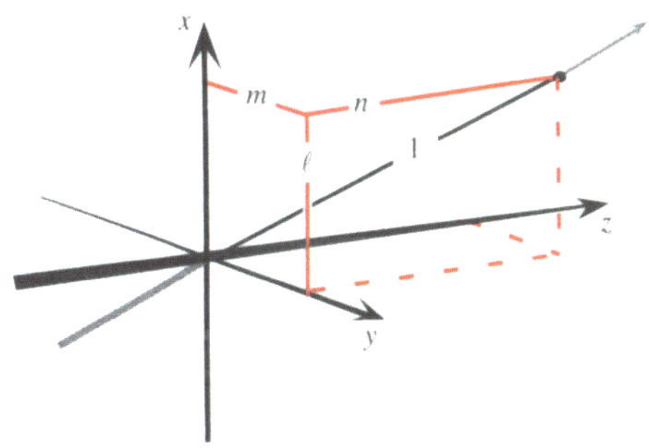

$$\ell_{\text{out}}=m\frac{\lambda_1}{\lambda_2}\left(\ell_{\text{obj}}-\ell_{\text{ref}}\right)+\ell_{\text{ill}},$$

$$m_{\text{out}}=m\frac{\lambda_1}{\lambda_2}\left(m_{\text{obj}}-m_{\text{ref}}\right)+m_{\text{ill}}, \text{ and}$$

$$n_{\text{out}}=\sqrt{1-\ell_{\text{out}}^2-m_{\text{out}}^2} \tag{30}$$

If m_{abc} is constrained to be zero, so that the abc-ray lies in the x-z plane, then ℓ_{abc}, which is always the cosine of the angle between the ray and the x-axis, becomes equal to the *sine* of the angle between the ray and the z-axis, the θ_{abc} as we have been defining it. If this is true for the obj, ref, and ill rays (and thus for all the out,*m* rays), then the equations we have been using are actually just half of the components of the full three-dimensional ray-tracing analysis. Our simplified approach could be extended whenever desired to handle the other relevant components also. However, we will continue as we have been doing, pointing out this interesting connection only in passing.

Conclusions

Ray tracing is highly accurate but computationally intensive and barren of physical insight. Simple approximations yield workable formulae that are handy for the purposes of designing systems and testing ideas. Analogies with conventional refracting optics can be drawn, although holograms are seen to be many lenses in one. Fully-three-dimensional ray tracing is seen to be possible with fairly straightforward extensions from techniques we have limited to the x-z plane.

References

i. Willebrord Snell van Roijen (1591–1626), professor at Leyden, formulated the law of refraction in 1621, though the version we give here using sines was published by René Descartes (1596–1630), who rarely gets credit for it.

ii. Welford, W. T., (1975). "A Vector Raytracing Equation for Hologram Lenses of Arbitrary Shape," *Optics Communications,* **14**, 3, pp. 322–323.

CHAPTER 9

Holographic Lenses and In-Line "Gabor" Holography

Introduction

In this chapter we will re-examine the interference pattern formed by two in-line point sources from a different point of view, and re-examine diffraction by that pattern in wavefront terms, confirming the behavior indicated by the ray-tracing results. The pattern is found to have many of the properties of conventional glass refracting lenses, so we will take a minute to review those properties for a moment too. Then, we will use these elements as a way of describing the operation of Gabor's original in-line type of transmission hologram, along with all its shortcomings.

Transition to Wavefront Curvature

Up to now, we have usually been referring to the phase-footprint of a source as being a function of the "location" of the source, the distance and angle of the source from the observation plane. Beginning now, we will emphasize a subtle change and usually refer instead to the same phase footprint as being a function of the *curvature* and *inclination* of the wavefront as seen at the observation plane. Instead of referring to the source location, which gave a sense of physical concreteness to the discussions, we will refer to the properties of the wavefront itself, which is all we can measure at the observation plane, after all. This "disembodiment" of the waves, as it were, will help us to avoid becoming dependent on a particular coordinate system choice, and will also make it easier to discuss wavefronts that might not correspond to single point sources or images. However, in most cases the correspondence is close enough that we will be able to oscillate between the phase footprint, the wavefront curvature and inclination, and the source/image location without any effort. Indeed, we have been doing so all along without commenting on it!

Definitions of "inclination" and "radius of curvature"

If we take a snapshot of a wavefront just as it comes to the observation surface, we can (in principle) make the physical measurements needed to characterize it with a ruler and a protractor. First, we construct a plane tangent to the wavefront at the (0, 0) location, the center of our local system of coordinates (typically the plane of the hologram). The perpendicular to that tangent plane defines the inclination, and presumably lies in the x-z plane, so only one angle is sufficient to define it, the angle θ that we have referred to. We might need to resort to direction cosines if the perpendicular sticks out of the plane, so that $\sin\theta$ becomes ℓ, and so forth.

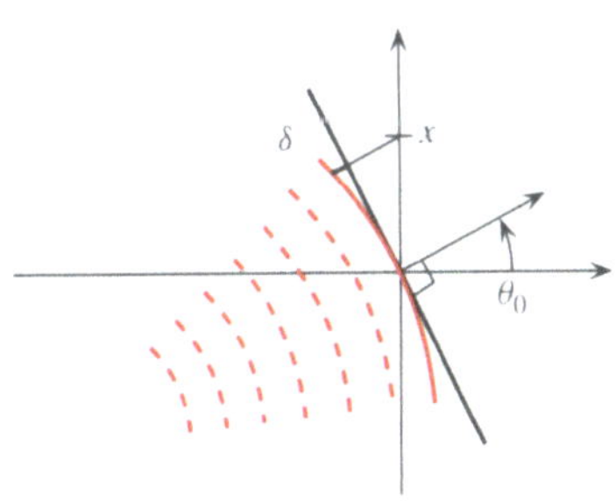

Unless it is a plane wave, the wavefront surface will separate from the tangent plane as we move in either the x or y direction, and the rate of separation increases as we move further away. Which is to say that the wavefront arrives at x later than it would if it were a plane wave, because of its curvature toward the left. In general, we describe the separation by a distance $\delta(x, y)$ where

$$\delta(x,y) = \frac{1}{2}\frac{1}{R_x}\left(x\cos\theta_0\right)^2 + \frac{1}{2}\frac{1}{R_y}y^2 \tag{1}$$

Note that $(x \cos\theta_0)$ is the distance along the tangent plane for an inclined wave. In some cases, the wavefront may have different curvatures in the x- and y-directions (that is, have a cylindrical or astigmatic component), depending on how it was generated. If the curvatures are equal, which means the wave came from a point source or is focused from/to a well-defined point virtual/real image, we will not use an x- or y-subscript, and save subscripts to denote the identity of the wave (object, reference, illumination, m-th order output, etc.).

Diverging waves** will always have **positive radii of curvature**, regardless of what direction they are traveling in, while **converging waves** will always have **negative radii of curvature.

Positive and negative = diverging and converging
Our definition of δ includes a sign; it is positive when the wave is bulging outward, or is *convex*, as it travels forward. That is, when the wave is *diverging*. When the wave is curved inwards, or is *concave*, it is said to be *converging*. This happens when a wave is being focused toward a point real image in space a distance R from the hologram. In this case, the wave at x arrives sooner than for a plane wave, and the δ is said to be negative. We describe the wave mathematically as having a *negative* radius of curvature, and again perhaps with different radii in the x- and y-directions.

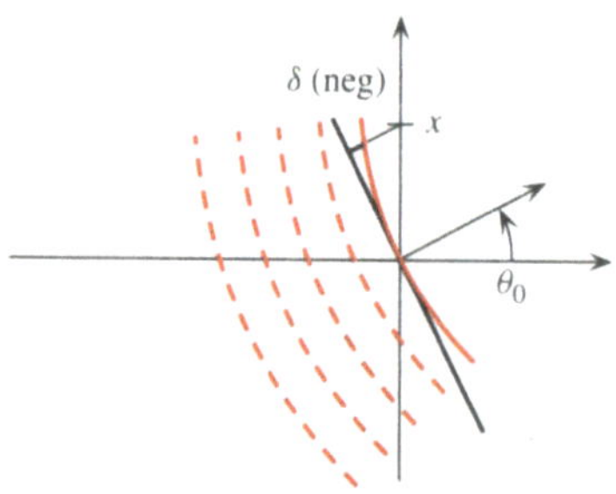

This picture is easy to understand if the wave is traveling from left to right, as light usually has been doing in our optical diagrams. However, we will soon have to deal with situations where we must let the light travel from right to left instead. A snapshot of a wave converging from right to left looks just like a wave diverging from left to right (we have included dashed lines here as a hint of the past, but these will not always be available), but the δ of the two cases have opposite signs, regardless. What we are really interested in is the phase footprint, and the phase is proportional to the δ with its correct sign. That is, the margin of a diverging spherical wave always arrives later than the equally-inclined plane wave, no matter what direction it is headed in! Which is to say that diverging waves will always have positive radii of curvature, regardless of what direction they are traveling in. Likewise, converging waves will always have negative radii of curvature.

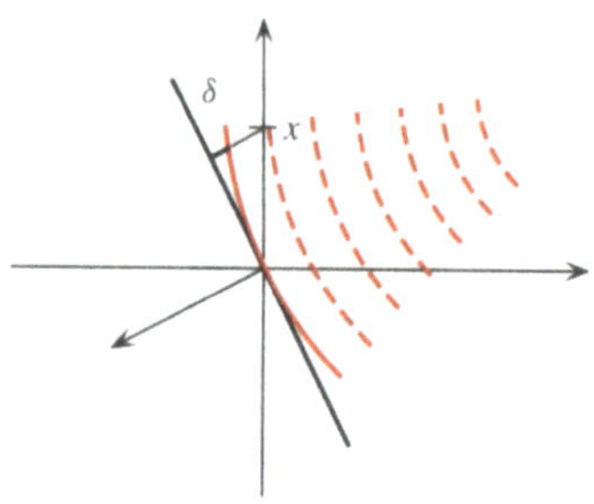

Phase Footprints, Again

From this geometrical discussion, we hope to re-derive the "phase footprints" of our rogue's gallery of simple wavefronts. The equation above is really the most general case we will need, so it suffices to work backward from that! The phase delay follows directly from $\phi(x, y) = (2\pi/\lambda)\delta(x, y)$.

Inclined plane wave
The increase in distance from an inclined plane is a linear function of distance up the x-axis, and we get the familiar phase footprint

$$\phi_{\substack{\text{plane}\\\text{wave}}}(x,y) = \phi_0 + \frac{2\pi}{\lambda}\sin\theta_0 x \tag{2}$$

General case

To find the phase footprint of an inclined spherical wave, it is only necessary to add the extra phase term due to the extra distance, the δ found above, giving

$$\phi_{\substack{\text{off-axis}\\\text{spherical}\\\text{wave}}}(x,y) = \phi_0 + \frac{2\pi}{\lambda}\sin\theta_0\, x + \frac{\pi}{\lambda}\left(\frac{\cos^2\theta_0}{R_x}x^2 + \frac{1}{R_y}y^2\right) \tag{3}$$

Further terms

A more complete expansion would include higher-order terms such as those below, which we will ignore except when they are needed for discussions of aberrations (especially spherical aberration and coma) as special topics. We'll return to a discussion of aberrations from time to time in this book, as needed to characterize the behavior of various diffractive things.

$$\begin{aligned}\phi_{\text{higher}}(x,y) = &-\frac{\pi}{\lambda}\frac{\cos^2\theta_0\,\sin\theta_0}{R_x^2}x^3 \\ &-\frac{\pi}{\lambda}\frac{\sin\theta_0}{R_xR_y}x\,y^2 \\ &+\frac{\pi}{8\lambda}\frac{\cos^2\theta_0\left(3-5\cos 2\theta_0\right)}{R_x^3}x^4 \\ &+\frac{\pi}{4\lambda}\frac{1-3\cos^2\theta_0}{R_x^2\,R_y}x^2\,y^2 \\ &-\frac{\pi}{4\lambda}\frac{1}{R_y^3}y^4 \\ &+O[x^5,y^6]\end{aligned} \tag{4}$$

On-axis spherical wave

Reducing θ_0 to zero brings us back to where we started, to the phase footprint of an on-axis spherical wave from a source a distance R from the hologram plane, except that now we are also prepared to deal with non-spherical or *astigmatic* waves that have different radii of curvature in perpendicular directions.

$$\phi_{\substack{\text{on-axis}\\\text{spherical}\\\text{wave}}}(x,y) = \phi_0 + \frac{\pi}{\lambda}\left(\frac{1}{R_x}x^2 + \frac{1}{R_y}y^2\right) \tag{5}$$

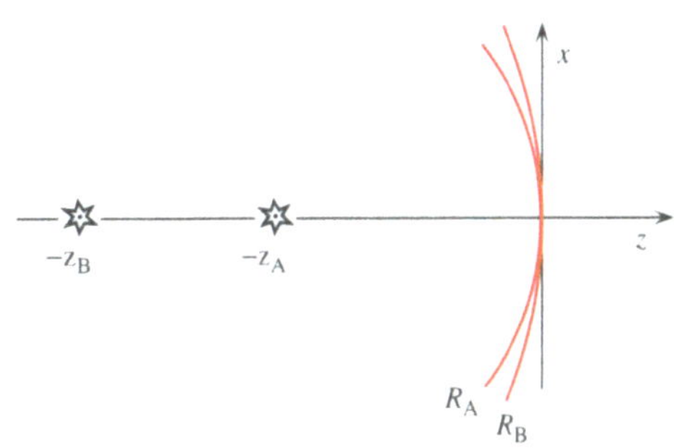

In-Line Interference, Again

Revisiting our familiar in-line interference case again in the new terminology, we must begin with the same old interference equation, where now we assume that both of the spherical (stigmatic) waves have an intensity of 0.25 at the hologram plane, but different radii of curvature (due to the difference in distance of their sources, which we are not supposed to know directly, but of course $R_i = -z_i$):

$$I_{\text{total}} = I_A + I_B + 2\sqrt{I_A \cdot I_B}\cos(\phi_A - \phi_B) \qquad (6)$$

The phase footprints, where we now introduce λ_1 to represent the recording wavelength, are given by

$$\begin{aligned} \phi_A(x,y) &= \phi_A + \frac{\pi}{\lambda_1 R_A}\left(x^2+y^2\right) \\ \phi_B(x,y) &= \phi_B + \frac{\pi}{\lambda_1 R_B}\left(x^2+y^2\right) \end{aligned} \qquad (7)$$

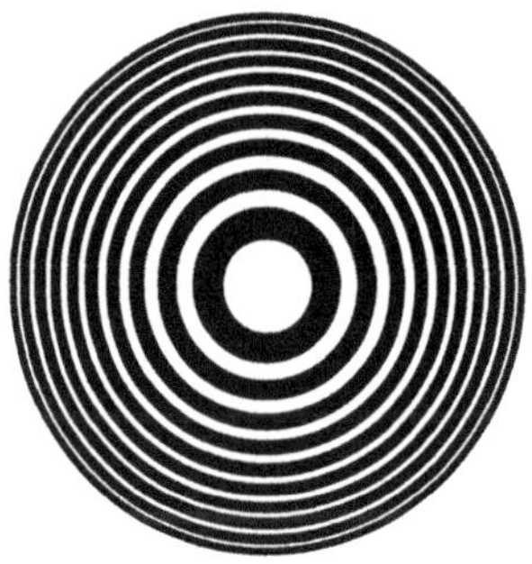

so that the exposing intensity pattern becomes the familiar Gabor zone plate:

$$I_{\text{total}}(x,y) = 0.5 + 0.5\cos\left(\frac{\pi}{\lambda_1}\left(\frac{1}{R_A} - \frac{1}{R_B}\right)\left(x^2+y^2\right)\right) \qquad (8)$$

Now we expose and process a holographic material to this pattern to produce a transmittance pattern that is its exact replica. That is, the transmittance pattern has the form

$$t_{\text{amp}}(x,y) = 0.5 + 0.5\cos\left(\frac{\pi}{\lambda_1}\left(\frac{1}{R_A} - \frac{1}{R_B}\right)\left(x^2+y^2\right)\right) \qquad (9)$$

Transmittance Proof of the Focus Equation

Next, we illuminate this transmittance pattern with yet a third spherical wavefront, the illumination wave, with a curvature R_c described by

$$E_{\text{ill}}(x,y,t) = \sin\left(2\pi\nu_2 t - \frac{\pi}{\lambda_2 R_c}\left(x^2+y^2\right)\right) \qquad (10)$$

The output wave is then given by

$$\begin{aligned} E_{\text{out}}(x,y,t) &= t_{\text{ampl}}(x,y)\cdot E_{\text{ill}}(x,y,t) \\ &= \frac{1}{2}\left(1+\cos\left(\frac{\pi}{\lambda_1}\left(\frac{1}{R_A}-\frac{1}{R_B}\right)\left(x^2+y^2\right)\right)\right) \\ &\quad \times \sin\left(2\pi\nu_2 t - \frac{\pi}{\lambda_2 R_c}\left(x^2+y^2\right)\right) \\ &= \frac{1}{2}\sin\left(2\pi\nu_2 t - \frac{\pi}{\lambda_2 R_c}\left(x^2+y^2\right)\right) \\ &\quad + \frac{1}{2}\sin\left(2\pi\nu_2 t - \frac{\pi}{\lambda_2}\left(\frac{\lambda_2}{\lambda_1}\left(\frac{1}{R_A}-\frac{1}{R_B}\right)+\frac{1}{R_c}\right)\left(x^2+y^2\right)\right) \\ &\quad + \frac{1}{2}\sin\left(2\pi\nu_2 t - \frac{\pi}{\lambda_2}\left(-\frac{\lambda_2}{\lambda_1}\left(\frac{1}{R_A}-\frac{1}{R_B}\right)+\frac{1}{R_c}\right)\left(x^2+y^2\right)\right) \end{aligned} \qquad (11)$$

These terms represent three spherical waves with three different curvatures, which we will designate as the $m = 0, +1$, and -1 diffraction orders. The locations of the corresponding foci follow from $R_i = -z_i$.

Things get a little more complicated if any of the waves are off-axis, but the principles are the same.

The three output wave curvatures can be represented by a single formula,

$$\frac{1}{R_m} = m\frac{\lambda_2}{\lambda_1}\left(\frac{1}{R_A} - \frac{1}{R_B}\right) + \frac{1}{R_C} \tag{12}$$

$$\frac{1}{R_m} = m\frac{\lambda_2}{\lambda_1}\left(\frac{1}{R_A} - \frac{1}{R_B}\right) + \frac{1}{R_C}$$

The "One-Over-R" equation

which is the same as the first term of the paraxial expansion of Chapter 7, and also follows the *focus law* for refractive lenses. Thus we see (yet again) that a simple holographic lens may be represented as paired positive and negative lenses, with equal and opposite focal lengths, and perhaps higher-order pairs with one-half and one-third the focal length, etc. We will usually call it the "One-Over-*R*" equation in our discussions. Thus our expectations from local ray-tracing analyses are confirmed by a global transmittance or "linear systems theoretic" analysis.

Off-axis illumination

If the illumination of Eq. (10) includes an off-axis term,

$$E_{\text{ill}}(x,y,t) = \sin\left(2\pi\nu_2 t - \frac{2\pi}{\lambda_2}\sin\theta_c\, x - \frac{\pi}{\lambda_2 R_c}\left(\cos^2\theta_c\, x^2 + y^2\right)\right) \tag{13}$$

then the transmitted wavefronts will have the same linear phase term imposed upon them, and the radius of curvature of the wavefront (at least in the y-direction, as we shall see later) stays the same while it gains an overall tip in the x-direction given by

$$\sin\theta_m = m\frac{\lambda_2}{\lambda_1}\left(\sin\theta_A - \sin\theta_B\right) + \sin\theta_C \tag{14}$$

$$\sin\theta_m = m\frac{\lambda_2}{\lambda_1}\left(\sin\theta_A - \sin\theta_B\right) + \sin\theta_C$$

The "Sine-Theta" equation

where we have also allowed for a tip in the object and/or reference beams. We will usually call this the "sinθ" equation in discussions to follow. Now the focus locations follow from

$$\begin{aligned} x_m &= -R_m \sin\theta_m \\ z_m &= -R_m \cos\theta_m \end{aligned} \tag{15}$$

With these generalized properties of the Gabor zone plate in hand, we can go ahead to apply them to help us understand the imaging properties of an early and simple type of hologram, the "in-line" or "Gabor" hologram.

In-Line (Gabor) Holograms

Now the object consists of a multiplicity of point sources arrayed near, but not entirely on, the z-axis so as to represent a three-dimensional object that reflects light that is coherent with the distant and on-axis reference point source. There are, as you might imagine, some practical problems in getting all the beams to the hologram without shadowing, but let's accept this simple picture for now. Each of the image points produces its own interference pattern that adds in intensity with those from the other points, plus some cross terms, to produce a superposition of Gabor zone plates in the hologram. These, in turn, produce a multiplicity of spherical waves, each in

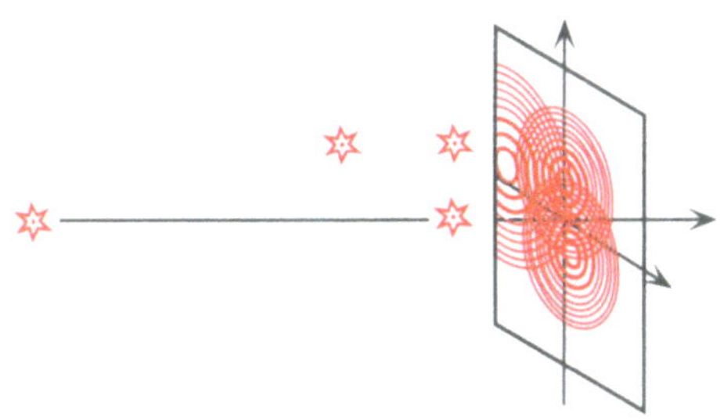

several orders, that combine to produce arrays of point images that replicate the three-dimensionality of the object.

Multiple image points

To defend this simple superposition principle, we need to look at the mathematics of interference again briefly. Let the reference and object waves be represented by completely general phase footprints (which means that this concept will apply to off-axis holograms too), where the reference beam intensity is K times that of the combined object beams. Just to set the stage, we will assume that there are N uniform object beams, all of equal (say, one over N, so they total to unity) intensity, and a uniform reference beam of intensity K (which is the beam ratio). Then we can think of the total wave as given by

$$\begin{aligned}
E_{\text{total}}(x,y,0,t) &= \left(\sqrt{\frac{2}{\varepsilon_0 c}}\right)\sqrt{K}\cos\big(2\pi\nu t-\phi_{\text{ref}}(x,y)\big) \\
&\quad+\left(\sqrt{\frac{2}{\varepsilon_0 c}}\right)\frac{1}{\sqrt{N}}\sum_{i=1}^{N}\cos\big(2\pi\nu t-\phi_i(x,y)\big) \\
&=\left(\sqrt{\frac{2}{\varepsilon_0 c}}\right)\sqrt{K}\cos\big(2\pi\nu t-\phi_{\text{ref}}(x,y)\big) \\
&\quad+\left(\sqrt{\frac{2}{\varepsilon_0 c}}\right)\frac{1}{\sqrt{N}}\cos\big(2\pi\nu t-\phi_1(x,y)\big) \\
&\quad+\left(\sqrt{\frac{2}{\varepsilon_0 c}}\right)\frac{1}{\sqrt{N}}\cos\big(2\pi\nu t-\phi_2(x,y)\big)+\cdots
\end{aligned} \tag{16}$$

Now, the total intensity is given by the time-average of the square of this expression, which we can look at term by term as

$$\begin{aligned}
I_{\text{total}}(x,y) &= \begin{pmatrix}\sqrt{K}\cos\big(2\pi\nu t-\phi_{\text{ref}}(x,y)\big)+\dfrac{1}{\sqrt{N}}\cos\big(2\pi\nu t-\phi_1(x,y)\big) \\ +\dfrac{1}{\sqrt{N}}\cos\big(2\pi\nu t-\phi_2(x,y)\big)+\cdots\end{pmatrix} \\
&\quad\times\begin{pmatrix}\sqrt{K}\cos\big(2\pi\nu t-\phi_{\text{ref}}(x,y)\big)+\dfrac{1}{\sqrt{N}}\cos\big(2\pi\nu t-\phi_1(x,y)\big) \\ +\dfrac{1}{\sqrt{N}}\cos\big(2\pi\nu t-\phi_2(x,y)\big)+\cdots\end{pmatrix} \\
&= K+\frac{1}{N}+\frac{1}{N}+\frac{1}{N}+\cdots \\
&\quad+2\sqrt{\frac{K}{N}}\cos\big(\phi_1(x,y)-\phi_{\text{ref}}(x,y)\big) \\
&\quad+2\sqrt{\frac{K}{N}}\cos\big(\phi_2(x,y)-\phi_{\text{ref}}(x,y)\big)+\cdots \\
&\quad+2\frac{1}{N}\cos\big(\phi_1(x,y)-\phi_2(x,y)\big)+2\frac{1}{N}\cos\big(\phi_1(x,y)-\phi_3(x,y)\big)+\cdots
\end{aligned} \tag{17}$$

= intensity of the beams by themselves ($N+1$ terms)

+ holographic cross terms (object × reference, N terms)

+ object – object cross terms (object × object, $N(N-1)/2$ terms)

The new class of interference terms is the third one, the "object-object" cross terms, which will have a total diffraction efficiency of roughly $(1/K)$ times the sum of the second or "holographic" cross terms. The second class of terms is the superposition set (or image set), where the contribution of each from each of the points of the object appears in a simple sum with the others. This constitutes the demonstration of "superposition" of holographic waves, which allows us to decompose a dimensional object into a 3-D array of points, separated by the resolving power of the optical system, and to trace the optical fate of each point separately, as if it alone existed in the exposure and reconstruction steps. In fact, we will trace the fates of only a few key points to obtain a representation of the spatial imaging parameters of the system, such as the 3-D location and magnification of the images.

Object self-interference terms

Note that the third class, or object-object self-interference terms, depend on only the differences between the phases of the two object points concerned. If these points are roughly at the same distance from the hologram, then the resulting pattern will consist of parallel fringes of roughly constant spatial frequency–an example of interference from side-by-side points. The resulting transmittance pattern will have $N(N-1)/2$ of these terms, with those of highest spatial frequency corresponding to interference between points at the extreme opposite boundaries of the object. The resulting transmittance pattern and image reconstruction is often termed "object self-interference noise," or "object shape-dependent noise," or "intermodulation noise." We shall see that it introduces a noisome fourth component into the array of output wave components of the hologram.

Multiple point images

We will begin by discussing a very simple image consisting of three points: A, B, and C. The central object point, A (located at $(x, z) = (0, z_{obj})$), will serve to establish the central locations of the resulting images, and the longitudinally displaced point, B (located at $z_B = z_{obj} + \Delta z$)) will serve to establish the longitudinal magnification of the images, M_{long}. The laterally displaced point, C (located at $(x, z) = (\Delta x, z_{obj})$), will then serve to establish their lateral magnifications, $M_{lateral}$.

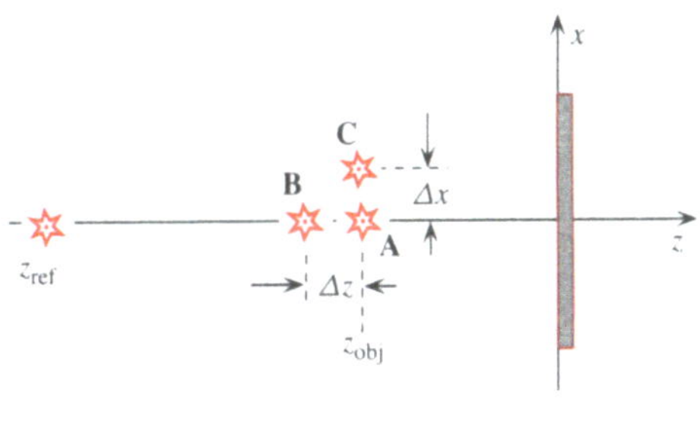

Location of the virtual image

To find the $m = +1$ or "virtual" image of the central point, A, which in more general terms we should call the "true" image, we resort to the focusing or "$1/R$" equation, Eq. (12), with the variables adapted for this occasion:

$$\frac{1}{\mathbf{R}_m} = m\frac{\lambda_2}{\lambda_1}\left(\frac{1}{R_{\text{obj}}} - \frac{1}{R_{\text{ref}}}\right) + \frac{1}{R_{\text{ill}}} \tag{18}$$

where the wavefront curvatures, the R_i, are related to the locations, the z_i, by $R_i = -z_i$. For the $m = +1$ case we then have:

$$\frac{1}{R_{+1 \atop \text{virtual}}} = \frac{\lambda_2}{\lambda_1}\left(\frac{1}{R_{\text{obj}}} - \frac{1}{R_{\text{ref}}}\right) + \frac{1}{R_{\text{ill}}} \tag{19}$$

Consider the specific case of an object point 300 mm from the plate, and the reference point 1000 mm from the plate. Recording is at 633 nm, and reconstruction is at 543 nm with a point source 1200 mm from the hologram. Cranking through the $1/R$ equation gives the $m = +1$ image location as $(x, z) = (0, -353)$ (a virtual image).

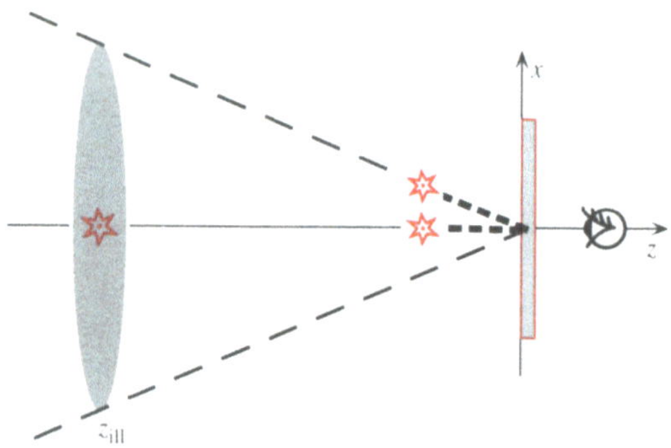

Location of the conjugate image

The m = −1 image is usually referred to as the "conjugate" image (because it and the "true" image are paired), or the "real" image (because it is often focused "downstream" of the hologram plane). Again, it is only necessary to plug the familiar terms into the $1/R$ equation, remembering to get the sign of m correct and to interpret a negative radius of curvature properly (both of which have proven over the years to be bugaboos to students!). Under the same conditions as above, the conjugate image will be found at $(x, z) = (0, 856)$ (a real image). Note that if the object is far enough away, relative to the reference source, the conjugate image will become a virtual image instead. If both the reference and illumination beams were collimated ($R_{\text{ref}} = R_{\text{ill}} = \infty$), then the real and conjugate images would be at equal distances from the hologram, but on opposite sides!

Higher-order image locations

No mysteries here: just plug in $m = 2, -2, 3, -3, 4, -4$ and so forth into the $1/R$ equation, and figure out the corresponding image locations. The positive-m images will all lie between the true image and the hologram, and the negative-m images will be between the conjugate image and the hologram (if the conjugate image is a real image, that is). Under the same conditions as above, those higher-order images will be found at $x = 0$, $z = -206, 315, -146, 193, -113, 139$, respectively.

If the reference and illumination sources are far enough away (*i.e.*, much farther than the object), we can approximate the $1/R$ equation (12 and 18) as

$$\begin{aligned} R_m &\approx \frac{R_{\text{obj}}}{m} + \frac{R_{\text{obj}}^2}{R_{ref}} - \frac{R_{\text{obj}}^2}{m\,R_{ill}} \\ &\approx \frac{R_{\text{obj}}}{m} \end{aligned} \tag{20}$$

That is, the higher-order images are at distances that are at integer fractions of the object distance. As the reference or illumination sources move toward the plate (compared to the object) that simple approximation starts to fall apart.

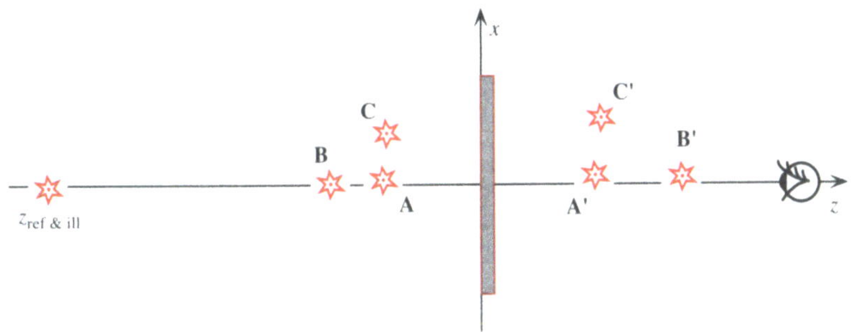

Object self-interference noise image

We said before that there would be components of the hologram transmittance pattern that would be diffraction gratings with a maximum spatial frequency determined by interference between points at the extremes of the object boundaries. For smaller spatial frequencies, there will be many possible pairs contributing grating components, and the density of those gratings becomes greater as the resulting spatial frequency approaches zero. Because these gratings are very nearly parallel-fringed with constant spatial frequency, the images they produce will appear at roughly the same plane as the illumination source. Because there are so many gratings, the images appear as a diffuse "halo" around the illumination source. For small object and illumination, we can say that the angle between the illumination source and the edge of the halo, as seen from the hologram, is the same as the angle between the extremes of the object in the same azimuth. The brightness of the halo is not very great, and tapers from a maximum near the illumination source to zero near the edge, often with a nearly-linear falloff. A careful mathematical analysis of the halo pattern reveals that its intensity is proportional to the "autocorrelation function" of the object intensity pattern, and it can have some shape of its own if the object shape is suitably complex.

Longitudinal magnification

If the object point moves to distance $z + \Delta z$, let's say away from the plate, then the image point will move out to $z_{+1} + \Delta z'$, where $\Delta z'$ is given by

$$\Delta z' = \frac{dR_{out}}{dR_{obj}} \Delta z_{obj} \tag{21}$$

and the longitudinal magnification follows from differentiation of the $1/R$ equation as

$$MAG_{long} = \frac{\Delta z'}{\Delta z_{obj}} = m \frac{\lambda_2}{\lambda_1} \left(\frac{R_m}{R_{obj}} \right)^2 \tag{22}$$

Note that the longitudinal magnification will be negative for negative m, which is to say that parts of the objects that are closer to the plate will be imaged closer to the hologram, no matter on which side of the hologram they are focused.

Pseudoscopic images

An implication of a negative longitudinal magnification is that the depth of the image will be reversed, as seen from the observer's downstream location. Such an image is said to be "pseudoscopic," and corresponds to what you see at the 3-D movies when you have the polarized glasses on upside-down (so that the right eye sees the left eye image, etc.; just putting the glasses on backwards doesn't cause this effect with polarized 3-D, by the way, though it does with red-green 3-D). Later, we will see that shaded surfaces give conflicting cues to the depth, but in simple Gabor holograms the inversion of depth is often easy to perceive, and to demonstrate with real images focused on cardboard or ground glass screens.

Lateral magnification

The side-to-side or lateral magnification of a hologram image can be determined from fairly simple geometrical considerations for the in-line Gabor hologram, and from a slightly more generalized view that extends to off-axis holograms too. We will pursue both in this section.

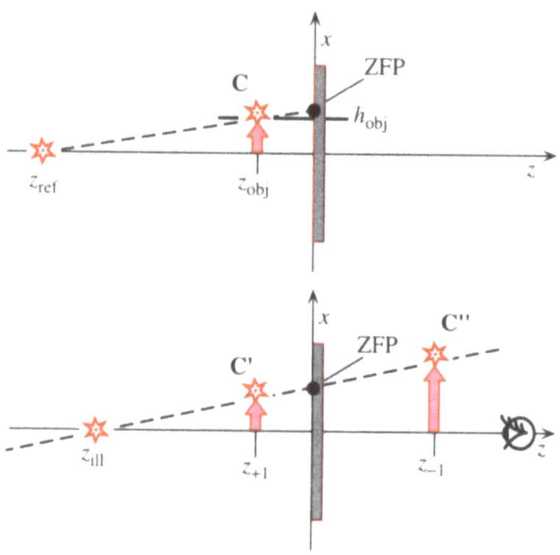

"Zero-frequency-point" geometrical analysis

Consider that point C is at the tip of an arrow, standing erect on the z-axis. The hologram of the tip of the arrow will be a Gabor zone plate centered on the ZFP or "zero-frequency point," which is defined by the intersection with the recording plate of a line drawn from the reference source to the tip, C, and extended to the plane of the plate.

Upon illumination of that Gabor zone plate with a point source, *all* images of the arrow tip will be formed somewhere along the "central ray," an infinitely-long line that passes through both the illumination source point and the hologram ZFP point. For all values of m and for all wavelengths, λ_2, the tips of the images must lie on that central ray. If we know the z_m, we can easily find the height of the image (the height of the ray at that point), and hence the lateral magnification of the image. While this approach lends itself well to a graphical solution, it also yields useful analytical results.

First we find the height of the zero-frequency point, h_{ZFP}, from similar triangles:

$$\frac{h_{ZFP}}{-z_{ref}} = \frac{h_{obj}}{-z_{ref} + z_{obj}} \tag{23}$$

The height of the image, h_{image}, is similarly determined by

$$\frac{h_{image}}{z_{image} - z_{ill}} = \frac{h_{ZFP}}{-z_{ill}} \tag{24}$$

Combining terms gives the image height as

$$h_{image} = \frac{z_{image} - z_{ill}}{-z_{ill}} \frac{-z_{ref}}{-z_{ref} + z_{obj}} h_{obj} \tag{25}$$

The lateral magnification is defined as the ratio of the image height to the object height,

$$MAG_{\text{lateral}} \equiv \frac{h_{\text{image}}}{h_{\text{obj}}} = \frac{\dfrac{z_{\text{image}}}{z_{\text{ill}}} - 1}{\dfrac{z_{\text{obj}}}{z_{\text{ref}}} - 1} \tag{26}$$

And with a few substitutions, invoking Eq. (18), we find that

$$MAG_{\text{lateral}} = m \frac{\lambda_2}{\lambda_1} \frac{R_{\text{image}}}{R_{\text{obj}}} \tag{27}$$

The lateral magnification will be positive, and the image will be erect, except when an odd-m image is virtual. Two special cases (which can be combined) are of interest: 1) if the illumination is collimated, the image size will be independent of the reconstructing wavelength; 2) if the reference beam is also collimated, the magnification will be unity, regardless of object distance.

Comparing the longitudinal and lateral magnifications, Eqs. (22) and (27), we find that

$$MAG_{\text{long}} = \frac{\lambda_1}{m\lambda_2} M^2_{\text{lateral}} \tag{28}$$

In conventional optics, the longitudinal magnification is always the square of the lateral magnification; in holography, we have the opportunity to change the wavelength after recording, and to observe a higher-order image, if that is useful. There is also a possibility of scaling the hologram pattern up or down, which we will not analyze here. Of course, the images will usually have higher-order aberrations that will obscure their details (except for the $m = +1$ image when the illumination is an exact replica of the reference beam), but these simple rules will give the locations and sizes of the images with good accuracy.

Angular subtense method

A different way to think about lateral magnification is based on considering the angle subtended by the object as seen from the center of the hologram. Call this angle Ω_1 (capital "omega", shown here as negative), given by

$$\Omega_1 = \frac{h_{\text{obj}}}{-z_{\text{obj}}} \tag{29}$$

Interference between the top and bottom points of the object produces a grating with a spatial frequency, f, that is given by

$$f = \sin\Omega_1 / \lambda_1 \approx \Omega_1 / \lambda_1 \tag{30}$$

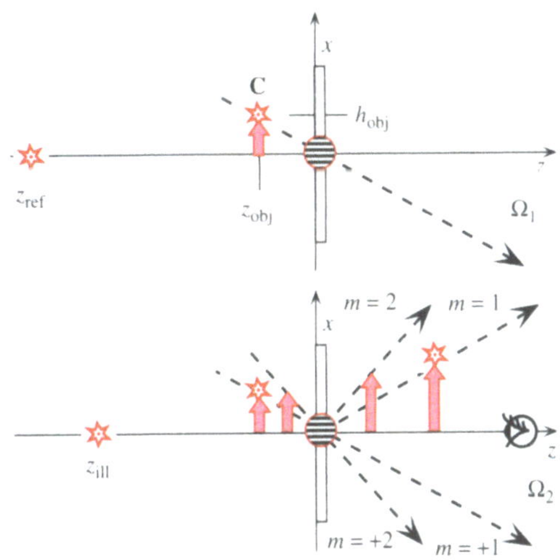

Upon illumination by an on-axis source, the output light will be diffracted by the spatial frequency, f, through an angle Ω_2, which might be different from Ω_1 if the order is different than +1, or if the wavelength changes:

$$\Omega_2 = \sin^{-1}(\lambda_2 m f) \approx \lambda_2 m f \approx m \frac{\lambda_2}{\lambda_1} \Omega_1 \tag{31}$$

Wherever the output images are, their end points must lie on the rays defined by this equation. By plugging in the z-axis locations found before (or the radii of curvature of the output wavefronts), we can find the height of any of the output images. The lateral magnification is then given by

$$MAG_{\text{lateral}} \equiv \frac{h_{\text{image}}}{h_{\text{obj}}} = \frac{z_{\text{image}} \cdot \Omega_2}{z_{\text{obj}} \cdot \Omega_1} = m\frac{\lambda_2}{\lambda_1}\frac{R_{\text{image}}}{R_{\text{obj}}} \tag{32}$$

which is what we wanted to show, as the two points of view must produce equivalent results.

Off-axis holograms

So far, we have assumed that the reference and illumination beams are directly on axis. For off-axis holograms, we would look instead at the difference in spatial frequencies of the gratings corresponding to the object points separated by Ω_1. That difference of spatial frequencies would give rise to a difference of diffracted angles, which would again define rays along which the top and bottom of all images must lie. Invoking simple differential calculus (as infrequently as possible), and letting the Ω angles become vanishingly small, we describe the small difference of spatial frequency as δf, where

$$\delta f = \frac{1}{\lambda_1}\left[\left(\sin\left(\theta_{\text{obj}} + \Omega_1\right) - \sin\theta_{\text{ref}}\right) - \left(\sin\theta_{\text{obj}} - \sin\theta_{\text{ref}}\right)\right] = \frac{\Omega_1}{\lambda_1}\cos\theta_{\text{obj}} \tag{33}$$

The two gratings are superimposed during the exposure, and produce separate transmittance terms in the final hologram. Illumination by an off-axis illumination beam produces output beams that must satisfy a similar relationship,

$$\delta f = \frac{1}{\lambda_2}\left[\left(\sin\left(\theta_{\text{out}} + \Omega_2\right) - \sin\theta_{\text{ill}}\right) - \left(\sin\theta_{\text{out}} - \sin\theta_{\text{ill}}\right)\right] = \frac{\Omega_2}{\lambda_2}\cos\theta_{\text{out}} \tag{34}$$

The lateral magnification is then given by

$$MAG_{\text{lateral}} \equiv \frac{h_{\text{image}}}{h_{\text{obj}}} = \frac{z_{\text{image}} \cdot \Omega_2}{z_{\text{obj}} \cdot \Omega_1} = m\frac{\lambda_2}{\lambda_1}\frac{R_{\text{image}}}{R_{\text{obj}}}\frac{\cos\theta_{\text{obj}}}{\cos\theta_{\text{image}}} \tag{35}$$

Whenever the central object and output beam angles are equal (which is usually the case for $m = +1$) we find that the lateral magnification is again given by the in-line hologram result, shown in Eqs. (27) and (32).

As for the longitudinal magnification, there was nothing in the derivation of Eq. (22) that depended on the angles of any of the rays, so that result applies to off-axis as well as on-axis holograms. However, because the 1/R equation will later be seen to apply only to horizontally focused light, Eq. (22) predicts the longitudinal magnification only for vertical line features (about which we will see much more later on).

The two ways of thinking about image formation, via the "ZFP and central ray method" and via the "central angles, Ω_1 and Ω_2, method" produce identical results. It is a little early to know which you will find easier to remember and to use as a problem-solving tool—you will probably just plug in the formulas, after all! We will soon move into the domain of off-axis holograms, where the ZFP approach becomes only approximate but the central angle approach extends accurately. So, make your peace with both before we have to move along!

What's wrong with "in-liners"?

The in-line hologram broke new intellectual ground in 1948. Nobody believed that it was possible to reproduce the phases as well as the amplitude of a wavefront, but Gabor showed photographs that proved it was true. The Nobel Prize in Physics for 1971 finally made the significance of these ideas clear to everyone, once the laser had made some important improvements in holographic technique possible. We have had to move beyond Gabor's configuration because of visual problems with the images it produced. Here is a brief catalog, which you can probably add to:

Glare of illumination: The fact that the desired image is directly in line with the illumination source can make for some very uncomfortable viewing, unless the image has somehow been made very bright. There ought to be a way to block the zero order light using polarizers, but this would require a special needle-like photographic grain to be invented.

Visibility of the conjugate image: This is a more profound problem: the oppositely diffracted or "conjugate" image is also directly in line with the desired image! Although it is far enough from the desired image to be substantially out of focus, it still provides a noisy coherent background, which especially degrades the resolution at the edges of the object.

Back-lit objects: In Gabor's configuration, the only light available for the object comes from scattered reference beam light. This means that a) there isn't much light, b) only forward-scattering and translucent objects can be used, and c) the reference beam will have holes in it, the shadows of the objects. We can get around all of these problems by using a beam-splitter to bring reflected object light into line with a reference beam, but this defeats some of the other advantages of Gabor holography.

"Halo" noise: The halo noise due to object self-interference, or intermodulation, is also centered around the illumination beam, so even those parts of the image not in line with the source, and not obscured by the conjugate image, have a contrast-reducing flare light behind them!

What's okay about "in-liners?"

In spite of the shortcomings of in-line transmission holograms mentioned above, a few are still made each year to take advantage of a few of the characteristics of this hologram type.

Low stability required: Because the path length difference between the reference and scattered object beam is relatively insensitive to object motion, the exposure system need not be as mechanically stable as for more advanced hologram types. Also, because the fringe spatial frequencies are typically quite low, the stability of the film holder need not be as high as usual.

Holograms can be very big: Very few optical parts are needed, and the low stability allows large pieces of film to be spread out and exposed with a minimum of precautions.

Coarse fringes = high-speed film: Because highest-resolution emulsions are not needed, coarser-grained higher speed emulsions may be used (*e.g.,* A-G 10E75 instead of 8E75, a 20× speed gain). This allows much shorter exposure times, again reducing the stability requirements, or the use of a smaller laser or larger film areas. On the other hand, it is only the finest-grained emulsions that produce bright and clear bleached holograms.

Only low-coherence light needed: The path length variations between the reference beam and the scattered object light are much smaller than for off-axis holograms. Thus the etalon may usually be taken out of an ion laser, and more power obtained, without degrading the image. Very small holograms can even be made without a laser, using something like a mercury arc source (as Gabor did).

Wide-band illumination works: As we will see, the spectral smear caused by white light illumination "points" in the direction of the light source. The closer we are to looking directly into the light source, the more nearly "end on" we are seeing the smear, and the less visible it becomes. Instead, the various colors stack on top of each other, and an achromatic image appears. As we move to the side of the hologram, color fringing does become apparent, however.

Overall, the disadvantages have vastly outweighed the advantages, and holography has moved on to off-axis techniques to separate the various image components, and allow holography to broaden the scope of its imaging capabilities.

Conclusions

In-line transmission holography quite literally involves a combination of diffraction gratings to change the direction of light, and holographic lenses to focus the light in three-dimensional space, and form images. The first of these elements, diffraction gratings, has been covered in the chapter on diffraction. Holographic focusing is more subtle, and we have looked at it in two ways, by ray-tracing, and by wavefront curvatures. The second of these gives us the tools to record and reconstruct spherical and astigmatic wavefronts without regard to actual or virtual source locations or image foci. Finally, we showed how the simple imaging properties of several holographic lenses, in the form of Gabor zone plates, can be combined to predict the three-dimensional imaging properties of in-line holograms of extended objects.

This is nearly the end of the road as far as learning new optical concepts goes; only two big new ideas lie ahead. From here, we take these building-block ideas and make lots of new kinds of holograms, exploring the properties of each as we go. That is a little like saying that Ohm's law and the Kirchhoff equations can explain all of electronics, but it is more than a little bit true! The "sinθ" and "One-Over-R" equations will serve similar roles in explaining what we see as we move ahead.

CHAPTER 10

Off-Axis "Leith & Upatnieks" Holography

Introduction

Many of the shortcomings of in-line "Gabor" holograms have been overcome by going to an off-axis geometry that allows the various image components to be separated, and also allowed opaque subjects to be front-illuminated. These discoveries were made by Emmett Leith and Juris Upatnieks, working at the Radar and Optics Lab of the University of Michigan's Willow Run Laboratories. They were working on optical data processing for a highly secret new form of side-looking radar when they found that their images were three-dimensional; they had rediscovered Gabor's ideas about holography, as they quickly realized. Around 1962, the first commercial helium-neon lasers became available and Leith and Upatnieks started making more ambitious holograms, slowly moving the reference beam off to the side and dividing the laser beam to illuminate the object.[i,ii] Finally, they made some holograms big enough (100 mm x 125 mm) to be visible with both eyes, and astonished everyone at the annual meeting of the Optical Society of America in 1964 with an incredibly vivid hologram of a brass model of a steam locomotive.[iii]

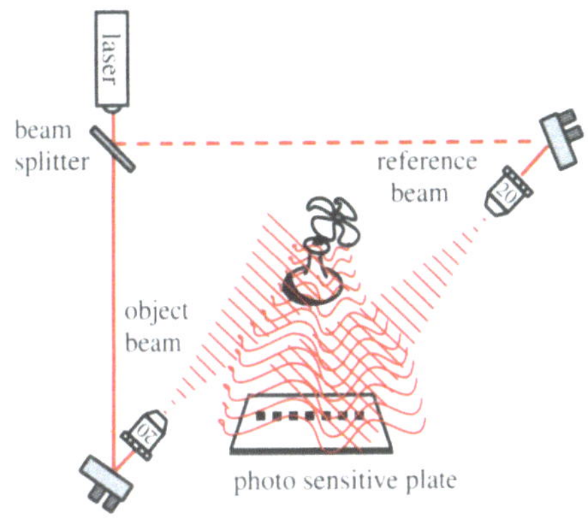

A typical setup is as shown in the margin. Most of the light goes through the beam-splitter to illuminate the object, and the diffusely reflected light, the "object beam," strikes the photo-sensitive plate. If that were all there were to it, we would just get a fogged plate. However, a relatively small amount of laser light is reflected off to be expanded to form the "reference beam," which overlaps the object beam at the plate to produce the holographic interference pattern. After exposure and processing, the plate (now called the "hologram") is put back in place, and illuminated with expanded laser light, usually with the same angle and divergence as the reference beam. Diffraction of the illumination beam produces several wavefront components, including one that reproduces the waves from the object—whence the 3-D image reconstruction. The various components are now separated by angles comparable to the reference beam angle, so that they no longer overlap and a clear window-like view of the scene is available.

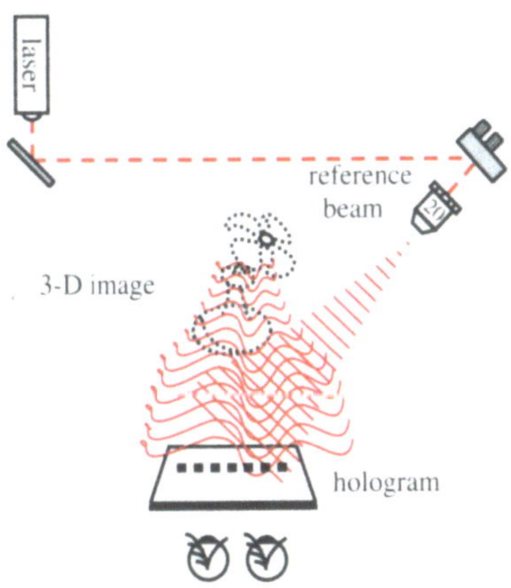

Implications of Off-Axis Holography

The dramatic increase of the angle between the reference and object beams has several important consequences:

Separation of image terms: Because there is a fairly large angle between the object and reference beam, the conjugate image will be well-separated from the true image, and may even be evanescent. Also, the straight-through beam, the zero-order component, will probably not fall into the viewer's eyes. The ability to see clearly a high-contrast, high-resolution image in vivid 3-D changed people's interest in holography literally overnight.

Much finer fringes: The large average angle means that the interference fringes will be much finer, typically more than 1000 fringes/mm^2. A typical photographic film can resolve details up to around 100 cy/mm^2, so ultra-fine-grained films are required for holography. Typical holographic materials have grains averaging 35 nm in diameter, compared to 1000 nm for conventional photo films (a volume ratio of one to 23,000!). Unfortunately, the sensitivity of emulsions drops quickly with decreasing grain size, and the equivalent ASA rating of the *8E75-HD* emulsion commonly used is about 0.001. That means that the exposure times will be quite long, usually up to ten seconds and sometimes much longer.

Another result is that the fringes will be closer to each other (a micron or so apart) than the emulsion layer is thick (five to seven microns, typically), so that volume diffraction effects can become noticeable. For the most part, this amounts to a modest sensitivity to the direction of illumination, but it also allows higher diffraction efficiencies to be reached with proper processing. At the same time, small defects in processing (especially during drying) become apparent if they cause mechanical shearing in the emulsion, and a distortion of the venetian-blind-like fringe structures.

Greater exposure stability required: The finer fringes mean that the recording material must stand still to within much higher tolerances during the exposure. And the lower sensitivity (compared to lower resolution emulsions) means that those exposures will be fairly long. In addition, because the beam paths are separated by the beamsplitter, vibrations of the mirrors are not canceled out in the two beams, so that the setup is more vulnerable to noise and shocks. Also, any element that is reflecting a beam (including the object!) need move only one-quarter wavelength to produce a shift of fringe position of one-half cycle, which washes out the fringes during exposure.

Frontal illumination of objects: Two more issues come up because we are reflecting light from fairly deep groups of ordinary diffusely-reflecting objects. If the lengths of the object and reference beam paths are matched for light reflecting from the front of the object, they will be mismatched for light from the rear by double the depth of the scene. This distance may be greater than the coherence length of the light from the particular laser used, which may be only a centimeter or two. Also, the steep reference beam angle means that the length of the reference beam will also vary across the width of the plate.

Phenomena related to polarization can also cause us some trouble. Interference happens only between similarly-polarized beams; the electric fields have to be parallel in order to add or subtract. Diffuse reflection (such as from matte paint) "scrambles" the polarization of a beam so that half of the object light simply fogs the plate, and is lost to the holographic exposure.

Beam ratio effects: Because we can usually adjust the reflection:transmission "split ratio" of the beamsplitter, we can adjust the ratio of the reference-to-object beam intensities, K, to any number we desire. This allows us to increase the diffraction efficiency of the hologram (the brightness of the image) more or less at will, up to the maximum allowed by the plate and processing. Typically, we will use a K of between 5 and 10. This will produce diffraction efficiencies of up to 20% with "bleach" processing. However, as the object beam intensity is raised relative to the reference beam (the K is lowered), additional noise terms arise caused by object self-interference. They grow as the third power of the diffraction efficiency, and reduction of the image contrast is often the practical limit on reducing K. Also, because only a small fraction of the object light is captured by the plate, increasing the beam split to the object increases light wastage, and thereby increases the exposure time significantly. Long exposure times often produce dim holograms, due to mechanical "creep" in the system, which defeats the purpose of decreasing the K.

Higher illumination bandwidth sensitivity: Although going off-axis increases the sensitivity to source spectral bandwidth (because we are seeing the spectral blur more nearly sideways), it also *decreases* the sensitivity to vertical source size—a feature that will become useful with white-light viewed holograms. However, it is only a $\cos\theta$ effect, which is not very strong.

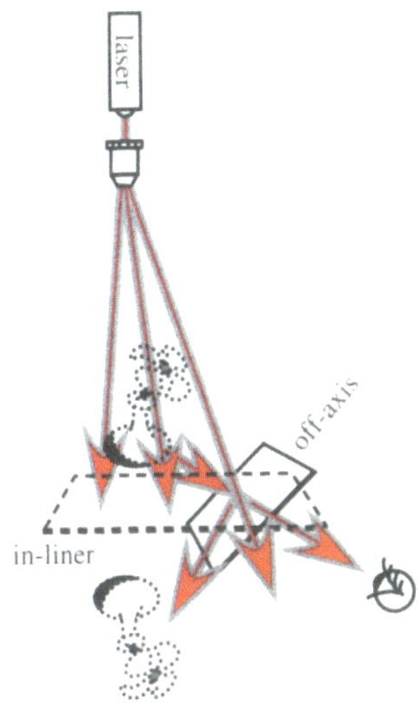

Interference and Diffraction in Off-Axis Holograms

If you get the chance to examine an in-line Gabor hologram, you may notice some of the features of off-axis transmission holography near the edges. At the edges of the plate, the angle between beams from the objects and the unscattered reference beam is large enough to separate the various other real and/or virtual images so that each may be seen more or less individually. If you imagine tilting a plate that is far from the zero-frequency point of a Gabor hologram, you have an off-axis hologram (except for the beam-split separation of the reference beam and object illumination beam). So, there really are no new physics concepts involved here, but their implications become quite different.

We might start the analysis by going through the same process that we did for the on-axis hologram: examine the phase footprints of the two waves involved (with a single point serving as a "stand-in" for the 3-D object), consider the interference pattern and the transmittance, add the illumination, and examine the output terms for likely suspects. Instead, we will invoke the master phase equation of holography as a short-cut.

We begin by defining terms. The reference beam comes in at some angle (positive in this example, for convenience), and the object beam will be on axis. As a rule, the radius of curvature of the reference beam will be much larger than that of the object beam, but this need not necessarily be the case as long as the intensity of the reference beam is fairly uniform across the plate.

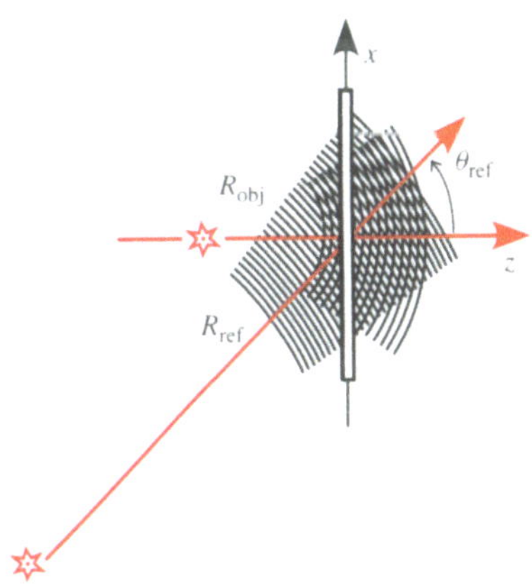

Phase footprint of the output waves

The phase footprint (the first few terms, anyway) of an off-axis spherical wave was described in Chapter 9, Eq. (3), and in the current situation becomes:

$$\phi_{ref}(x,y) = \phi_0 + \frac{2\pi}{\lambda_1}\sin\theta_{ref}\, x + \frac{\pi}{\lambda_1}\left(\frac{\cos^2\theta_{ref}}{R_{ref}}x^2 + \frac{1}{R_{ref}}y^2\right) \quad (1)$$

By comparison, the phase footprint of an on-axis point object wave should look familiar by now (note that $\cos^2\theta_{obj} = 1$):

$$\phi_{obj}(x,y) = \phi_1 + \frac{\pi}{\lambda_1 R_{obj}}\left(x^2 + y^2\right) \tag{2}$$

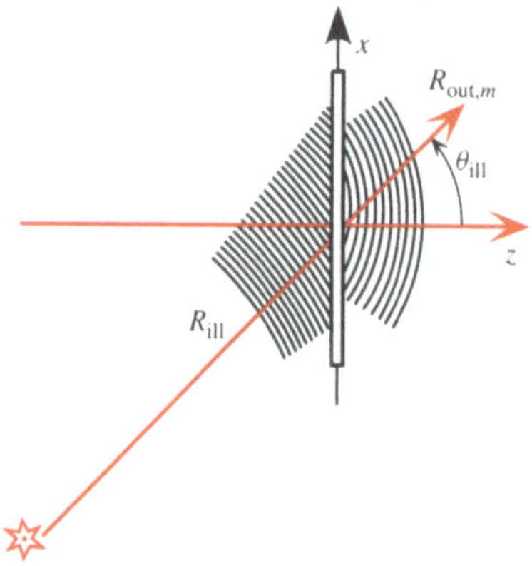

All that we lack is the illumination beam, which will again be an off-axis spherical wave, with a phase footprint of the same general form as the reference wave:

$$\phi_{ill}(x,y) = \phi_2 + \frac{2\pi}{\lambda_2}\sin\theta_{ill}\ x + \frac{\pi}{\lambda_2}\left(\frac{\cos^2\theta_{ill}}{R_{ill}}x^2 + \frac{1}{R_{ill}}y^2\right) \tag{3}$$

Now we will invoke the fundamental phase-addition law of holography, first revealed in Chapter 7 ("Platonic Holography"):

$$\phi_{out,m}(x,y) = m\left(\phi_{obj}(x,y) - \phi_{ref}(x,y)\right) + \phi_{ill}(x,y) \tag{4}$$

where each of the output waves has its own angle of inclination and radius of curvature,

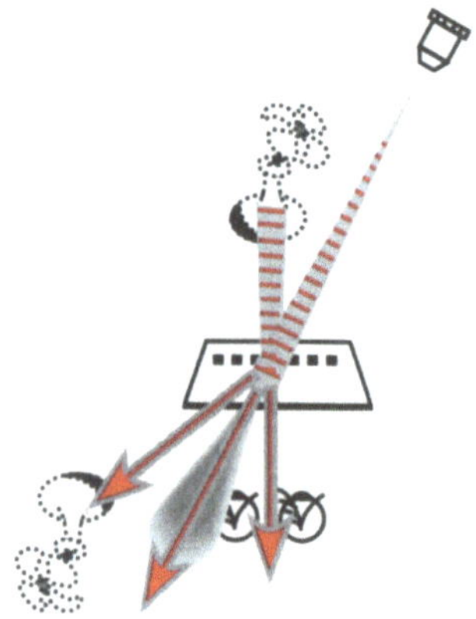

$$\phi_{out,m}(x,y) = \phi_3 + \frac{2\pi}{\lambda_2}\sin\theta_{out}\ x + \frac{\pi}{\lambda_2}\left(\frac{\cos^2\theta_{out}}{R_{out,x}}x^2 + \frac{1}{R_{out,y}}y^2\right) \tag{5}$$

Now it is only necessary separately to match the coefficients of the linear terms in x, and the quadratic terms in x and y (we do not bother with the constant phase terms, of course). This produces the results that characterize the output wave:

$$\sin\theta_{out,m} = m\frac{\lambda_2}{\lambda_1}\left(\left(\sin\theta_{obj} = 0\right) - \sin\theta_{ref}\right) + \sin\theta_{ill} \tag{6}$$

$$\frac{\cos^2\theta_{out,m}}{\mathrm{R}_{out,m,x}} = m\frac{\lambda_2}{\lambda_1}\left(\frac{1}{\mathrm{R}_{obj}} - \frac{\cos^2\theta_{ref}}{\mathrm{R}_{ref}}\right) + \frac{\cos^2\theta_{ill}}{\mathrm{R}_{ill}} \tag{7}$$

$$\frac{1}{\mathrm{R}_{out,m,y}} = m\frac{\lambda_2}{\lambda_1}\left(\frac{1}{\mathrm{R}_{obj}} - \frac{1}{\mathrm{R}_{ref}}\right) + \frac{1}{\mathrm{R}_{ill}} \tag{8}$$

$$\frac{\cos^2\theta_{out,m}}{\mathrm{R}_{out,m,x}} = m\frac{\lambda_2}{\lambda_1}\left(\frac{1}{\mathrm{R}_{obj}} - \frac{\cos^2\theta_{ref}}{\mathrm{R}_{ref}}\right) + \frac{\cos^2\theta_{ill}}{\mathrm{R}_{ill}}$$

The "$\cos^2/R$" equation

Note that these are just our familiar "$\sin\theta$" and "$1/R$" equations, plus a new addition, the "cosine-squared (over R)" equation for the radius of curvature of the output wave in the x-direction.

Perfect reconstruction

Note that if we again have $\lambda_2 = \lambda_1$, $R_{ill} = R_{ref}$, and $m = +1$, we achieve "perfect reconstruction" in that $\theta_{out} = 0°$, and $R_{out,x} = R_{out,y} = R_{obj}$. That is, the image will be located at the same place as the object, which will be true for every point in the object.

The conjugate image

Let's leave everything about the illumination the same, but examine the $m = -1$ or "conjugate" image for a moment. Note that the output beam angle is now

$$\theta_{-1} = \sin^{-1}\left(2\sin\theta_{ref}\right) \tag{9}$$

and does not exist if the reference beam angle is 30° or more (*i.e.*, the wave will be evanescent). This is the usual case in off-axis holography, as typical reference beam angles are 45° or 60°. We might deliberately make some shallow-reference-angle holograms just to make the conjugate image easier to see. Instead, we usually display the conjugate image by illuminating the hologram from the other side of the z-axis, with $\theta_{ill} \approx -\theta_{ref}$ (so that the conjugate image comes on-axis), or by more often by illuminating through the back of the plate, with $\theta_{ill} \approx \pi + \theta_{ref}$, (about which much more will be said in later chapters).

If the conjugate image exists at all, it is very likely to be a real image. Consider first the y-curvature (letting $\lambda_2 = \lambda_1$ and $R_{ill} = R_{ref}$ and $\theta_{ill} = \pm\theta_{ref}$ for simplicity):

$$R_{-1,y} = -R_{obj}\left(\frac{\dfrac{R_{ref}}{R_{obj}}}{\dfrac{R_{ref}}{R_{obj}} - 2}\right) \tag{10}$$

As long as the reference point is more than twice as far away as the object, the conjugate image will be real. Otherwise, it will be a virtual image, appearing beyond the illumination source. But consider now the x-curvature:

$$R_{-1,x} = -\cos^2\theta_{-1}\, R_{obj}\left(\frac{\dfrac{R_{ref}}{R_{obj}}}{\dfrac{R_{ref}}{R_{obj}} - 2\cos^2\theta_{ref}}\right) \tag{11}$$

Note that it is, in general, very different from the y-curvature. It may even have a different sign! This is our first real taste of the dreaded *astigmatism*, which will plague us for the rest of our holographic careers. It means that the rays that are converging to the point-like real-image focus will cross first in the x-direction, and later in the y-direction (as a rule). In general, we will have to treat the x- and y-focusing of the hologram separately at each step. Because the x-direction will often be vertical, we will call it the vertically-focused image (or *tangential* focus, in conventional lens-design terms). The y-focus is then the horizontally-focused image (or *sagittal* focus).

Higher-order images

Note that, if the $m = -1$ term is evanescent, the $m = +3$ term will usually be evanescent too, and all the higher-order terms (assuming that $\theta_{out,+1} \approx 0°$). Some of those higher order terms can be brought into view by manipulating the illumination angle and/or wavelength. They will be formed closer to the hologram, just as for the in-line hologram, and follow the same rules (for the wavefront y-curvature, anyway).

Imperfect reconstruction and astigmatism

Considering again the $m = +1$ or "true" image, note that if the illumination wave is not a perfect replica of the reference wave (*i.e.*, it has a different wavelength, angle, or divergence), the output wave will not be a perfect replica of the spherical wave created by the point object. In fact, it will probably not even be a spherical wave!

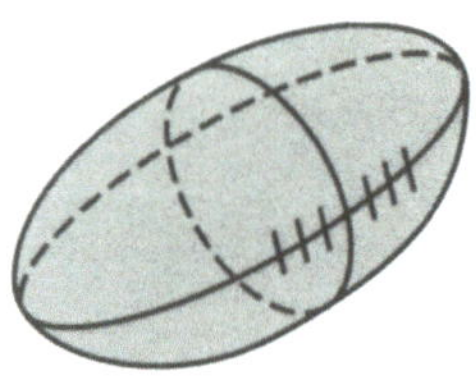

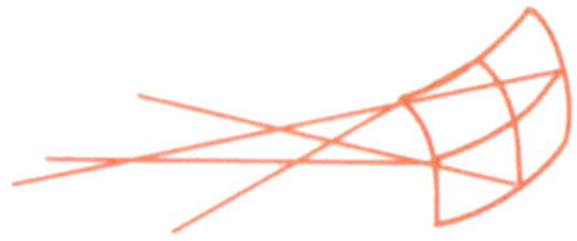

For "imperfect" reconstructions, the radii of curvature in the x- and y-directions, given by Eqs. (7) and (8), will be different, often significantly so.

It is difficult to get used to thinking about astigmatic wavefronts and astigmatic ray bundles, and we will make several tries at making it clear. A wavefront with different curvatures in two perpendicular directions has a shape like that of the surface of an American football where a passer usually grabs it (near the stitching). It has a *small* radius of curvature around the waist of the ball, and a long radius of curvature from end to end. If you try to focus such a wave onto a card to see what kind of source produced it, you would first see a vertical line, then a round circle, and then a horizontal line as you passed the card from the first center of curvature to the second. Many people have astigmatism in their eyes (usually from a cornea that is non-spherical) and have a cylindrical lens component in their prescription to allow a sharp focus to be restored.

Thinking about it in ray terms, a point source produces a *stigmatic* ray bundle (from the Greek for pin-prick or tattoo mark), a bundle of rays that seem to have passed through a single point in space. Instead, an *astigmatic* (non-stigmatic) ray bundle seems to have passed through two crossed slits that are somewhat separated. The curvature in each of the two directions is equal to the distance to the perpendicular slit, and the rays have no common origin point. In addition to blurring a focused image, the usual visual effect is that the distance of an image seems to be different depending on the direction we move our head (side-to-side versus up-to-down).

Interestingly, there are some conditions of imperfect illumination that do not produce astigmatism. One condition that is easy to derive is obtained if the object and image are perpendicular to the plate and if

$$\frac{\sin^2\theta_{ill}}{R_{ill}} = \frac{\sin^2\theta_{ref}}{R_{ref}} \tag{12}$$

Another case, of some practical interest later on, occurs when only the distance to the illumination source changes. If the object and image angles are equal and opposite to the reference and illumination angles (also equal), then there will be no astigmatism for any pair of reference and illumination distances. That is to say, all of the $\cos^2$ terms in Eq. (7) are equal, and so divide out.

If you are a photographer, you may also have come across lenses called *anastigmats*. That name comes from the Greek for "again" and "pin-prick" or "point-like," which is only to say that the lenses claim to produce a particularly sharp spherical-wave focus.

Astigmatism will be a much stronger effect when we deal with real image projection in a few chapters, and we will be studying it in some detail. For the time being, we will be content with the examples at the end of the chapter. Its effects in virtual image reconstruction are usually so weak as to be almost invisible, but it is important to understand astigmatism in principle, even now. Strangely, it is a subject that is not much discussed or appreciated in the holography literature, although researchers noted its existence early in the history of the field.[iv,v]

Models for Off-Axis Holograms

The three equations that describe image formation by an off-axis hologram seem pretty opaque at first glance, although they will gradually become more familiar as we gain experience. In the meantime, it is tempting to draw some simple physical models to describe the optical properties of off-axis holograms. We will look at two such models; the first is a deliberate "straw man," appealingly simple but hopelessly inaccurate. It can be used only for a very rough first judgement of physical reasonability.

Off-axis zone plate

We have seen that the off-axis hologram can be considered as an extreme case of an on-axis hologram, at least conceptually. Why, then, can't we apply the same model of a Gabor zone plate, using simple raytracing through key landmarks, such as the zero-frequency point, the ZFP? Such a model might look like the sketch, which shows a collimated illumination beam at 20°, which is presumably the same angle as the reference beam. If the object was 100 mm from the plate, the ZFP is 36.4 mm above the axis. The distance from the hologram to the real and virtual foci should be equal in collimated illumination, so the real image location is predicted to be $(x, z) = (72.8, 100)$. The more carefully calculated location is (68.4, 72.9), significantly different! What is the problem with the Gabor zone plate model now? Recall that our analysis assumed that the rays of interest travel close to and at small angles to the optical axis of the zone plate, what we called a "paraxial" analysis. But for an off-axis hologram, the rays of interest pass through the center of the hologram, which is far from the ZFP and the optical axis of the zone plate. The off-axis and large-angle aberrations have become too large to accurately predict anything but the location of the virtual image in near-perfect reconstruction.

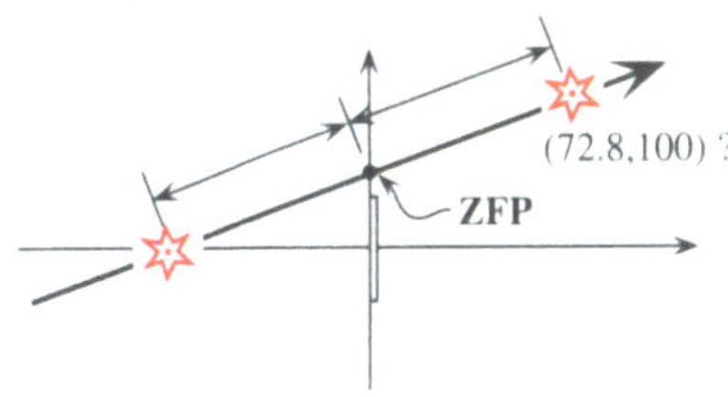

Prism + lens (grating + zone plate) model

What the "sinθ" and "$1/R$" equations for the $m = +1$ image are telling us is that the light turns upon reaching the hologram, as though deflected by a diffraction grating (or its refractive equivalent, a base-down prism), and then is focused (well, diverged) by an on-axis Gabor zone plate (or its equivalent, a negative or double-concave lens). On the other hand, the $m = -1$ image is deflected the opposite way (the opposite order of the image, or a base-up prism) and focused by the opposite power of the zone plate (or its equivalent, a positive or double-convex lens). Higher order images are generated by prisms and lenses, each having multiples of the base power, always paired. Refracting elements seem to be more photogenic than their diffractive equivalents, so we often sketch an off-axis hologram as a combination of two lens-prism pairs (in idealized optics, it doesn't matter which comes first). Upon examination of the transmittance pattern, we find a constant spatial frequency term plus a term with a linearly varying frequency, which can be interpreted as two diffractive elements in tandem, exactly as suggested by these sketches. Thus this model brings us quite close to the mathematical as well as physical reality of off-axis holograms.

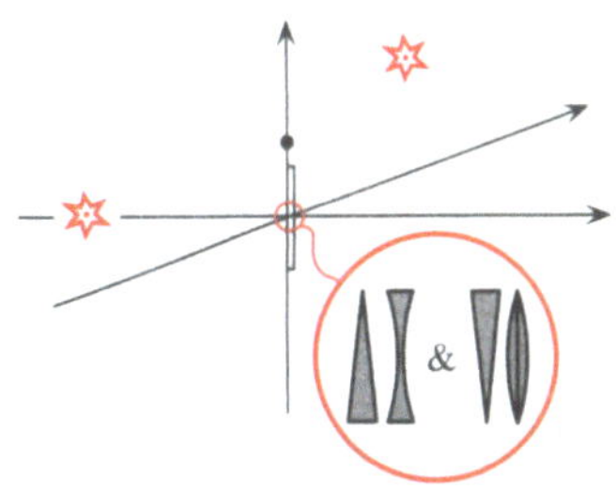

The focus in the x-direction is a little different, as there is some coupling between the power of the equivalent lens and the equivalent prism, so that the lens itself has different curvatures in the two directions, as would a lens designed to correct astigmatic vision. The appearance of an astigmatically focused image is difficult to describe. For an image focused on a card, vertical and horizontal lines will come into sharp focus at slightly different distances. An aerial image—viewed in space by an eye—may seem to have different magnifications in the two directions. The implications will be context-specific, so we will explore them as they arise in holographic imaging systems.

The "sinθ" equation is exact; it is a raytracing equation after all. But the focusing equations are valid only for small variations of angle or location around the accurately raytraced component. We call this a "parabasal" type of analysis, one that is valid only in the mathematical vicinity of the "basal ray" that is traced through the part of the hologram of interest, even though that ray strays far from the z-axis and has several large-angle bends.

Image Magnification

Now that we have found the image locations fairly accurately, all that remain to be found are the magnifications of the images to finish our characterization of off-axis holograms as 3-D imaging systems.

Longitudinal magnification

Note that the "1/R" equation is the same for off-axis and on-axis holograms, and recall that this is the equation that governs longitudinal magnification. Thus the same equation (which followed from the derivative of the R_{out}) applies, but now re-stated in terms of wavefront curvatures:

$$MAG_{\text{long}} = \frac{\Delta R_{\text{out},m}}{\Delta R_{\text{obj}}} = m\frac{\lambda_2}{\lambda_1}\left(\frac{R_{\text{out},m}}{R_{\text{obj}}}\right)^2 \tag{13}$$

We have only to point out that the radii are now measured along a line through the center of the hologram and the center of the object, which may be at a large angle to the z-axis.

The x-focus or "cos^2" equation (7) moves the images around and changes their magnification.

Lateral magnification

The angular subtense approach is the only workable handle on lateral magnification in this case, as the "ZFP & central ray" method is no longer applicable. Considering the interference patterns caused by light from the top and bottom of an arrow some distance from the hologram, we can see thatthe marked tilt of the reference beam causes these two object beams to generate slightly different spatial frequencies. The subtense of the output rays is then determined by the difference in the output angles for those same frequencies. Recalling the discussion that led up to the final equation (35) of the previous chapter, we have the lateral magnification expressed as

$$MAG_{\text{lateral},x} = m\frac{\lambda_2}{\lambda_1}\frac{\cos\theta_{\text{obj}}}{\cos\theta_{\text{out},m}}\frac{R_{\text{out},m,x}}{R_{\text{obj}}} \tag{14}$$

This is the magnification in the x-direction, and requires knowledge of the corresponding image distance (or wavefront curvature). Diffraction in the y-direction is less clearly analyzed in our terms, but the angular subtense does not depend on the angles involved, so the corresponding equations follow as:

$$\begin{aligned}\frac{\Delta\theta_{\text{out},m,y}}{\Delta\theta_{\text{obj}}} &= m\frac{\lambda_2}{\lambda_1}\\ MAG_{\text{lateral},y} &= m\frac{\lambda_2}{\lambda_1}\frac{R_{\text{out},m,y}}{R_{\text{obj}}}\end{aligned} \tag{15}$$

An example: horizontal (y) focus

This magnification business may become a lot clearer if we work through a specific case. Consider exposing a hologram at 633 nm and later illuminating the finished hologram at 543 nm. The basic equations that tell us where various orders of images will show up are

$$\begin{aligned}\frac{\sin\theta_{\text{out},m,y} - \sin\theta_{\text{ill}}}{\lambda_2} &= m\frac{\sin\theta_{\text{obj}} - \sin\theta_{\text{ref}}}{\lambda_1}, \qquad m = 0, \pm 1, \pm 2, \ldots\\ \frac{1}{\lambda_2}\left(\frac{1}{R_{\text{out},m,y}} - \frac{1}{R_{\text{ill}}}\right) &= m\frac{1}{\lambda_1}\left(\frac{1}{R_{\text{obj}}} - \frac{1}{R_{\text{ref}}}\right)\end{aligned} \tag{16}$$

so if we plug in a few numbers:

$\theta_{\text{obj}} = 10°, \theta_{\text{ref}} = 45°, R_{\text{obj}} = 150\,\text{mm}, R_{\text{ref}} = 2000\,\text{mm}, \theta_{\text{ill}} = 60°,$

$R_{\text{ill}} = 1000\,\text{mm}$

we find that the location and angle of the $m = +1$ image is

$$\theta_{\text{out},m,x} = 24.1°, R_{\text{out},m,x} = 149\,\text{mm}$$

Now, what's the magnification? The relevant equations here are

$$\begin{aligned}MAG_{\text{lateral},m,y} &= \frac{width_{\text{image}}}{width_{\text{object}}} = m\frac{R_{\text{out},m,y}}{R_{\text{obj}}}\frac{\lambda_2}{\lambda_1}\\ MAG_{\text{longitudinal},m,y} &= \frac{depth_{\text{image}}}{depth_{\text{object}}} = m\frac{\lambda_2}{\lambda_1}\left(\frac{R_{\text{out},m,y}}{R_{\text{obj}}}\right)^2 = \frac{1}{m}\frac{\lambda_1}{\lambda_2}MAG^2_{\text{lateral},m,y}\end{aligned} \tag{17}$$

and the resulting magnifications are 91% lateral and 96% longitudinal.

The same example, but vertical (x) focus

In the same case as above, what happens along the other axis? Now we have a different set of equations telling us where things go:

$$\frac{\sin\theta_{\text{out},m,x} - \sin\theta_{\text{ill}}}{\lambda_2} = m\frac{\sin\theta_{\text{obj}} - \sin\theta_{\text{ref}}}{\lambda_1}, \qquad m = 0, \pm 1, \pm 2, \ldots$$

$$\frac{1}{\lambda_2}\left(\frac{\cos^2\theta_{\text{out},m,x}}{R_{\text{out},m,x}} - \frac{\cos^2\theta_{\text{ill}}}{R_{\text{ill}}}\right) = m\frac{1}{\lambda_1}\left(\frac{\cos^2\theta_{\text{obj}}}{R_{\text{obj}}} - \frac{\cos^2\theta_{\text{ref}}}{R_{\text{ref}}}\right) \tag{18}$$

and the $m = +1$ image is at

$$\theta_{\text{out},m,x} = 24.1°, R_{\text{out},m,x} = 149\,\text{mm}$$

The magnification equations here are

$$MAG_{\text{lateral},m,x} = \frac{height_{\text{image}}}{height_{\text{object}}} = m\frac{\lambda_2}{\lambda_1}\frac{\cos\theta_{\text{obj}}}{\cos\theta_{\text{out},m,x}}\frac{R_{\text{out},m,x}}{R_{\text{obj}}}$$

$$MAG_{\text{longitudinal},m,x} = \frac{depth_{\text{image}}}{depth_{\text{object}}} = m\frac{\lambda_2}{\lambda_1}\left(\frac{\cos\theta_{\text{obj}}}{\cos\theta_{\text{out},m,x}}\right)^2\left(\frac{R_{\text{out},m,x}}{R_{\text{obj}}}\right)^2 = \frac{1}{m}\frac{\lambda_1}{\lambda_2}MAG^2_{\text{lateral},m,x} \tag{19}$$

and work out to be 92% lateral and 99% longitudinal.

Intermodulation Noise

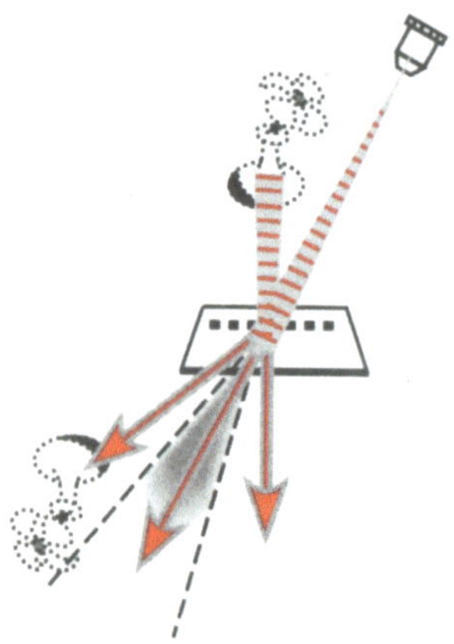

Another component of the light is what we have been calling "halo light," which is also called "intermodulation noise" and "object shape dependent noise." It produces a diffuse fan of light around the zero-order beam, the attenuated straight-through illumination beam. If the non-linearities in the emulsion response are very strong, it also causes diffuse light to appear in and around the image, but here we will concentrate on the halo of light around the zero-order beam, and find the conditions that will keep it from overlapping the image light. The key question is "what is the angle of the halo fan?"

The halo is caused by the interference of light from points on the object. We have been considering the hologram as though there were only one object point at a time. When there are many points (the usual case), coarse interference fringes arise from interference among them. Because the object points are all at roughly the same distance from the hologram, the gratings that "intra-object" interference produces are of approximately constant spatial frequency across the hologram. To find the limits of the fan of halo light, we only need to consider interference between the most widely spread object points (which will create the highest spatial frequency pattern). We designate the angle subtended by the object as $\Delta\theta_{\text{obj}}$. The maximum spatial frequency of the intra-object interference grating (f_{IMN}) is then, assuming that the center of the object is perpendicular to the plate,

$$f_{\text{IMN}} = \frac{2\sin\left(\frac{\Delta\theta_{\text{obj}}}{2}\right)}{\lambda_1} \tag{20}$$

To avoid overlap of the halo light and the image light, it is only necessary that the minimum spatial frequency of the image gratings be greater than f_{IMN}. This relationship is expressed as

$$f_{\text{obj-min}} = \frac{\sin\theta_{\text{ref}} - \sin\left(\frac{\Delta\theta_{\text{obj}}}{2}\right)}{\lambda_1}$$

$$\geq \frac{2\sin\left(\frac{\Delta\theta_{\text{obj}}}{2}\right)}{\lambda_1}, \quad \text{or} \tag{21}$$

$$\sin\theta_{\text{ref}} \geq 3\sin\left(\frac{\Delta\theta_{\text{obj}}}{2}\right)$$

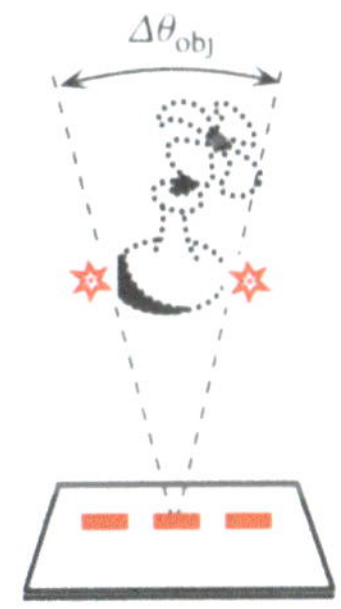

Thus the size, or rather the angular subtense, of an object is limited by the choice of reference beam angle, if the overlap of halo light is to be avoided. If the object has an angular subtense of 30°, for example, then the reference beam angle must be at least 51°. The intensity of halo light drops off smoothly from the center to the edges of the fan, so these limitations can be stretched a bit before much image degradation is visible. However, there are several other sources of scatter that can send illumination beam light into the image area, so that controlling halo is only one issue to pay attention to.

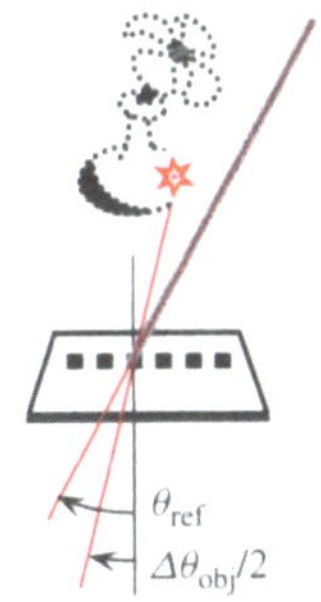

Conclusions

Off-axis holograms may require three times as many equations as diffraction gratings, but they involve the same physical principles and fit in the same logic that we started developing several weeks ago. Compared to in-line holograms, they require one new equation, the "cos-squared" focusing law that describes the *astigmatism* of off-axis holographic imaging. Astigmatism has minimal implications for virtual images, but will soon have to be dealt with very carefully for real images. In exchange for this mathematical complexity, we have moved into the domain of holograms that produce truly impressive three-dimensional images!

References

i. Leith, E. N. and J. Upatnieks, (1962). "Reconstructed Wavefronts and Communication Theory," *J. Opt. Soc. Amer.*, **52**, pp. 1123-1130.

ii. Leith, E. N. and J. Upatnieks, (1963). "Wavefront Reconstruction with Continuous-Tone Objects," *J. Opt. Soc. Amer.*, **53**, pp. 1377-1381.

iii. Leith, E. N. and J. Upatnieks, (1964) "Wavefront Reconstruction with Diffused Illumination and Three-Dimensional Objects," *J. Opt. Soc. Amer.*, **54**, pp. 1295-1301 (1964). This famous "Train and Bird" hologram is on display at the MIT Museum.

iv. Meier, R. W. (1965) "Magnification and Third-Order Aberrations in Holography," *J. Opt. Soc. Amer.*, **55,** pp. 987-992 (1965).

v. Ward, A. A. and L. Solymar, (1986). "Image Distortions in Display Holograms," *J. Photog. Sci.*, **24**, pp. 62-76.

CHAPTER 11

Non-Laser Illumination of Holograms

Introduction

Laser-lit off-axis transmission holograms remain the "holographer's holograms," even today. The images are razor sharp, and can reach from the tip of your nose to the horizon (we will not dwell on the drawbacks of "laser speckle" at this point). But laser illumination of holograms presents some serious practical problems in many image display environments. High-powered laser light is still expensive, in terms of dollars per lumen, and uncommon—compared to sunlight, for example. Some kinds of laser light can impractical: most big gas lasers have a specific startup procedure that must be followed, take a while to warm up, require cooling water, and have various subsystems that may go sour. Also, the beam itself is typically expanded and cleaned with a spatial filter that needs routine cleaning and tweaking for best performance. And, various government agencies seem to regard almost all lasers bigger than laser-pointer size as "death rays." The amount of paperwork required to provide laser illumination for large-scale holograms is incredible—for instance, the state of New York requires a "Mobile Laser Operator's Certificate of Competence" for everyone who would plug in and turn on a laser in a public place. The desire to bring holograms out of darkened basement laboratories and into the public's awareness has motivated several approaches to white-light viewable holograms, resulting in the development of the white-light transmission or "rainbow" hologram, and the white-light reflection hologram. We emphasize that these holograms must still be made with lasers; they are specially designed to be viewable in white light. Even that light must have spatial coherence, coming from something approximating a point source (a line source, in some cases). In this chapter, we will look at the problems of trying to view ordinary holograms with coherent and incoherent sources, and examine a few preliminary solutions to the problems involved.

Problems with Laser Illumination

Let's look again at the laser illumination situation. The most common gas lasers, by far, are helium-neon gas lasers, our beloved He-Ne type, which are readily available at modest cost for powers up to 5 milliwatts or so. We will see below that 5 mW is just enough to illuminate a 4" × 5" hologram under ordinary conditions, which is why we speak of laser light as expensive. 5 mW is also about the power output at which one should start thinking seriously about eye-safety issues (at about 500 mW, it's time to start worrying about setting things on fire, too!) Low powers are adequate for darkened laboratory conditions, but we have to be thinking of bringing holograms out into the real world at every opportunity, and then the stakes go up considerably. For large holograms, on the order of a meter square, ion lasers are the only choices as powers on the order of a watt are required. These lasers are expensive to own and to operate (10 kW

3-phase input power, plus cooling water). Diode lasers offer a reasonable alternative to He-Ne lasers, and as of this writing lab-grade red 635 nm diode lasers are available up to tens of milliwatts at somewhat lower prices than He-Ne lasers. The output beams of common diode lasers are markedly elliptical, requiring special optics for shaping into more useful round beams, and the temperature has to be controlled. Diode lasers are also available (though not cheap) at blue and violet wavelengths like 475 nm and 405 nm. To get green, it's necessary to take advantage of the fact that some materials respond to high light intensities in a nonlinear fashion (generating second harmonics—what you would call harmonic distortion if your audio amplifier did it). Diode-pumped frequency-doubled lasers typically start with an infrared diode laser at 1064 nm whose output is converted to a beautiful 532 nm green output with powers up to a few hundred milliwatts, though at the high end these are as costly as ion lasers. And none of these systems addresses the chronic problems of laser speckle and safety registration that are bound to arise.

Photometry: how to get 20 ft-L

Let's take a moment to examine the question of "how much laser illumination power is needed to provide a reasonably bright image?" First, our conventional method of measuring laser power, in milliwatts, tells us how much heat such a beam can deliver, but doesn't tell us how bright the beam will appear to be (it could be invisible, *e.g.,* if it were infra-red or ultra-violet light). As we saw in Chapter 2, Eq. (15), the conversion from watts to lumens (or "visually apparent power") is via the CIE Eye Response Curve, established in the 1930s and regularly refined ever since. At the peak of the eye's sensitivity, in the green area of the spectrum, one watt of radiation produces 683 lumens. That is our central calibration point. Second, we need to know what a "reasonably bright image" is supposed to be. Here, we can simply rely on long experience with color television sets, where a peak white is expected to have a luminance of 20 foot-Lamberts (in the old notation), or 70 lumens per square meter per steradian (expressed in SI units, also 70 "nits" or candles/m^2). This is the measure that describes how bright a diffusely illuminated surface will seem to be; as strange as its units may be, it works pretty well. We would have to take a longer detour through radiometry and photometry than seems justified to explain much more, but at this point we are simply ready to plug in some numbers and look at the results.

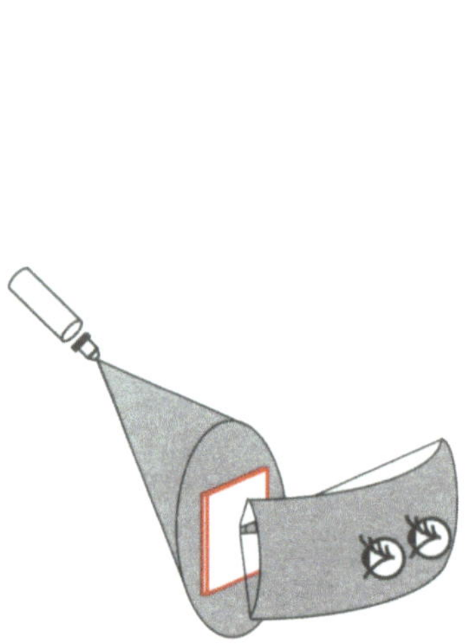

4" × 5" hologram

Let's start with a 100 mm x 125 mm (4" × 5") hologram; for simplicity, lets assume we have 1 mW of He-Ne illumination, or 0.17 lumens. Let's then assume that the illumination overfills the hologram by a factor of two (to provide more uniform lighting, or because the aspect ratio at 45° is unfavorable; 2× is optimistic, by the way), so that the illuminating intensity averages to be 6.85 lumens/m^2. Let the average diffraction efficiency be 20%, which is plausible for a well-bleached hologram, and let the viewing zone be 60° wide and 30° high (or 0.55 steradians, generous but plausible for a laser transmission hologram). Now we can obtain an overall luminance of 2.5 nits. However, the peak extended-white luminance of an "average" scene is typically five times the average, so we could expect a white surface in the scene to have a luminance of 12.5 nits. To bring this up to the hoped-for 70 nits, we obviously need to increase the illumination five-fold, to about 5.6 mW—not a small He-Ne laser!

This number may come as a surprise to those of us who are used to peering into dim holograms lit by laser pointers or weak He-Ne lasers, but that usually happens in a darkened basement lab, and we are talking here about images bright enough to be seen in ordinary *room light situations*, alongside televisions and other everyday imaging devices. The power scales with the area of the hologram, so that a 12" × 16" hologram would require about 60 mW, which is getting close to the largest He-Ne lasers made (the venerable Spectra-Physics 125). What are we to do? Some folks have made an art form of limiting the view-zone angle to the minimum acceptable, perhaps 15° across and 10° high, which provides an "antenna gain" of brightness of more than 10×, but limits "look-around," one of the most charming features of holographic 3-D (limited vertical viewing zones are, however, one of the keys to the brightness of rainbow holograms). Note too that as the image approaches the hologram plane, the brightness of an extended-white area is limited by the diffraction efficiency at that area, and we lose the 5× advantage due to averaging a peak luminance over the entire hologram area. Thus the prospects for bright laser-illuminated holograms are dim indeed!

Of course, higher-power lasers are readily available, but only at substantial expense, and with the other problems mentioned above. Practical holographic display has turned instead to non-laser, or "thermal" sources of illumination. The problem is that the output power of such sources increases as their source area and spectrum width increase, both of which cause a degradation in the sharpness of the resulting image. Thus there is an inevitable tradeoff between image brightness and sharpness that is determined by the quality of the illumination. This chapter will examine first the sensitivity of holograms to color blur and source size blur, and then re-examine the qualities of candidate light sources.

Sources of Image Blur

Our discussion of holography, including image reconstruction, has presumed the use of sources of arbitrarily high spatial and temporal coherence. Which is to say that the sources were perfectly point-like, and monochromatic (indeed, single-frequency). That will typically still be the case for image recording (some relaxations are possible, which we will not have time to describe here), but the fact is that hologram viewing is possible under some circumstances with highly incoherent sources, resulting in only mild blurring of the resulting images. The two dimensions of coherence, temporal and spatial, correspond to independent contributions to image blur, and we will discuss them one at a time.

Color blur

As in many of our hologram analyses, we will consider the angle and distance issues in sequence, corresponding to the grating and Gabor zone plate components of the holographic fringes. As a simplification, we will assume at the end that the object beam angle and central output beam angles are perpendicular to the plate, as is generally the case in display applications. Then, we can expand the output angle as a function of illumination wavelength as

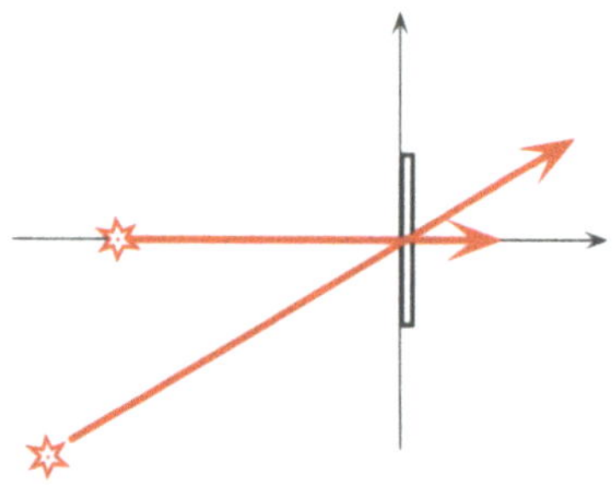

$$\frac{d(\sin\theta_{\text{out}})}{d\lambda_2} = \cos\theta_{\text{out}}\frac{d\theta_{\text{out}}}{d\lambda_2} = m\frac{1}{\lambda_1}\left(\sin\theta_{\text{obj}} - \sin\theta_{\text{ref}}\right)$$
$$\frac{\sin\theta_{\text{ref}}}{\lambda_1} = \frac{\sin\theta_{\text{ill}}}{\lambda_2},\ \text{if}\ \theta_{\text{out,central}} = \theta_{\text{obj}} = 0,\ \text{so} \tag{1}$$
$$\Delta\theta_{\text{out}} = -\sin\theta_{\text{ill}}\frac{\Delta\lambda_2}{\lambda_{2,0}}$$

where $\lambda_{2,0}$ is the central wavelength of the illumination, or the filter over the white light source in the sketch. If we consider the angular resolution of the unaided eye to be about one minute of arc (290 microradians), then a hologram illuminated at 45° with 540 nm light would need a spectrum width of less than 0.22 nm to avoid noticeable color blur, a very narrow spectrum indeed! Note that on-axis holograms, with very small θ_{ill}, have much lower color blur than off-axis holograms for the same source bandwidths, as observed earlier.

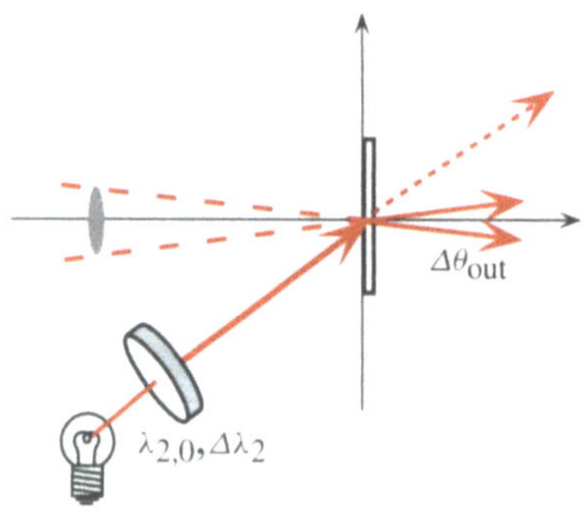

This analysis gives us the *angular subtense* of the color blur as seen from the hologram plane. But to gauge the blur at other viewing distances, we will need to know the *location* of the color blur too. In fact, the color-blurred image of a point source will be tipped at an angle of special interest, which we will come to know as the "achromatic angle," designated by α.

The central location of the image is given by the same "$\cos^2\theta$" equation as before (note that we are concerned with the vertically focused blur in this case—there is no horizontal component to color blur). Thus

$$\frac{1}{R_{\text{out}}} = (1)\frac{\lambda_{2,0}}{\lambda_1}\left(\frac{1}{R_{\text{obj}}} - \frac{\cos^2\theta_{\text{ref}}}{R_{\text{ref}}}\right) + \frac{\cos^2\theta_{\text{ill}}}{R_{\text{ill}}} \tag{2}$$

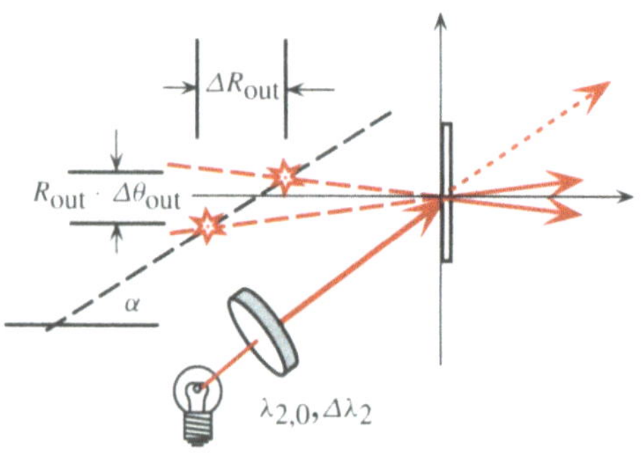

and the vertical extent of the color blur of the image, or its height, is given by

$$h_{\text{color blur}} = R_{\text{out}} \cdot \Delta\theta_{\text{out,color}} = R_{\text{out}} \cdot \sin\theta_{\text{ill}} \cdot \frac{\Delta\lambda_2}{\lambda_{2,0}} \tag{3}$$

The detailed variation of the image distance with wavelength is given by

$$\frac{d}{d\lambda_2}\left(\frac{1}{R_{\text{out}}}\right) = -\frac{1}{R_{\text{out}}^2}\frac{dR_{\text{out}}}{d\lambda_2} = \frac{1}{\lambda_1}\left(\frac{1}{R_{\text{obj}}} - \frac{\cos^2\theta_{\text{ref}}}{R_{\text{ref}}}\right)$$
$$\frac{dR_{\text{out}}}{d\lambda_2} = -\frac{R_{\text{out}}^2}{\lambda_2}\left(\frac{1}{R_{\text{out}}} - \frac{\cos^2\theta_{\text{ill}}}{R_{\text{ill}}}\right) \tag{4}$$
$$\Delta R_{\text{out}} = -\left(R_{\text{out}} - R_{\text{out}}^2\frac{\cos^2\theta_{\text{ill}}}{R_{\text{ill}}}\right)\frac{\Delta\lambda_2}{\lambda_2}$$

The tangent of the angle of the blur image is then given by

$$\tan\alpha = \frac{R_{\text{out}} \cdot \Delta\theta_{\text{out}}}{\Delta R_{\text{out}}}$$

$$= \sin\theta_{\text{ill}} \frac{1}{1 - \frac{R_{\text{out}} \cdot \cos^2\theta_{\text{ill}}}{R_{\text{ill}}}} \quad (5)$$

As a rule, the illumination distance is much greater than the image distance, so that the second term can be ignored and the generally useful relationship becomes,

θ_{ill}	α
30°	26.6°
45°	35.3°
60°	40.9°

$$\tan\alpha = \sin\theta_{\text{ill}} \quad (6)$$

A simple off-center zone plate model of an off-axis hologram would predict, for collimated illumination, that the achromatic angle would be exactly equal to the illumination angle. However, this chapter's more careful consideration shows that the color blur is actually tilted significantly more toward the *z*-axis. The difference is large enough to decrease markedly the color blurring in holograms that properly use the achromatic angle concept to help correct color blur (discussed in subsequent chapters).

Source size blur

We have seen before, for on-axis holograms, that motion of the illumination source away from the axis moves the image off the axis too. For off-axis holograms, the relationship is only slightly different: a vertical motion of the light source produces a vertical motion of the image through an angle that is proportional to the cosine of the illumination angle:

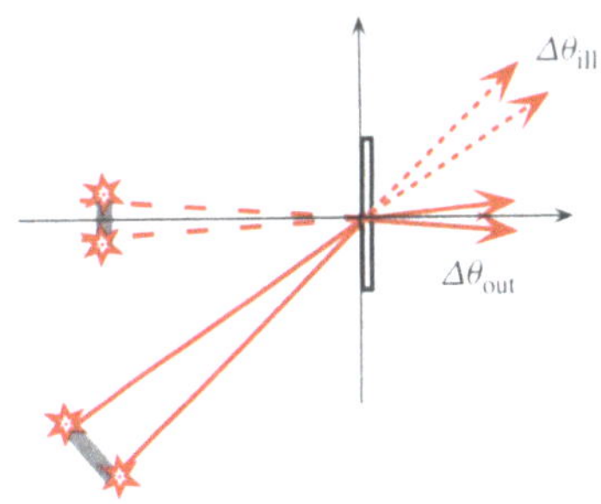

$$\frac{d(\sin\theta_{\text{out}})}{d\theta_{\text{ill}}} = \cos\theta_{\text{out}} \frac{d\theta_{\text{out}}}{d\theta_{\text{ill}}} = \frac{d(\sin\theta_{\text{ill}})}{d\theta_{\text{ill}}} = \cos\theta_{\text{ill}}$$

$$\Delta\theta_{\text{out}} = \cos\theta_{\text{ill}}\, \Delta\theta_{\text{ill}} \quad (7)$$

There is also a small variation of the vertically-focused image distance, which we shall consider to be negligible:

$$\frac{d}{d\theta_{\text{ill}}}\left(\frac{1}{R_{\text{out}}}\right) = -\frac{1}{R_{\text{out}}^2}\frac{dR_{\text{out}}}{\theta_{\text{ill}}} = \frac{1}{R_{\text{ill}}}\frac{d}{d\theta_{\text{ill}}}\left(\cos^2\theta_{\text{ill}}\right)$$

$$\Delta R_{\text{out}} = \frac{R_{\text{out}}^2}{R_{\text{ill}}} \sin 2\theta_{\text{ill}}\, \Delta\theta_{\text{ill}} \approx 0 \quad (8)$$

To consider the effect of a finite source size, we simply imagine that, instead of moving a single source point over the angle $\Delta\theta_{\text{ill}}$, all those points are present simultaneously, and that they are incoherent with each other so that the images they produce all add in intensity. The result is an enlarged blur spot, with the height of the source-size blur given by

$$h_{\text{source-size blur}} = R_{\text{out}} \cdot \Delta\theta_{\text{out, source}} = R_{\text{out}} \cdot \cos\theta_{\text{ill}} \cdot \frac{\Phi_{\text{ill}}}{R_{\text{ill}}} \quad (9)$$

where we signify the diameter of the light source by Φ_{ill}.

Motion of the source from side to side gives a variation of output angle equal to the variation of illumination angle (no cosine effect), so that the width of the source-size blur is simply

$$w_{\text{source-size blur}} = R_{\text{out}} \cdot \Delta\theta_{\text{ill}} = R_{\text{out}} \cdot \frac{\Phi_{\text{ill}}}{R_{\text{ill}}} \tag{10}$$

Thus, to keep the blur due to source size below the perceptible limit of the human eye, the source must subtend an angle of less than one minute in width, and a little over a minute in height. This is about the angular subtense of a US quarter-dollar coin at a distance of 82 meters (270 feet)!

The angular sensitivity of the eye puts stringent limitations on the spectrum width and area of conventional "thermal" or non-laser sources of illumination. Virtually monochromatic and point-like sources are needed in the general case. We will go on to consider some of the ways around this apparent dilemma.

Narrow-Band Illumination

Various sorts of more-or-less monochromatic light sources are available to us, and we will consider a few of them in this section.

Mercury arc

There are a wide range of gases that, under stimulation by electrical currents, give rise to one or more relatively pure spectral lines. Neon is widely seen in advertising signs, and sodium makes an appearance in street lights. However, most of these sources are large-area or large-volume tubes, operating at relatively low pressures. Mercury-vapor lamps, operating at high pressure, are probably the best-known point-like gas sources of single-color light. The higher the pressure, the smaller the glowing part of the discharge becomes, but the wider the output spectrum becomes also, so there is some room for trade-offs. The most prominent mercury output spectral lines are "green" at 546 nm, with a line width of about 5 nm, and a "yellow" pair of lines at 577 and 579 nm, which are generally widened to a total of 7 nm. A one-hundred-watt lamp radiates about 150 mW/steradian (sr) in the 546 nm line, which comes from a bright spot near the cathode that is about a millimeter across.

Filtered incandescent light

An incandescent tungsten wire emits radiation over a wide spectral band, peaking in the infra-red, but with plenty of energy out to the deep blue. The higher the temperature of the tungsten, the greater the number of watts of light per square millimeter (and the more blue light, relative to red) are emitted. However, the lifetime of the wire drops quickly due to evaporation of tungsten from the hot surface. The addition of halogen gases (iodine, especially) and the use of a very hot lamp envelope (quartz, to stand the heat) recycles some of the evaporated tungsten back onto the wire, extending the lifetime of such lamps, especially at high temperatures. Even so, temperatures are limited to about 3400°K for practical lifetimes (a thousand hours or so).

Narrow-band interference filters can be put over the beam to select out a fairly narrow spectral band, but obviously the luminous flux decreases linearly with spectral width (or even faster, as the peak transmittance of such filters decreases for narrow filters; it is only 50% in the best of cases). Filter bandwidths 15 or 20 nm wide are typical of the narrower pass bands.

Light-emitting diodes
In recent years LEDs (which are the incoherent cousins of semiconductor lasers) have advanced from little indicator lights to serious illuminators;[ii] some high-output ones now even come with the same eye-safety warnings as lasers. LEDs have lifetimes in the tens or hundreds of thousands of hours, generate little heat, and are available in wavelengths throughout the visible spectrum (including some delightful and unexpected orange-yellows and blue-greens). Their spectral widths are at least ten times those of semiconductor lasers, though—commonly about 25–50 nm. Resonant-cavity LEDs (RCLEDs) are sort of half-way between LEDs and lasers, and have spectral widths of between 5 and 20 nm, but aren't yet available in the full palette of wavelengths of normal LEDs.

Holograms as narrow-band filters
Rather than put the narrow-band filter in the light source (which is difficult with sunlight, for example), it is possible to put it into the hologram. Very thick transmission holograms can have considerable angle and wavelength selectivity (the Bragg selectivity effects of volume holograms), but it is reflection holograms that offer the highest wavelength selectivity. We will see that their volume structure includes multiple layers of alternating high and low refractive indices, much like a vacuum-coated interference filter, and that they can self-select a narrow (approximately 15 nm wide) portion of the visible spectrum for their reconstruction. Of course, the narrower the reflection spectrum, the slimmer the "photon catch" and the dimmer the image becomes, even if sharper. Thus some reflection holograms are deliberately processed to widen their reflection spectrum to increase brightness at the expense of sharpness.

Point-Source White Illumination

If we have a hologram that can be illuminated with white light (about which we'll learn more in coming chapters), then we will need not just white light but a point source of it. Unfortunately, the ideal point source of white light simply doesn't exist. So-called "white" lasers put out three or so wavelengths, not a continuous spectrum (as the sun does). Arc lamps are the next best thing, and have a high enough luminance to be virtual point sources. After that come incandescent lamps. Far behind are fluorescent lamps, and then the holographer's nightmare, an overcast hazy day! "White" LEDs, incidentally, are usually blue LEDs exciting a phosphor that adds red and green light, and emit light from a fairly large and diffuse region.

The sun
The sun is just another incandescent source, although it has a higher temperature and a higher brightness (about 4000 lumens/m^2-sr). It subtends an angle of about 0.5°, or a solid angle of 76 microsteradians (almost exactly the same as the moon). The luminance of the surface of the sun, as seen directly overhead from the earth's surface on a clear day, is about 1.6×10^9 nits.

High-pressure arc (xenon)
Xenon lamps (possibly containing a small amount of mercury) can be made with a very short arc and thus create a small region of sunlight-like illumination. They also make a lot of ultraviolet light, as it isn't attenuated by the quartz envelope (and a secondary problem is that the ultraviolet light can ionize the oxygen around the lamp and generate substantial amounts of ozone). They are often used as light sources in projectors and as headlights in high-end automobiles. The discharge near the cathode of these lamps reaches a luminance of 1.8×10^8 nits.

Zirconium arc
Another arc lamp that can produce a sub-millimeter emission region of intense white light uses electrodes coated with zirconium oxide, and can reach a luminance of 4.5×10^7 nits, within a particularly well-defined area. These lamps are sometimes used as microscope illuminators, but they are becoming hard to find.

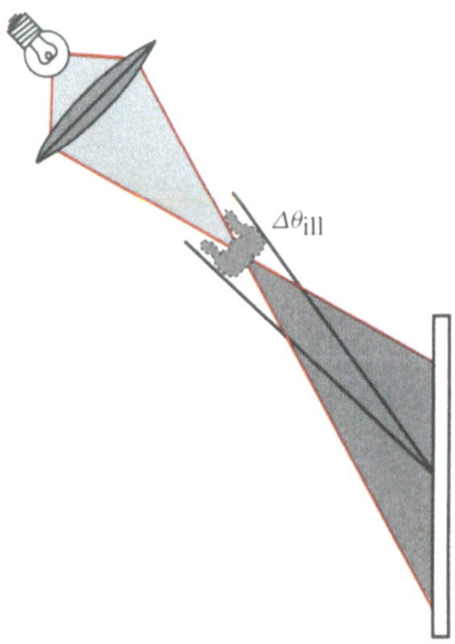

Quartz-halogen lamps
Rather than use a large filament to illuminate a hologram, we can use a small filament and focus the light onto the hologram with a lens or concave mirror. This produces a large-area image of the filament near the output aperture of the illuminator, and the same formulas apply if we use the area of this source image instead of the lamp itself. The "optical brightness" theorem says that the luminance of an extended surface stays constant as that surface is successively imaged through an optical system (assuming that the aperture on the measuring device is the limiting aperture—*e.g.*, the pupil of the human eye—which is usually the case). That is, the brightness of the image of the filament (in lumens/m^2-sr) is the same as the brightness of the filament itself. The hologram gets more light because the filament image fills a larger solid angle as seen from the hologram than the filament would at the same distance. There is no other way to get more flux to the hologram surface! The luminance of a tungsten filament reaches about 2.4×10^7 nits.

Image Depth Effects

So far, we have been thinking of the hologram as a window that the viewer looks through, almost with his or her nose pressed up to it. The height (and width) of the blur image increases linearly with image depth because the image blur subtends a constant angle, $\Delta\theta_{out}$, as seen from the plane of the hologram.

But if we let the viewer back away from the hologram, the angle subtended *at the viewer's eye* by the image blur decreases, and it is the blur angle at the eye that must be below one minute of arc for the image to appear sharp. If we designate the viewer's distance from the hologram as D_{view} and the depth of the image as D_{image} ($= R_{out}$), then the blur angle at the eye (call it ω_{blur}) becomes:

$$\omega_{blur} = \frac{\left(h_{blur} = R_{out} \cdot \Delta\theta_{out}\right)}{D_{view} + D_{image}} = \frac{1}{\left(1 + \dfrac{D_{view}}{D_{image}}\right)} \Delta\theta_{out} \tag{11}$$

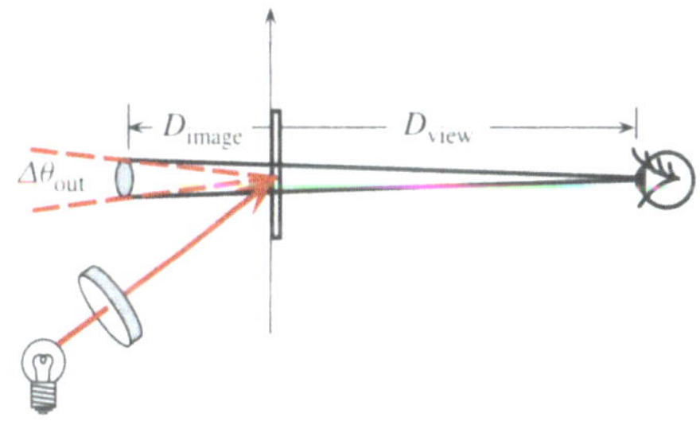

Thus, if the viewer's distance is much greater than the image depth, the visual blur decreases significantly. Or, considered differently, much larger sources may be used, allowing more illuminating flux to reach the hologram. For example, if a hologram is viewed at arms length (500 mm) under sunlight ($\theta_{ill} = 45°$, $\Delta\theta_{ill} = 0.5°$), then the image may be 25 mm deep (behind the hologram, or 23 mm in front) before any visual blurring is perceptible.

Standard-definition television offers a pixel size that is at least double the magic one minute of arc under good conditions. In practice, much greater amounts of blurring are tolerable in holograms too, as long as the visual center of interest is reasonably sharp. But the general trend has become clear: holograms have become things one looks *at* rather than *through*. They are held at arm's length, or viewed on a wall, considered more as a photograph than as a porthole into another spatial world.

Other Approaches

There are other things we can do besides simply trying to emulate the coherence of laser sources by brute force; we can try to design holograms and systems that are simply designed to work within the limitations of white-light illumination, or even to take advantage of some of its characteristics. Two approaches worth mentioning here are dispersion compensation and parallax limiting.

Dispersion compensation

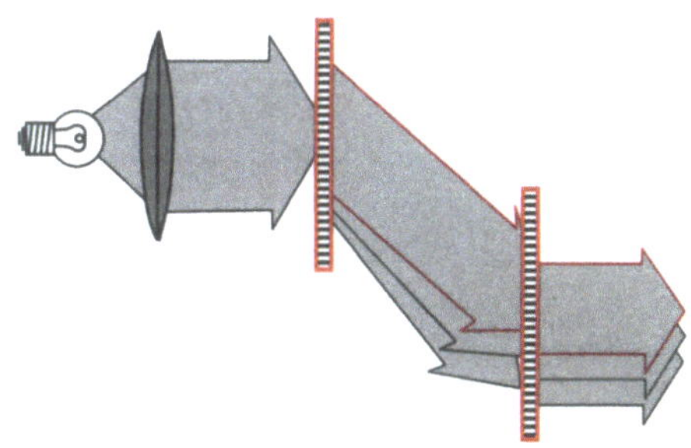

One way of partially overcoming the color blur of a wide-band illumination source is to pre-disperse the various colors from the illuminator so that they are incident on the hologram at angles that result in their all being diffracted at equal angles toward the viewer. This is generally done by diffraction with a pre-grating that has about the same spatial frequency as the hologram, so that rays that start parallel from the illuminator wind up parallel again after diffraction by both gratings. This means that R, G, and B images of a distant point would be superimposed, producing an image free of color blur! Different focal situations can achromatize on-axis images at any chosen depth, or from a viewpoint at a chosen on-axis distance. However, as the parts of the image move from the central location (or the viewer moves from that central location) the blurring (color fringing) will increase—it is not possible to achromatize a large volume of image space in this way, only specific points.

Parallax limiting

One way to think about vertical color fringing, which is the main effect of color blur, is that differing vertical perspectives are becoming mixed in different wavelengths. Objects at different depths shift by different amounts with respect to the hologram plane as the wavelength changes; they rotate and move up or down. If we see more than one wavelength at a time, we see more than one image at a time, and the difference of their locations causes a blurring (we will revisit this point of view in a later chapter). So, one way of eliminating this source of blur is to eliminate all but one perspective view in the vertical direction such that no rotation or shift is possible, and hence no blurring. This is one way of looking at the principle of white-light transmission "rainbow" holograms, which we will be talking about in detail fairly soon.

Conclusions

Most people who have seen holograms have never seen one illuminated with a laser. When we holographers get the chance to do so, we should remind ourselves what a rare pleasure it is, one that few "civilians" will ever enjoy. Laser-illuminated holographic images can extend from the tips of our noses to the far horizon, with exquisite sharpness at every depth. But from here, we will be moving toward white-light viewed images, for which we must give up those extravagant vistas for a "space in a box" that we can hold in our hands. It will be impressively and realistically deep, and perhaps provoke new kinds of spatial thinking, but don't forget to drag out your laser from time to time for a look at "the real thing!"

References

i. A newer sort of diode laser called a vertical cavity surface-emitting laser (VCSEL) makes a perfectly circular beam, though the available selection of these devices is still somewhat limited.

ii. Baba, J., A. Yaeda, H. Asakawa, T. Shibuya, and M. Wakaki (2007). "Development of Lighting System for Hologram Using High Power LEDs," *Proc. SPIE Practical Holography XXI,* **6488**, 64880Z.

CHAPTER 12

Phase Conjugation and Real Image Projection

Real Image Projection Techniques

In the preceding chapter, we saw that having an image be much closer to the plate—compared to the viewer's distance—allows the use of larger-area and wider-band sources (*i.e.*, brighter sources, if we're making them by filtering white light) without blurring the image noticeably. But this is not as simple as just putting the object closer to the hologram plate—there it will usually block parts of the reference beam! And it is difficult to arrange for attractive object illumination if things get too close to the plate. Thus there has been a lot of interest in techniques for optically "relaying" an image of a remote object, and then letting that relayed image serve as the subject of the hologram. Here, we will look briefly at some "conventional" techniques for image relaying, also called *real image projection*, and then concentrate on the more widely used holographic method.

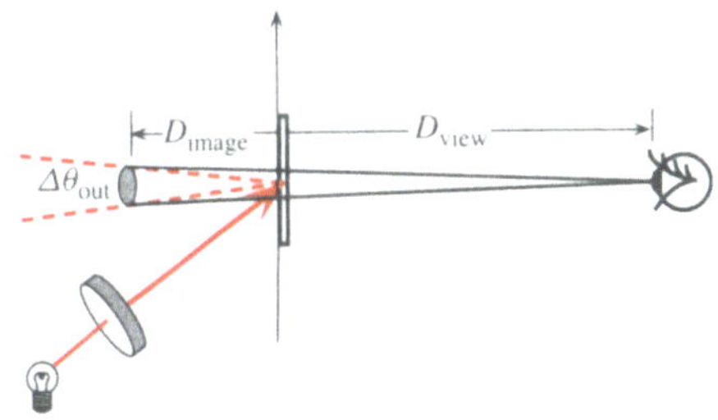

Positive lens

A positive lens can form either a real or a virtual image of a scene. Here, we will consider the $2F{:}2F$ geometry, in which the lens forms a real image, same-sized but upside-down, on the right-hand side. There, it can serve as an "optical object" for the hologram. We let the image straddle the hologram plane, with half of its depth on one side and half on the other, minimizing the maximum distance of the any part of the image from the hologram plate.

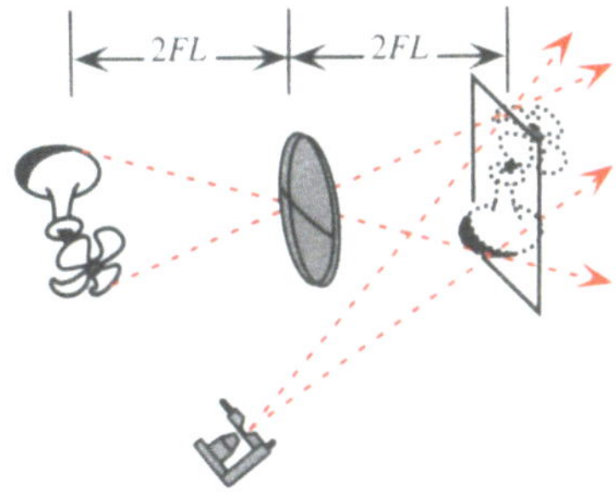

A major weakness of this approach is *vignetting*, a porthole effect on what the viewer sees caused by the limited width of the lens. Because the viewer can see only those parts of the image that have open lens area behind them, only the central area of the image appears. As the viewer moves from side to side or up and down, different parts of the image become visible as they are "back lit" by the lens. Getting around this problem requires using lenses that are much larger than the object, perhaps twice its width, which become very expensive (if even practical—the example shown here is already an $f/1$ lens, which is very unusual!).

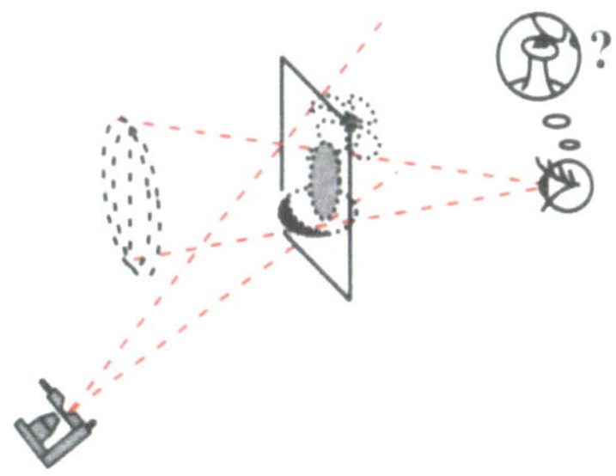

A secondary effect is that the image becomes distorted due to non-uniform magnification for those object parts not exactly $2F$ from the lens.

More complex setups use a second lens at the hologram plane to overcome the vignetting and non-uniform magnification, but then it is the viewing area that becomes limited.[i] The "bottom line" is that very few holo-cameras have been built that use lenses as the main imaging elements.

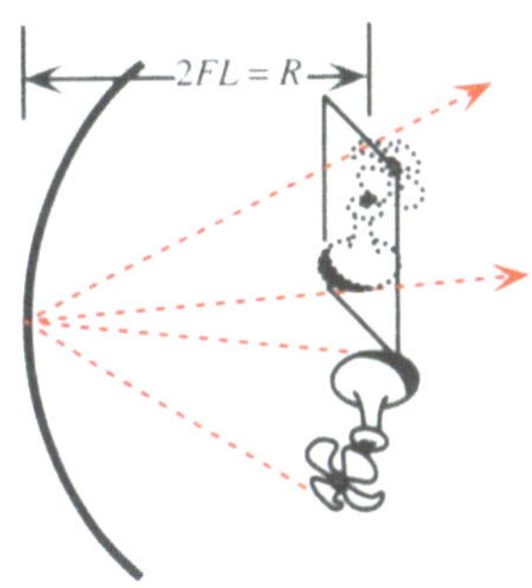

Concave mirror

Much larger diameter-to-focal-length ratios are possible with concave mirrors, of which only spherical mirrors are really practical to fabricate. Most science museums have displays of real images produced by such mirrors, usually as part of a magical illusion (a common one involves coins appearing to be within reach of the viewer but in fact some distance away). Vignetting is not so much of a problem in this case, but the non-uniform magnification becomes even worse at large viewing angles.

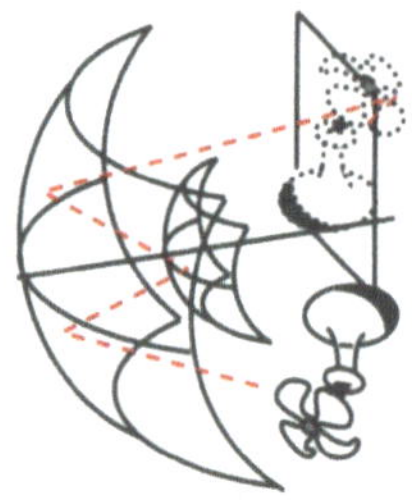

Multiple mirror systems (like the one sketched in the margin) have been described that correct for many of these distortions, but no such systems have yet found their way into practical use.[ii]

Two-step holography

Holograms have slowly emerged as the optics of choice for real image projection, taking advantage of their conjugate-image projection properties. They can be made almost arbitrarily wide relative to the object, thus affording an unvignetted image over quite a wide angle of view. Of course, this becomes a non-real-time or two-step process: the first hologram has to be exposed and processed, and only then illuminated to provide the image for the second or final hologram. With two holograms to deal with, we have to adopt a naming strategy to avoid confusion. The first hologram is often called the "master" hologram or the "H1." The second hologram is generally called the "transfer" hologram or the "H2." We will arbitrarily adopt H1 and H2 as the designators for most of our discussions here.

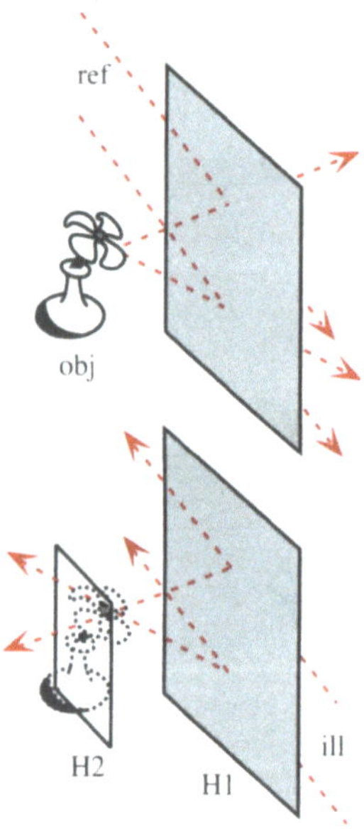

The second sketch shows the H1 being used in a way that is brand new for us—it is being illuminated through its back. The image it produces presents a real image to the H2 all right, and it is properly described as an $m = -1$ or *conjugate* diffracted order, but it has several properties that we will have to explore fairly carefully. This chapter will concentrate on this new type of conjugate image projection before going ahead to consider the resulting H2 and its properties.

Phase Conjugation—a Descriptive Approach

Holograms have a property that no other optical device has: the ability to generate a so-called "phase conjugate" image, one that behaves as though the waves from an object were literally traveling backwards in time to generate an image of that object focused in space. The image is called "phase conjugate" because the sign of the phase of its wavefront, as generated by the hologram, is exactly the opposite of that of the "true" or virtual image wave. It is the same conjugate reconstruction term that we associate with the $m = -1$ order, except that the illumination is now traveling in the direction opposite (typically right-to-left in diagrams) to the reference beam. Other descriptions of this kind of reconstruction are "reverse ray tracing," "time-reverse waves," and similar-sounding terms. Optical devices such as retroreflectors ("Scotchlite"™ for example) approximately conjugate the wave from a point source, sending the light roughly back in the direction that it came from. Photorefractive and nonlinear optical materials can be used in the four-wave mixing mode, which is also called "real-time holography," and produces an exactly phase-conjugated wavefront. But in this book we will limit our attention to the two-step holographic, recording and reconstruction, type of phase conjugation.

Thick hologram—general conceptual approach

We can understand the central concepts of wavefront phase conjugation with a fairly simple geometrical example. These are concepts that apply to thick as well as thin holograms, and this "proof" will include both cases. The basic idea is that the exposure of a holographic plate is a summation of energy over time, and that once the exposure is finished, the plate has no idea of whether time was running forward or backward. Consider holograms made in two different ways: by two diverging waves traveling from the left, as in "A" in the margin, and by two converging waves traveling from the right, as in "B." The curvatures and angles of the two reference waves (and also the object waves) are the same, but the waves are traveling in opposite directions, as though a high-speed movie of the "A" waves were being played backward in "B." That is, reference beam B is a converging beam, focused to the location of the point source for reference beam "A" (and object beam "B" is similarly focused to location of the object point source "A"). The identification of the reference and object beams is left deliberately ambiguous, as the result doesn't depend on which is which, but you might think of the upper arrow in "A" as the object beam, and the lower as the reference beam.

A

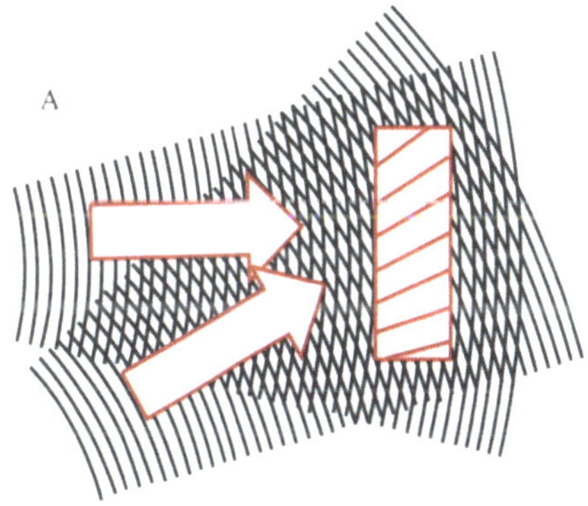

B

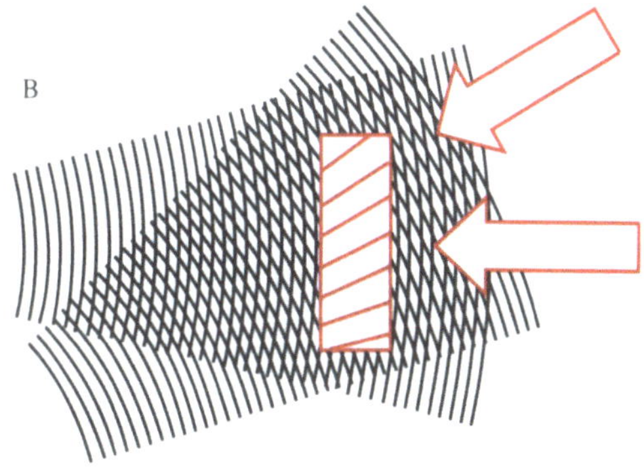

It might already be clear that the holograms/gratings produced in these two cases are identical! The exposure doesn't distinguish between waves traveling toward the right and toward the left. But let's keep each with its intended illumination for just a minute more.

Certainly, if we illuminate the hologram from "A," which we will call H_A in "C," with a replica of its reference beam, it will reconstruct a perfect replica of the diverging object wave, producing a virtual image of a point at the location of the source for "A." Likewise, if the hologram from "B," which we will call H_B, is illuminated with a replica of its reference wave as in "D," it will reconstruct a perfect replica of its converging object wave, producing a real image of a point, also at the location of the source for "A."

C

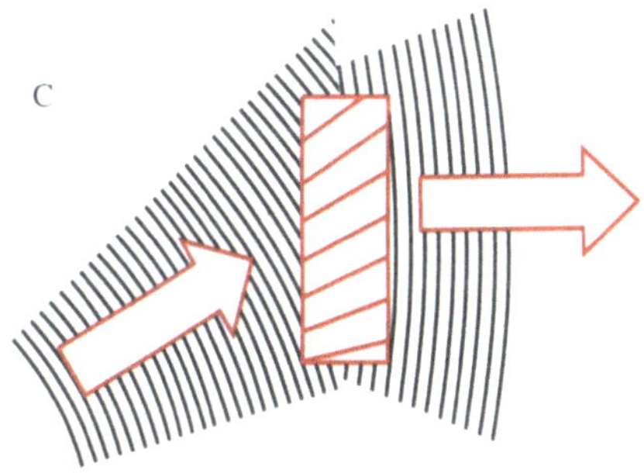

Now, the trick is to switch the two holograms, H_A and H_B, while nobody is looking. Because they are identical, no one will be able to tell the difference—each will reconstruct perfectly in the other's intended illumination. That is, a hologram of a point source (H_A) can produce a real image of a point simply by illuminating it through its back with a wave that has a particular relationship to the original reference wave—it must be its *phase conjugate* (it has the same shape, but is traveling as though reversed in time).

D

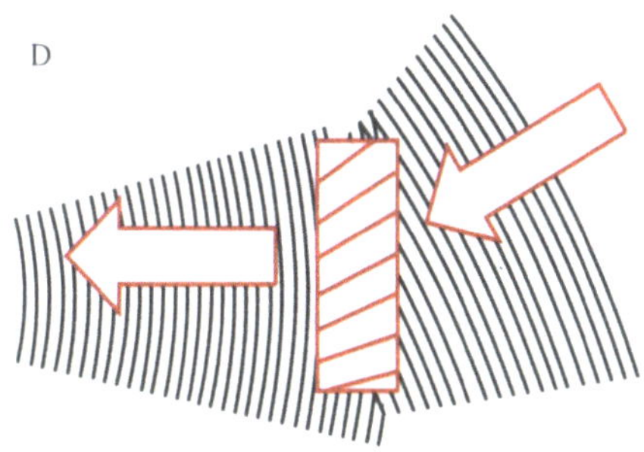

This generalizes to waveforms of arbitrary shape, as long as the amplitudes of the reference and illumination waves are uniform so as to produce an accurate replica of the object waves—no matter how complex the shape of the object waves. Thus, this principle applies to light diffusely reflected by a solid three-dimensional object, among many other things. For complex objects, which can be considered as collections of points arrayed in space, it becomes clear that the image produced by perfect phase conjugate illumination is also three-dimensional, perfectly undistorted and projected into

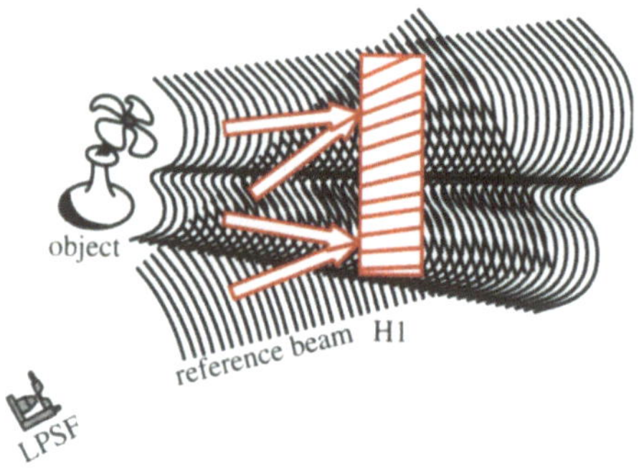

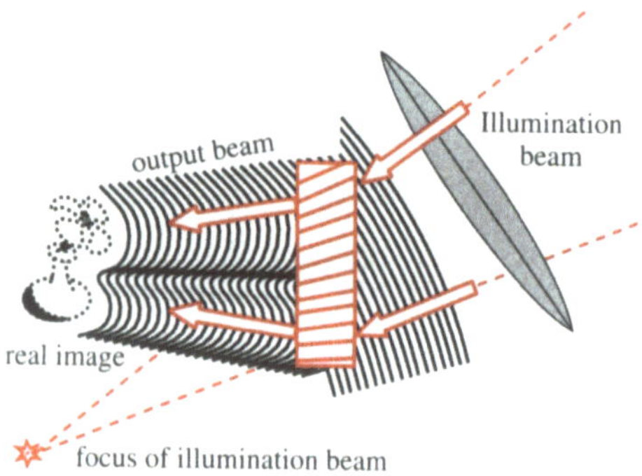

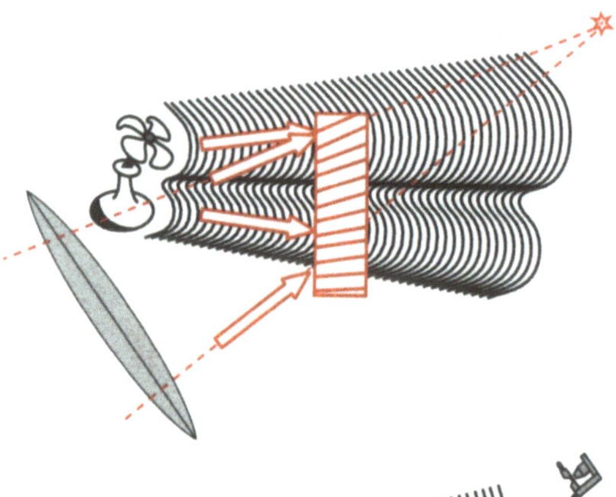

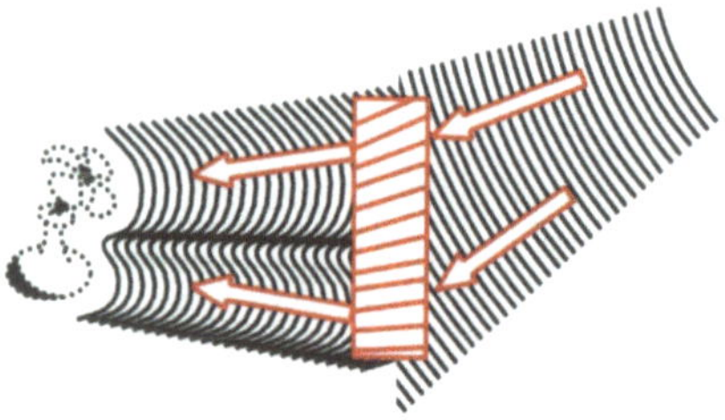

space, but with some peculiar properties that we will explore in just a minute.

Perfect Conjugate Illumination (Examples)

The accuracy of the 3-D reconstruction depends on the accuracy of the phase conjugated illumination, measured with respect to the reference wave. Thus we will briefly examine some practical implications of a few examples before continuing.

Diverging/converging
The first example in the margin shows a diverging reference beam, for which the phase conjugate is a converging reference beam, focused at the origin point of the reference beam. This illumination beam has to be converged by some optical device, typically a positive lens or a concave mirror, which has to produce an accurate point focus, without any aberrations. Note that, in general, the optic for producing such a beam has to be significantly larger than the hologram it is intended to illuminate. Because the cost of an optical element typically seems to grow roughly in proportion to the third power of its maximum diameter (or faster!), lens size is an important economic consideration.

Converging/diverging
The converging optic may also be used for the reference beam, which is handy for illumination situations where there is no opportunity to use extra optical elements. However, very short illumination beams require impractically "fast" reference beam converging lenses, compared to virtual or direct image projection. The reference beam lens must be as close as possible to the hologram to keep its diameter to a minimum, which also makes the setup awkward.

Plane/plane
The most generally useful configuration is one in which collimated light is used for both the reference and illumination beams. We should have qualms about using *any* optical element after the spatial filter in a reference beam, due to the magnification of the effects of dust and so forth, but given the lack of practical alternatives, this seems like a reasonable compromise. The collimators need be only as big as the hologram, although some extra size helps keep the beams free of "edge ringing" patterns, and they can be placed as far from the plate as is convenient, which helps simplify the exposure geometry. Later on we will make white-light viewable holograms, for which the sun is a handy illuminator and which produces collimated light, so that a collimated reference beam is actually an appropriate choice.

Collimator Choices

A holographer typically needs at least one collimator, and preferably two, for making high-quality holograms. They are inevitably large, expensive, heavy, and easily damaged items, which brings new aspects of thoughtful care to the laboratory. Holographic-grade collimators are not available "off the shelf" anywhere, and have to be custom made or adapted from available components. Let's stop for a moment's practical discussion of some of the options that confront this choice.

Refractive collimators (lenses)

A fairly simple positive lens can produce a collimated beam with acceptable accuracy. Ideally, one surface of the lens should have an aspheric shape (called a Rubé lens), but non-spherical surfaces are incredibly expensive to make and test.

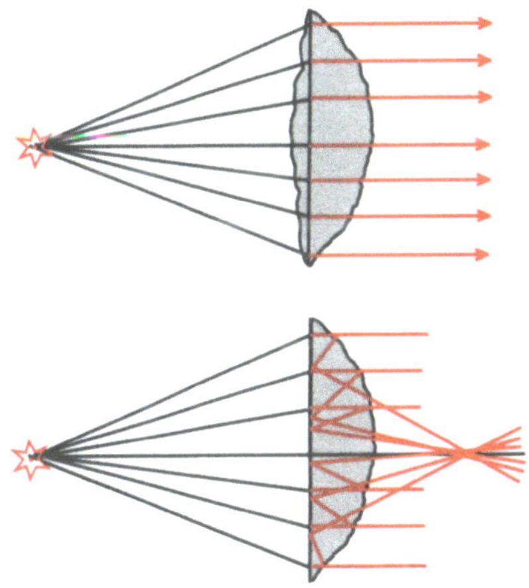

If the lens has spherical surfaces, the focal length at the edge will be slightly shorter than at the center, a phenomenon called "spherical aberration." This aberration can be minimized by making the surface of the lens facing the point source much less curved than the other side, in effect "bending" the lens. Using more than one element can further minimize the spherical aberration, and can also minimize the variation of focus with color, the chromatic aberration. But multiple surfaces can create serious multiple-reflection problems too, and even a single element can be very expensive in large sizes.

The optimum curvature of the lens surface facing the point source is about a tenth that of the other surface, but it is much cheaper to have a flat surface made than any spherical surface. There is very little glass to remove from the blank, and it is very easy to test (opticians usually have to make a testing element for every new surface curvature). Thus all of the curvature is usually put on one surface, the one facing away from the point source. The problem is that all of these compromises mean that the lens focal length has to be at least four times the diameter of the lens or the ray-pointing errors at the edge of the lens will become unacceptable. The exact criterion depends on the use the holograms will see, and the precision of the imaging. But limiting simple plano-convex collimators to $f/4$ or more (ratio of focal length to diameter) is a good rule of thumb.

Getting the overall surface curvature correct is something that most optical shops can do well. The only issue may be the maximum diameter of lens that their equipment can handle. However, variations of surface curvature and flatness over small distances are especially troublesome in laser applications. The glass surface often takes on a very shallow random waviness during polishing. When used as a collimator, the beam may look uniform at near distances, but will take on a mottled appearance a few meters downstream. This pattern resembles the roughness of the skin of an orange, and it is important to specify that there should be no "orange peel" on the lens surface.

Another issue is bubbles and "striae" in the glass. Normal specs for "bubble free" usually say "no bubbles bigger than a millimeter in diameter in the center third of the lens," or something similar. For this application we can tolerate no bubbles at all, which may limit our choices of glass types to the most popular varieties, such as BK-7. If the molten glass is improperly mixed before cooling, "ropes" of material of higher or lower index may form inside the glass. These produce "striae" (they look like streamers in the downstream light). This is often not a problem for conventional uses, but a disaster in a hologram reference beam. There are many front elements from big theatrical spotlights that would make wonderful collimators except for this usually-hidden defect.

A final issue is anti-reflection coatings. Without treatment, each naked glass surface reflects about 4% of the incident light, the Fresnel reflection. Enough light is doubly reflected to produce a point image two focal lengths in front of the collimator. A single layer of an evaporated material such as magnesium fluoride that is exactly a quarter of a wavelength thick can reduce the Fresnel reflection to under 1%. But because the point image concentrates the light, it can still be objectionably strong, and more elaborate coatings are usually needed. A three-layer coating can be designed to completely eliminate the reflection at a single wavelength (a "V-coat") or to reduce it to about 0.25% over the entire visible spectrum (a "BBAR-coat"). The choice is difficult because one always wants to keep open the option for full-color holography, even with the chromatic aberration of the lens in mind. Most of the lenses used in Benton's work have had BBAR coatings, and even so it's typical to move the collimator far enough away from the hologram to attenuate the effect of the weakened point image.

Holographers are occasionally tempted to have collimators made in acrylic plastic. A number of companies do very good work in plastic optics, and it grinds so much more quickly than glass that there are huge cost savings. However, it is very difficult to avoid an "orange peel" surface in polishing plastics, and the anti-reflection coatings available are not very durable (they have been compared to cake frosting by one dismayed holographer). And acrylic is soft and incredibly vulnerable to accidental damage during use.

Reflective collimators (mirrors)

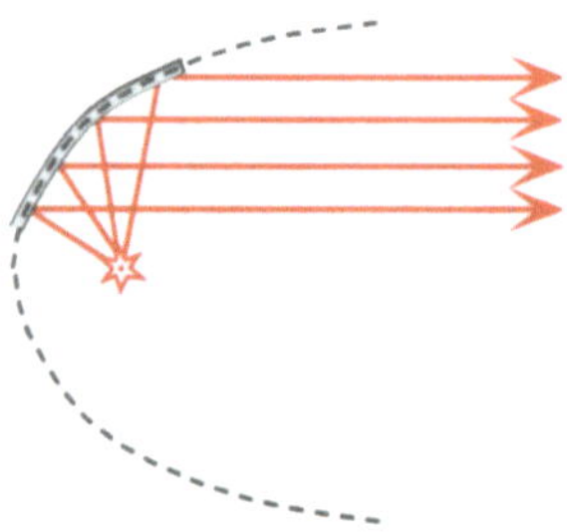

A realistic alternative to a collimating lens is a collimating mirror. Telescope mirrors, after all, produce point images of distant stars—the inverse of collimation! An ideal collimating mirror would have a parabolic shape, as most large telescope mirrors do. However, conventional telescopes put the pickup optics along the axis of the mirror, where they block the very center of the beam. This is not acceptable in holography, so the ideal mirror would be an off-axis section of a parabola. Unfortunately, only rotationally symmetric mirrors can be made with high accuracy, so a very large parabola would have to be generated, and one-sixth of it cut out for use—but the customer would have to pay for the fabrication of the other five-sixths as well.

Instead, holographers use spherical mirrors tipped off axis, and try to correct the resulting astigmatism by feeding the mirror through a lens that is also slightly tipped to produce the opposite astigmatism. Fairly inexpensive spherical mirrors are available from astronomy suppliers but typically have focal lengths at least ten times their diameters, which makes for setups that spread out over large distances.

"Orange peel" is also an issue for mirrors, and scratches and digs take the place of internal bubbles. Striae are not a problem, though, and neither is chromatic aberration. Holographers tend to be strongly partisan in their preference for refractive or reflective collimators, so be careful whom you ask about which type!

Diffractive collimators (gratings)
A third alternative would seem to be an ideal choice, but is not yet commercially available: a holographic collimator. If someone who had invested in a good collimator were willing to make holographic "clones" of it, he/she would find a ready market, even at several hundred dollars per hologram. Simply exposing a plate to a collimated beam and a steeply diverging beam, and mounting it on flat anti-reflective-coated glass would likely produce exactly what is needed. However, such diffractive optical elements are the most sensitive types of holograms to exposure and/or processing flaws, and collimators of useful quality are incredibly difficult to make. Nevertheless, perhaps some holo-entrepreneur will decide to take on this selfless task, and help put this important tool into the hands of even small-scale holographers.

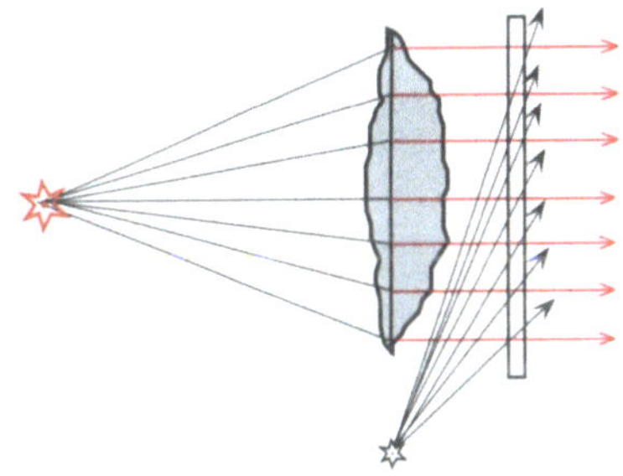

Perfect Conjugate Illumination (More Examples)

Now we return to more discussion of the peculiar real image that is projected by a phase-conjugate illumination beam. From the vantage point of an observer, who is down-stream in the now reverse-propagating optical beam, it presents an appearance never seen before!

Outside-in
Note that the hologram can reproduce information about only the parts of the object that it has "seen," which are those parts closest to the hologram plate (which obscure its view of the more distant parts). Thus only the "front" surface of the object (or right-hand surface in this sketch) is reproduced in space as a glowing texture—the back of the object simply does not appear. The observer is then seeing the front of the object through the back—the occlusion cues are just the opposite of what they should be! With practice, an observer can learn to see this image as an "outside-in" version of the object. A ball will appear as a cup, for example. The apparent depth of the object has been reversed with respect to its occlusion cues, as they would be for a stereoscope image if the right and left views were reversed. The descriptor for this reversed-depth type of image is *pseudostereoscopic*, or more commonly, *pseudoscopic*. For an untrained observer and a complex object, though, the occlusion cues usually dominate the parallax cues, and the image appears to have normal depth but rotates as the observer moves from side to side.

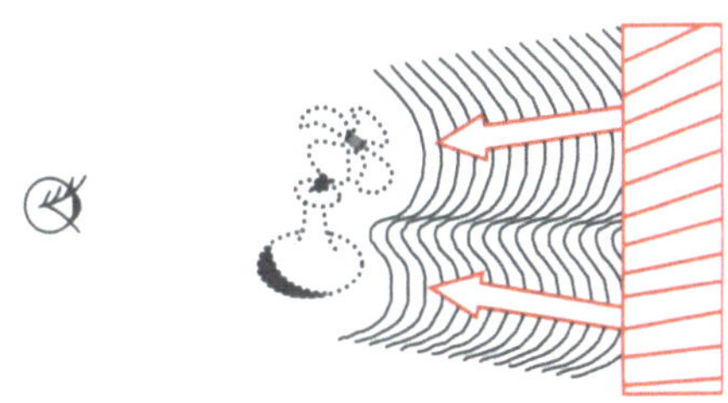

pseudoscopic image

Effects of Imperfect Conjugates

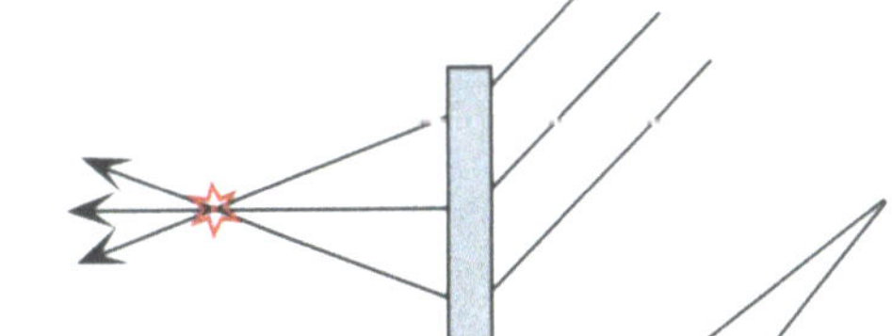

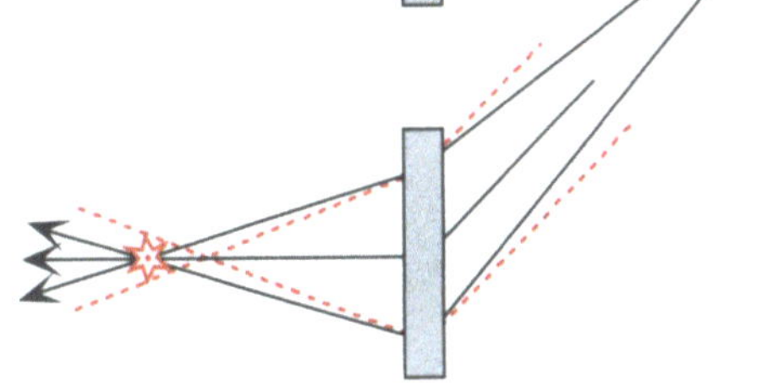

Nothing in our conceptual discussion prepares us to describe what happens when the illumination departs from being the *perfect* phase conjugate of the reference beam. For that, we need to develop some mathematical models to describe the behavior of the light waves. Also, we need to give up our ability to describe really thick holograms. As the local illumination angle rotates from that required for perfect phase conjugation, the output beam angle rotates in the same direction (although by a somewhat different amount, owing to the non-linearity of the sine function), and the amplitude of the beam decreases because the incoming and outgoing beams no longer satisfy the Bragg conditions for volume diffraction. We will assume that the hologram is thin enough that the Bragg angle mismatch problems are not very severe.

If we start by assuming that the perfect phase-conjugate illumination would be collimated, then reconstruction by a diverging illumination with the same central angle means that although the center of the plate is illuminated at the proper angle, the illumination at the top of the plate is rotated slightly to the right, and the illumination at the bottom of the plate is rotated slightly to the left. The output rays from the top and bottom are then also rotated slightly to right and left respectively, and cross the undeviated ray from the center somewhat further from the hologram than before. Thus the image distance will become greater as the radius of curvature of the illumination wave becomes shorter.

Image Location (Analytical)

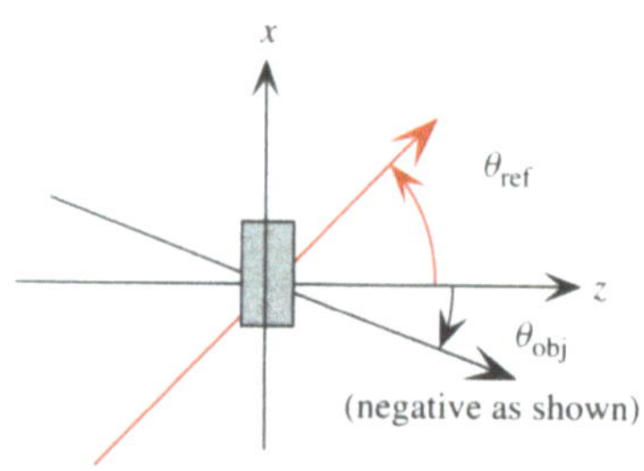

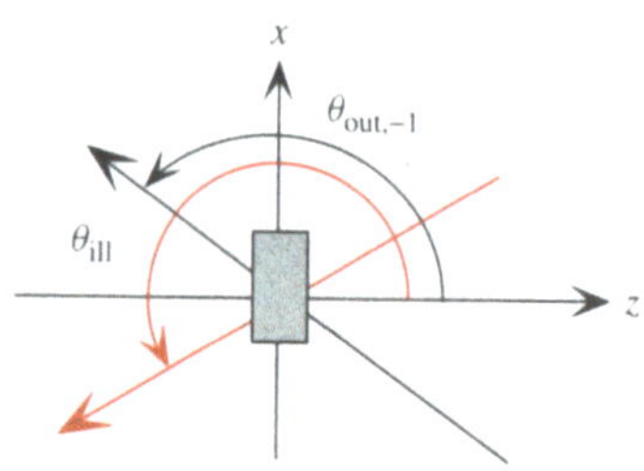

The relationship between the illumination and output angles is described by exactly the same equation that we have seen before, with the definition of angles increased explicitly to include angles greater than 90°, which must be measured "the long way around" from the plate perpendicular (which continues to be defined as coming out of the side of the plate *opposite* to that exposed to the object beam). Namely, the "sinθ" equation, Eq. (3) of Chapter 8, still applies,

$$\sin\theta_{\text{out},m} = m\frac{\lambda_2}{\lambda_1}\left(\sin\theta_{\text{obj}} - \sin\theta_{\text{ref}}\right) + \sin\theta_{\text{ill}} \tag{1}$$

And if we apply this equation over a small area and note where the output rays intersect (within the paraxial approximation), we find that the same focus-law relationships, the "1/R" and "cos-squared" equations (Eqs. (7) and (8) of Chapter 10), also apply to this case:

$$\frac{\cos^2\theta_{\text{out},m}}{R_{\text{out},m,x}} = m\frac{\lambda_2}{\lambda_1}\left(\frac{\cos^2\theta_{\text{obj}}}{R_{\text{obj}}} - \frac{\cos^2\theta_{\text{ref}}}{R_{\text{ref}}}\right) + \frac{\cos^2\theta_{\text{ill}}}{R_{\text{ill}}} \tag{2}$$

$$\frac{1}{R_{\text{out},m,y}} = m\frac{\lambda_2}{\lambda_1}\left(\frac{1}{R_{\text{obj}}} - \frac{1}{R_{\text{ref}}}\right) + \frac{1}{R_{\text{ill}}} \tag{3}$$

In other words, the same equations apply regardless of what direction the light is traveling in, as long as we are careful to define the angles and distances properly, especially by identifying converging waves with negative radii of curvature. In addition, the diffracted order of interest in phase conjugation is almost always the $m = -1$ order.

If we imagine that the holograms are perfectly thin, then the illumination and the output waves have the same phase patterns whether they are traveling from left or right, and the transmittance of the hologram operates the same way in both directions. Our preference for now using leftward-going illumination is to make a better match to the angle of thick holograms for bright reconstructions in the $m = -1$ order, and to make certain that the image reads properly top to bottom and left to right. Otherwise, it is important to realize that this is not a new hologram output term. It is the same conjugate term that we have seen before, except that the illumination has been angled to make it more accessible.

Image Magnification

Because the $\sin\theta$ equation still applies in phase conjugate reconstruction, all of our previous image formulas also apply—because they all followed from the application of the $\sin\theta$ equation in the direct or forward reconstruction context. Thus the magnification formulas still apply, again provided only that the usual case is $m = -1$ and all angles and distances/radii of curvature are measured correspondingly. In general, because the most usual case of imperfect conjugation is using an illumination wave that is diverging more than it should, the output waves will be diverging more than they should, and the image will focused further from the hologram than it should be.

For the image of an extended object, consisting of many points, the rays through the center of the hologram will be traveling in the same directions as for perfect conjugation, but the rays from the margins of the hologram will again cross the central rays further away from the hologram. The real image will therefore be magnified in the ratio of exposure to reconstruction distances as for the previous cases, and it will also suffer the same effects of change of the wavelength of the illumination.

Recalling our discussion of image magnifications from Chapter 10, we can calculate the image magnifications in this situation. The longitudinal magnifications are

$$
\begin{aligned}
MAG_{\text{long},x} &= \frac{\Delta R_{\text{out},m,x}}{\Delta R_{\text{obj}}} = m\frac{\lambda_2}{\lambda_1}\left(\frac{R_{\text{out},m,x}}{R_{\text{obj}}}\right)^2 \\
MAG_{\text{long},y} &= \frac{\Delta R_{\text{out},m,y}}{\Delta R_{\text{obj}}} = m\frac{\lambda_2}{\lambda_1}\left(\frac{R_{\text{out},m,y}}{R_{\text{obj}}}\right)^2
\end{aligned} \tag{4}
$$

The lateral magnifications are

$$
\begin{aligned}
MAG_{\text{lateral},x} &= m\frac{\lambda_2}{\lambda_1}\frac{\cos\theta_{\text{obj}}}{\cos\theta_{\text{out},m}}\frac{R_{\text{out},m,x}}{R_{\text{obj}}}, \\
MAG_{\text{lateral},y} &= m\frac{\lambda_2}{\lambda_1}\frac{R_{\text{out},m,y}}{R_{\text{obj}}}.
\end{aligned} \tag{5}
$$

Example: horizontal (y) focus

Let's re-do the calculations we looked at in chapter 10 (starting at Eq. (16)). Again we'll expose a hologram at 633 nm and later illuminate the finished hologram at 543 nm. The basic equations that tell us where various orders of images will show up are

$$
\begin{aligned}
&\frac{\sin\theta_{\text{out},m,y} - \sin\theta_{\text{ill}}}{\lambda_2} = m\frac{\sin\theta_{\text{obj}} - \sin\theta_{\text{ref}}}{\lambda_1}, \qquad m = 0, \pm1, \pm2, \ldots \\
&\frac{1}{\lambda_2}\left(\frac{1}{R_{\text{out},m,y}} - \frac{1}{R_{\text{ill}}}\right) = m\frac{1}{\lambda_1}\left(\frac{1}{R_{\text{obj}}} - \frac{1}{R_{\text{ref}}}\right)
\end{aligned} \tag{6}
$$

and we'll use the same numbers as before, except that we're looking for the $m = -1$ image, and we need to add 180° to the illumination angle:

$$\theta_{\text{obj}} = 10°, \theta_{\text{ref}} = 45°, R_{\text{obj}} = 150\,\text{mm}, R_{\text{ref}} = 2000\,\text{mm}, \theta_{\text{ill}} = 240°,$$

$$R_{\text{ill}} = 1000\,\text{mm},\ \text{m} = \text{-1}$$

we find that the location and angle of that image is

$$\theta_{\text{out},m,x} = 204.1°, R_{\text{out},m,x} = -233\,\text{mm}$$

When we apply the y magnification equations (4) and (5) above, we find that the resulting magnifications are 133% lateral and 207% longitudinal.

The same example, but vertical (x) focus

In the same case as above, but along the other axis, again we are looking for the location of the $m = -1$ image. Recall that here w have to use have a different set of equations:

$$\frac{\sin\theta_{\text{out},m,x} - \sin\theta_{\text{ill}}}{\lambda_2} = m\frac{\sin\theta_{\text{obj}} - \sin\theta_{\text{ref}}}{\lambda_1}, \qquad m = 0, \pm 1, \pm 2, \ldots$$

$$\frac{1}{\lambda_2}\left(\frac{\cos^2\theta_{\text{out},m,x}}{R_{\text{out},m,x}} - \frac{\cos^2\theta_{\text{ill}}}{R_{\text{ill}}}\right) = m\frac{1}{\lambda_1}\left(\frac{\cos^2\theta_{\text{obj}}}{R_{\text{obj}}} - \frac{\cos^2\theta_{\text{ref}}}{R_{\text{ref}}}\right) \tag{7}$$

and the $m = -1$ image is at

$$\theta_{\text{out},m,x} = 204.1°, R_{\text{out},m,x} = -164\,\text{mm}$$

The above magnification equations for x evaluate to 101% lateral and 119% longitudinal.

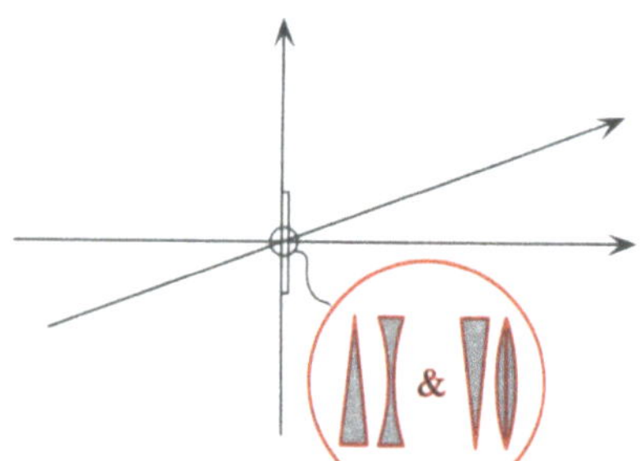

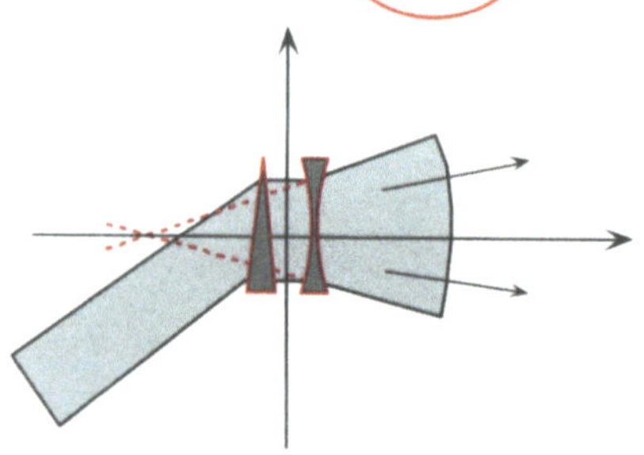

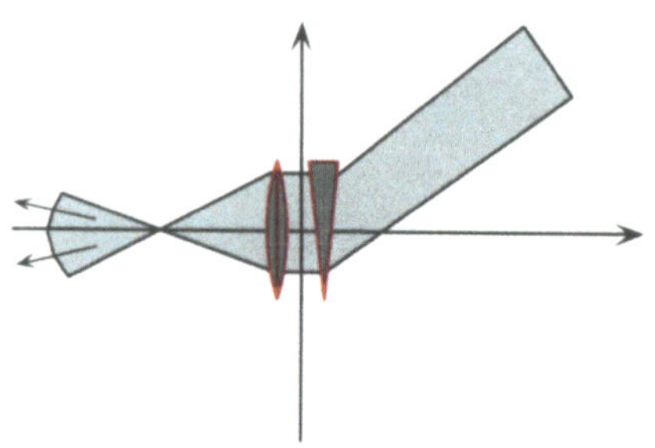

Relation to the Lens and Prism-Pair Model

Recall the model of an off-axis transmission of a single point that consisted of a base-down prism plus a negative lens, and that had an accompanying base-up prism and positive lens (each prism of the same but opposite deflecting power, and each lens of the same but oppositely signed focal length). It doesn't matter which element comes first in the beam, and rearranging slightly makes it easy to see how the "opposite set" (the conjugate order of the hologram) comes into play to deflect and focus the collimated illumination to produce a real-image focus at the location of the former virtual image. Each point of the object gives rise to such a pair of prism and lens sets, and thus gives rise to a corresponding real image point.

Image Aberrations—Astigmatism

When $m = -1$, the reference and illumination terms of the focusing equations (the one-over-R and cos-squared equations) add, and so any departures from ideal phase-conjugate illumination also add. Most typically, both the reference and illumination beam are weakly diverging, causing the image to be formed farther from the plate than the object location, and thus magnified. In addition, the wavefronts are likely to be astigmatic, so that no sharp focus can be obtained. Here the trick of splitting the reference and object beam angles (and later, the image and illumination angles) with the perpendicular to the plate comes in especially handy, as astigmatism is balanced out in this case. The table below shows the vertical and horizontal focal distances for the three exposure and reconstruction plate angles shown in the sketch.

	plate angle A	plate angle B	plate angle C
horizontal focus	312.5 mm	312.5 mm	312.5 mm
vertical focus	416.7 mm	312.5 mm	277.8 mm

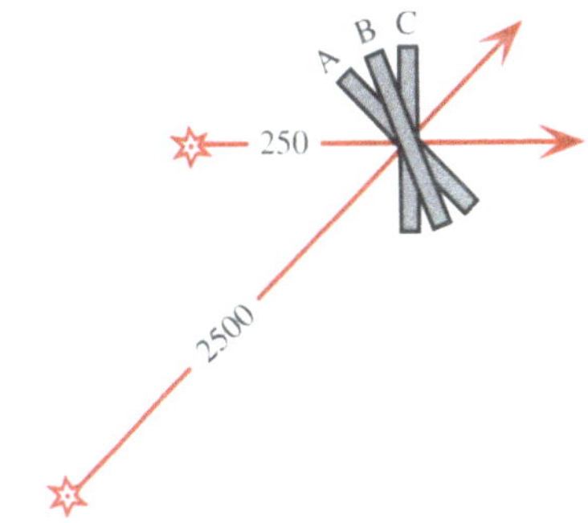

Of these, the two most common geometrics are "B," in which the plate normal/perpendicular bisects the object and reference beams, and for which there is no astigmatism (a very important consideration), and "C," in which the object and output are perpendicular to the plate, and for which the vertical focus is always closer to the plate than the horizontal focus.

Conclusions

Real-image projection by phase conjugate illumination will turn out to be one of the most powerful techniques we use in holographic imaging. It will certainly enable us to do some valuable things in some upcoming chapters! There are only three things to remember to do: 1) measure illumination and output angles "the long way around" from the perpendicular to the "back" of the plate, 2) let the order number, *m*, be negative one, and 3) remember that negative radii of curvatures signify converging waves (producing real images). This marks the end of the convention that light will be traveling from left to right, and that the front of the plate will be facing left. From now on, we will expose plates from whatever direction is convenient, and reconstruct them after moving them around. The local coordinate system will have to follow the plate accurately, which can get pretty confusing! So make sure that you understand what is happening here before we start tumbling around in holographic space.

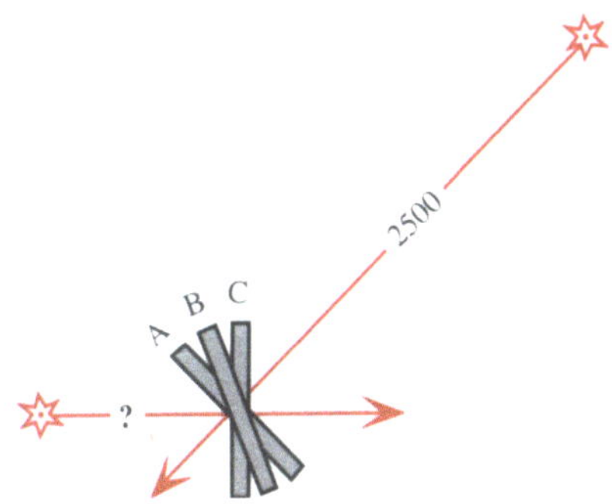

i. Benton, S. A., H. S. Mingace, Jr., and W. R. Walter (1980). "One-Step White-Light Transmission Holography," *Proc. SPIE Recent Advances in Holography,* **215**, pp. 156-161. A concise discussion of the relevant math based on the work of Benton and of S. St. Cyr appears as an appendix to Saxby, G. (2004). *Practical Holography,* Institute of Physics Publishing, Bristol UK.

ii. Steel, W. H. and C. H. Freund, (1984). "Single-Step Rainbow Holograms Without Distortion," *Opt. Comm.*, **51**, 6, pp. 368-370.

Dennis Gabor

Emmett Leith, Juris Upatnieks

Stephen A. Benton

Yurii Denisyuk

Train, Leith and Upatnieks, 1964

Chess Set, Stephen A. Benton, 1968

Holographic Filament, Yurii Denisyuk, 1958-1962

Rind II, Stephen A. Benton, 1977

Chess Pieces, Stephen A. Benton, 1979

Tricia, Stephen A. Benton, 1980

Martian Grove, MIT SPI Group, 1984

Leonardo's Vision, Polaroid/MIT SPI Group, 1985

Robie House, MIT SPI Group, 1986

Edge-lit Hologram, MIT SPI Group, 1988

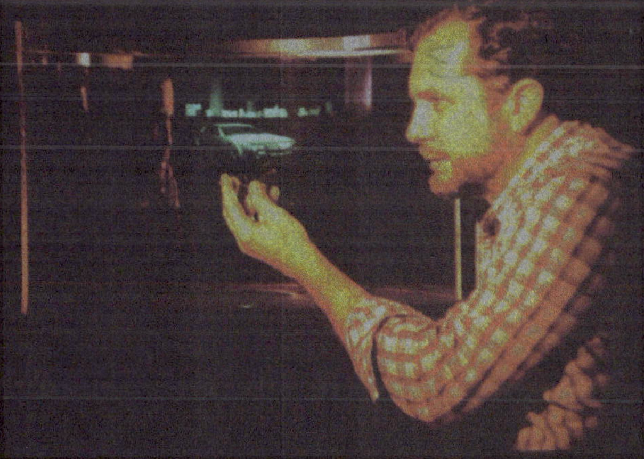

Alcove Hologram, MIT SPI Group, 1986

Teacup Photopolymer Ultragram, MIT SPI Group, 1993

Hubcap Ultragram, MIT SPI Group, 1990

Still Life, MIT SPI Group, 1988

Acura NSX, MIT SPI Group, 1991

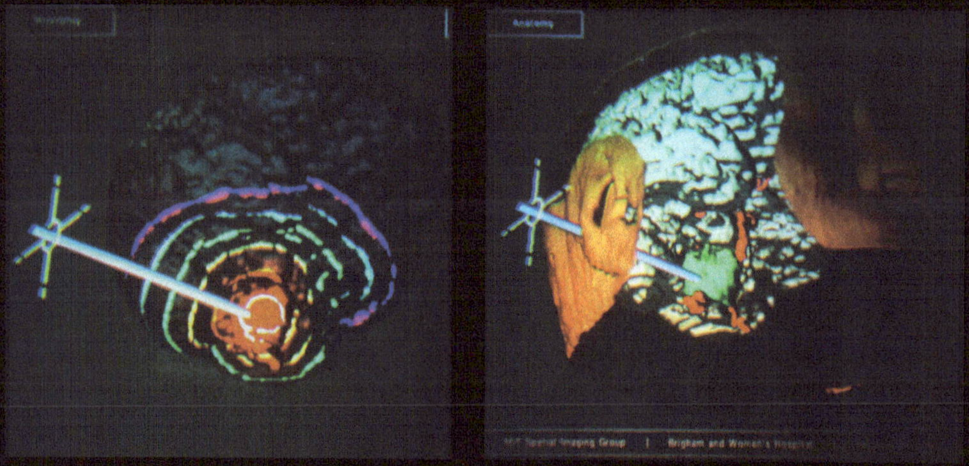

Radiation Treatment Planning Hologram, MIT SPI Group, 1991

World's Largest Hologram, Zebra Imaging, Inc., 1997

Holo-Video Flower Pot Mark I,
MIT SPI Group, 1990

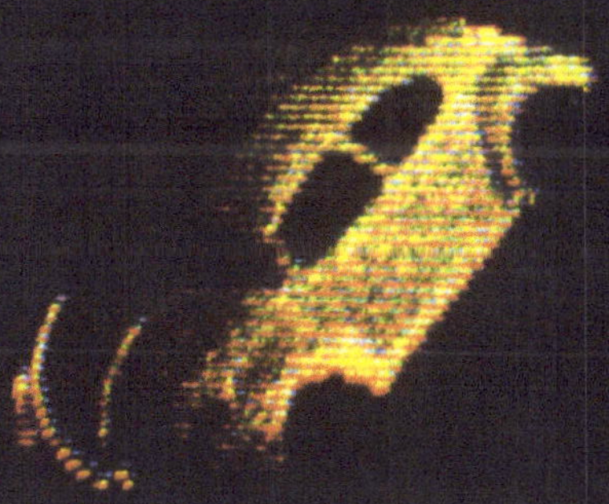

Holo-Video VW Mark I,
MIT SPI Group, 1991

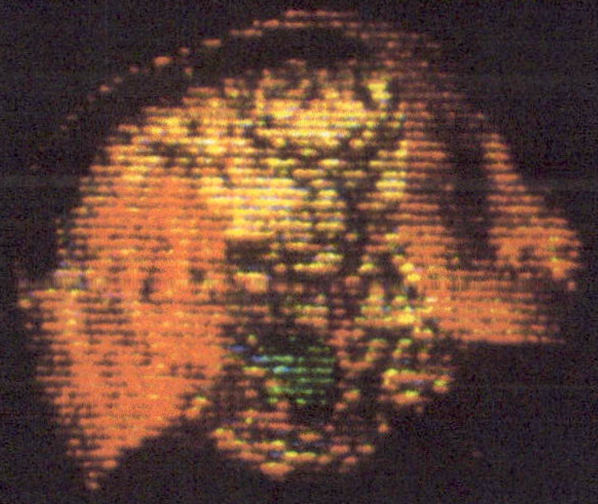

Holo-Video Medical Image
Mark I, MIT SPI Group, 1991

Holo-Video Mark I three color set-up, MIT SPI Group, 1991

Holo-Video Honda EPX Mark II, MIT SPI Group, 1992

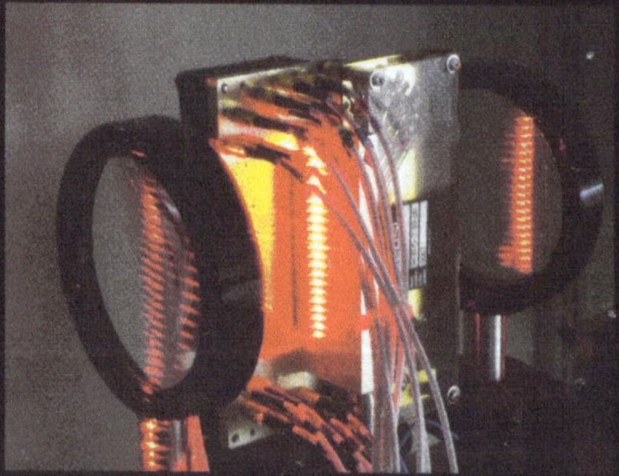

Holo-Video Mark II 18-channel light modulator, MIT SPI Group, 1992

Holo-Video RIP Lincoln Cube Mark II, MIT OBM Group, 2005

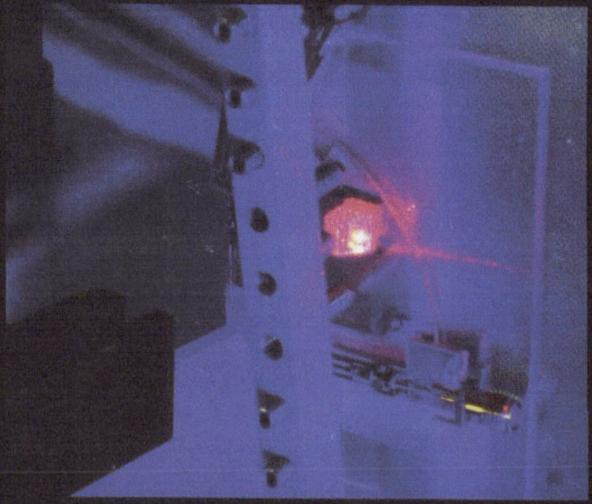

Holo-Video Mark III light modulator, MIT OBM Group, 2007

CHAPTER 13

Full-Aperture Transfer Holography

Full-Aperture Transfers

The preceding chapter discussed to the two-step, "master-transfer," or "H1-H2" method of making holograms, in which a first hologram is used to create a real image in space, which then becomes the object for a second hologram. Normally, the image is brought as close as possible to the plane of the second hologram, the H2, so as to minimize the sensitivity of the resulting image to the source-size and color blurs usually produced by an ordinary white-light source, such as a spotlight or the sun. As such, the resulting hologram was first popularly described as an "image-plane hologram," although technically the image has no plane because it has depth! We prefer to call these "open-aperture transfers," or "full-aperture transfers" for reasons that will become clear in the next chapter. It is true that the image is usually intended to straddle the hologram plane very carefully, to minimize the maximum depth of the image.

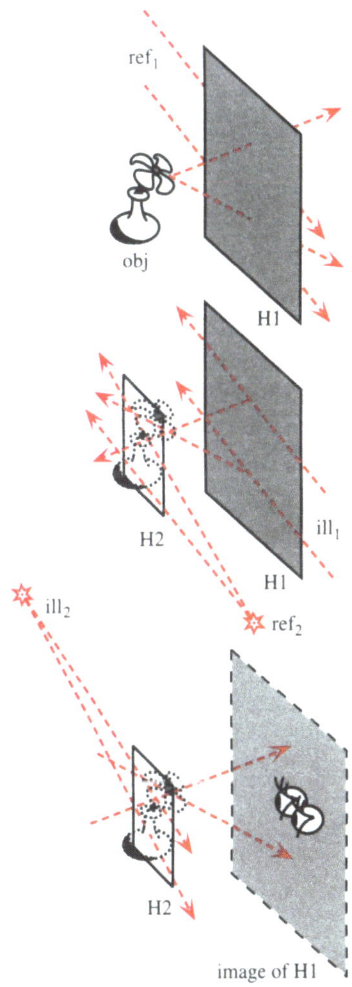

This chapter is the story of the H2 and its optics. The previous chapter tells us almost everything we need to know about the H1, and a great deal of what we will need to know about the H2 as well. Basically, the H2 is also recorded as an off-axis transmission hologram that is later reconstructed with phase-conjugate illumination. The pseudoscopic image that results is then a depth-reversed image of a projected real image that was itself pseudoscopic, or depth-reversed. "Two pseudos make an ortho," might be the rule—the final image reads with the correct depth compared to the original object. Thus we have produced a "right-reading" holographic image that is remarkably clear when viewed with ordinary light sources.

There are quite a few complicating factors that we have to take into account though. First, the coordinate system for the H2 is oriented differently from what we have been used to, and we have to agree upon a convention for the rotation and translation of local hologram coordinate systems in general. Second, the H2 is actually a hologram of two things at once: of the projected real image of the object, and of the H1 itself. Consideration of the second brings a new point of view to the imaging process. The exposure of the H2 is to a focused and nearly photographic-like real image, with large intensity variations over small distances, which makes the beam ratio more difficult to measure and adjust. And finally, because the H1 and H2 play very different roles in the imaging process, their exposure and processing should be separately optimized with quite different criteria in mind.

As impressive as a full-aperture transfer hologram is upon first sight, be cautioned that it is only a transitory state. The technique is of major importance for reflection holograms (serving as the H2), and we will revisit it later. But it will serve here mainly to frame the discussion in the following chapter about more advanced "rainbow" transmission holograms. There are quite a few concepts to layer on here before we are ready to go forward, and full-aperture transfers are wonderful tools for learning.

Further Discussion of H1-H2 Technique

The creation of images that came up to and through the hologram plane was a revolutionary step when introduced in 1966 by Rotz and Friesem, of the University of Michigan group.[i] Within a few years, it became the technique of choice for most display holograms.[ii] Although there were several attempts to produce image-plane holograms in a single optical step, by the use of large lenses and mirrors for example, the two-step hologram technique has come to be the generally accepted practice. It is a technique that requires an extra holographic step, which means separate setups for mastering and transferring, usually by tearing down the first and replacing it with the second. This makes the usual "cut and try" methods of holography impractical, as re-shooting the master becomes more and more time-consuming (except for those few with the luxury of two tables, lasers, and sets of gear!). The use of a few mathematical calculations makes it much easier to get it right, or nearly right, on the very first try, and the recognition of the utility of shop-math-based holography followed the emergence of these two-step techniques. They allow a degree of precision in "previsualization" that is necessary for efficient work, so that the holographer can judge with some confidence what is likely to appear in the final image.

At about the same time, it became generally realized that the illumination for holograms would have to come from above, as it does for most other display media, if holography were to become competitive. Looking into the beam of a side-lit hologram can be an uncomfortable, or at least worrying, experience. And rainbow holograms are absolutely going to require vertically-inclined (usually overhead) illumination, so here is where we start to deal with it in earnest. Although we can blithely sketch "underhead" reference beams to result in overhead illumination beams, it is difficult to bring such beams up through the solid tables we usually work with. Many holographers have come up with clever multi-mirror periscope schemes to allow vertically-inclined reference beams, but it is best to avoid mirrors in reference beams whenever possible. The easy way out is simply to turn everything on its side! This complicates things a bit when it comes time to describe the direction or location of this or that item, as the final hologram's frame of reference is turned 90° to the laser table's frame of reference. We will ignore this particular practical issue for most of this chapter's discussion, and will continue to sketch as though we had transparent tables to work upon.

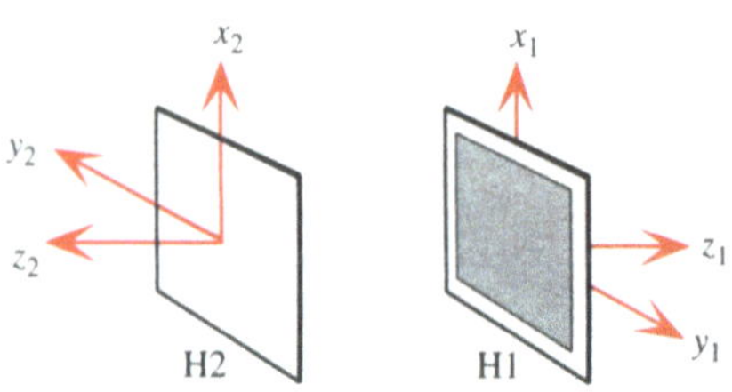

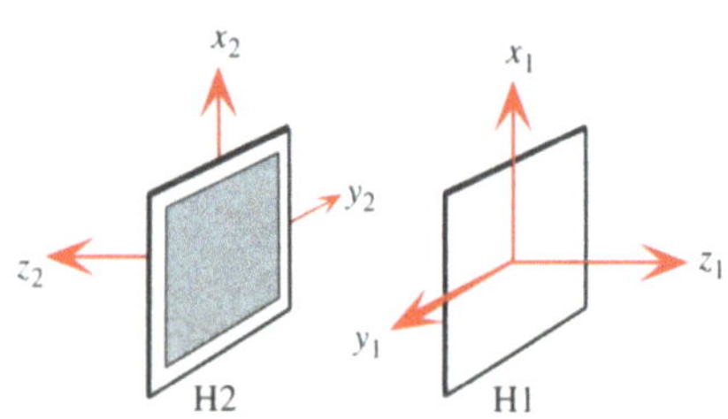

Holo-Centric Coordinate System

Up to now, the holograms have (very conveniently) been facing straight along the "minus z" direction, so that angles could be measured in the usual way: a positive angle is one that a clockwise rotation would bring to the z-axis that emerges from the back of the hologram. Now we will have to construct a small traveling coordinate system for each hologram, and our convention will be that the z_i-axis will be sticking out of the "back" of the i-th hologram, no matter what its orientation will be! And, we will define the "front" of the hologram as the face that receives the object-beam exposure (the reference beam also hits the "front" of the hologram, for transmission holograms). As a gesture of friendship, we will continue to show the master hologram, the H1, as facing in the general direction of "minus z," so that object and reference beam angles will usually be between plus and minus π (±180°) in the global coordinate system (they are always between plus and minus π in the holo-centric coordinate system). The transfer hologram, the H2, on the other hand, will generally be facing in the opposite direction, so that its "plus z_i" direction is roughly in the minus global-z direction. Assuming that the local coordinate system is "glued" to the i-th plate, there are two ways to go from what we have been using to what we need for the H2—by rotating "head over heels" with the horizontal y-axis used as an "axle," or by spinning around the vertical x-axis. We will choose the second or "spinning" method, so that the x_i-axis of the H2 will stay roughly vertical but the y_i-axis will now poke *into* the page, as shown in the approximately-isometric marginal sketches. For the transfer hologram, the H2, positive angles will be those for which a *counter-clockwise* rotation brings them into the z_i-axis.

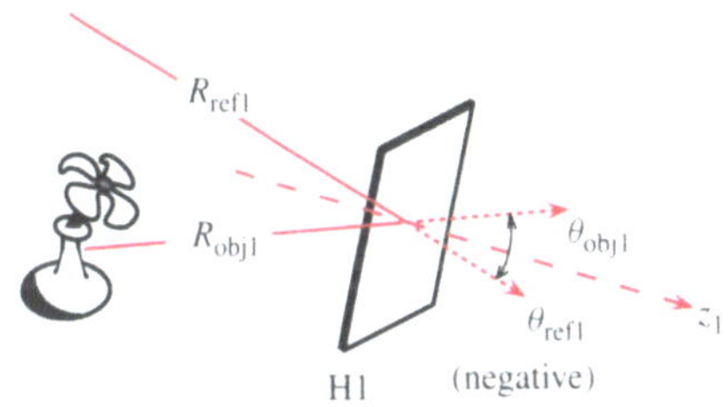

While angles may be a little difficult to keep track of, distances and curvatures are no different than before. A diverging wave will have a positive radius of curvature whether it is traveling from left to right, or from right to left (or from top to bottom, of course). And a negative radius of curvature will denote a converging wave, whatever its angle of propagation.

This can get a little confusing for spatially challenged thinkers. From time to time it is helpful to think of yourself as being in the center of the hologram, in order to see what it sees and to judge what kind of fringes and optical behavior might be produced. Now, you just have to add a spear sticking out of your back, denoting the positive z_i axis, with your right arm sticking straight up to be in the direction of the positive x_i axis, and your left arm pointed out sideways to indicate the positive y_i axis (a good old-fashioned right-handed coordinate system). Now just pivot around and shuffle about (like practicing Latin dance steps in your mind) to take on the orientations of the various plates in a two- or even three-step system, and you can readily gauge what angles and radii to plug into the three equations that we will continue to use.

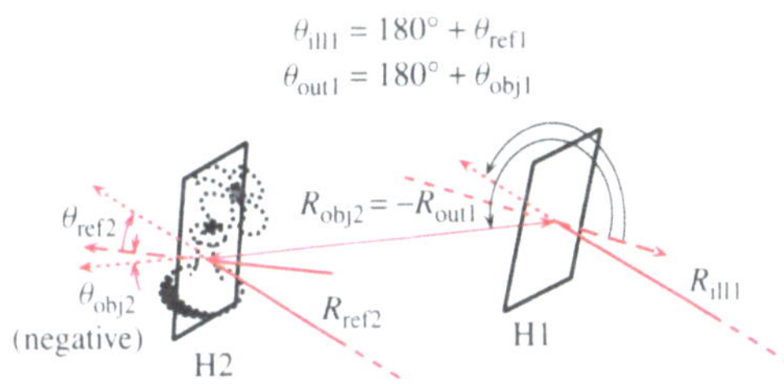

Example

As an exercise, let's just walk through a typical full-aperture geometry, as sketched alongside. In the first exposure, or "mastering" step, the object beam angle (θ_{obj1}) is positive because we are deliberately tipping the H1 back so as to make the interference fringes be perpendicular to the plate (for easier processing design). The reference beam is from "overhead" (in the hologram frame), and has a negative angle (θ_{ref1}). Both the object beam and reference beam are diverging in this example (R_{obj1} and R_{ref1} are positive).

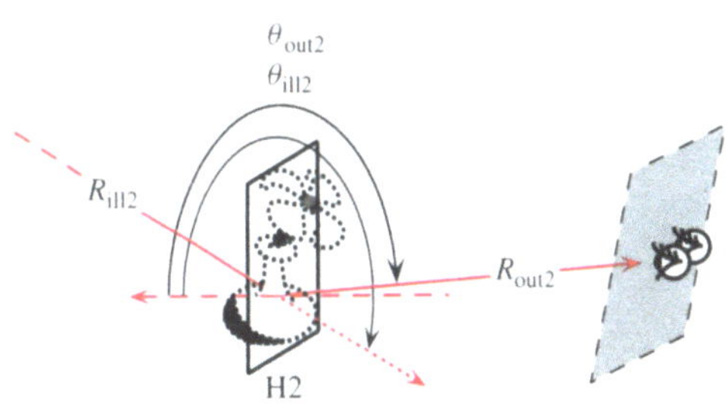

In the second exposure, or "transfer" step, the H1 is illuminated with a beam in the direction opposite to that of its reference beam ($\theta_{ill1} = 180° + \theta_{ref1}$), and the output beam is traveling in the direction opposite to that of the object beam (θ_{out1} = 180° + θ_{obj1}, assuming no change of wavelength between exposure and transfer). The output beam is a converging wave, focused at the distance of the H2 to produce an image straddling the hologram plane ($-R_{out1}$ = S (separation) = R_{obj2}). Both the reference and illumination beams for H1 are diverging, so that the output radius is larger than the object radius (though still negative), so that the image is farther away and magnified.

In the final "viewing" step, again assuming the same wavelength is used, the H2 illumination is angled in the direction opposite to that of the reference beam (θ_{ill2} = 180°+θ_{ref2}), and the output beam is traveling in the direction opposite to that of the H2's object beam (θ_{out2} = 180°+θ_{obj2}). Because both the reference and illumination beams are diverging, the output wavefront's radius of curvature is again larger than the exposing wavefront's radius of curvature, and the real image of the H1 is formed farther from the H2 than the H1 was during exposure.

At this point we have not yet discussed *astigmatic* focusing, although both H1 and H2 are clearly not enjoying perfect phase-conjugate illumination, and their output beams will be markedly astigmatic. For the purposes of side-to-side parallax and triangulation by the eyes, it is the horizontal or *y*-focus that matters, and the "1/*R*" equation is the relevant focusing law for placing the apparent distance of the image projected by the H1 exactly at the H2 plane. However, it is the *vertical* parallax between the image and the H2 that allows blur under white-light or extended-source illumination, so that the vertical focus or "$\cos^2\theta$" relationship is the relevant equation for the H1, if image sharpness is the main issue.[iii] The calculations are straightforward, if a little tedious to do by hand.

Separate Optimization of the H1 and H2

This is a good time at which to mention that the H1 and H2 will typically be very different types of holograms, as the exposure and processing of each is optimized for the characteristics most important at each step. In a teaching lab we are likely to use the same techniques for both, but in commercial practice they are usually very different. A general statement of the different roles is: the H2 needs to be bright above all else, and the H1 needs to produce a "clean" image above all else.

Master hologram issues

Let's discuss the issues for the H1 or "master" hologram first.

High contrast vs. brightness: The ratio of intensity between the whites and blacks in a glossy paper print is limited to about 50:1, and is commonly less than that for television screens (though often higher for projected images). Reaching a 50:1 matte white to shadow ratio in a holographic image is really quite difficult. It requires using a high beam ratio, between 30 and 50 typically, to keep intermodulation noise low. Even though this produces a fairly dim image, the low scattered light is more important. The hologram may also be exposed and processed without bleaching, as it is widely believed that bleaching lowers the contrast and degrades the archival stability of the hologram.[iv] And the master is typically recorded on a glass plate for flatness and durability, and used while index-matched to reduce scatter by any surface relief in the emulsion. The details of optimizing the contrast and brightness of a hologram image require a careful study of exposure, beam ratio, and processing effects for each recording material used. Phillips *et al.* (1980) report conducting several thousands of tests during the development of their reflection hologram processing techniques, for example.[v]

Split-angle recording: Obtaining clean and undistorted real image projections requires that the angles of the fringes recorded in the thickness of the hologram do not change between exposure and reconstruction. Most processing chemistries change the emulsion thickness by several percent (up to 20%, in principle), which would significantly change the angle of any fringes that are not vertical to the emulsion surface. Therefore, master holograms are usually tilted so that their perpendicular bisects the angle between the reference beam and the center of the object, to make the fringes as "vertical" as possible. In addition, "splitting the angle" makes the reconstruction free of astigmatism, if the illumination beam cannot be a perfect conjugate of the reference beam. However, this is not the configuration that minimizes "coma" in the image (for which the plate should be almost perpendicular to the object), and that may be a more important consideration, especially for rainbow holograms. It is easy to tell when the plate is bisecting the reference-object beam angle, by the way: the reference beam will reflect onto the object.

Transfer hologram issues

The final hologram, on the other hand, has almost the opposite qualities as desiderata. The image must be maximally bright, because without adequate luminance a holographic display is pointless. The contrast is usually degraded more by external light leaks than by intermodulation noise (although these can often be overcome by careful masking). Thus the hologram is usually exposed at a low beam ratio, and almost always bleached. It may or may not be laminated, which provides index matching as well as protection from the elements. The hologram must typically hang vertically, in order to be as inconspicuous as possible (and as much like a photograph as possible), thus tilted fringes are inevitable. Avoiding shrinkage effects means pre-compensating for them, or using only non-shrinking processing chemistries. Also, to reduce the cost, transfers are often recorded on flexible film base materials, which require care to keep acceptably flat.

We will revisit many of these issues when we spend more time talking about processing chemistries and techniques.

Another Point of View: H1 as Multi-Perspective Projector

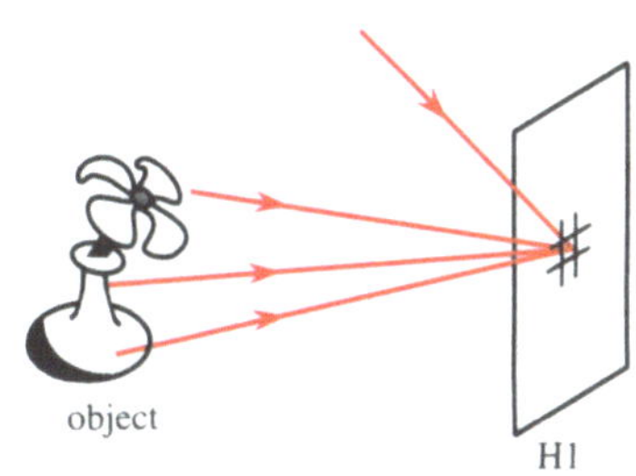

"All problems in optics are straightforward, if you look at them the right way," says the old maxim, and there usually are several points of view that can be tried for any particular question. In the present case, we have been thinking of the H2 as recording a hologram of the real image projected by the H1, just as though it were any ordinary object that happened to be able to straddle the hologram plane.

And that point of view explains a great deal of what happens when we make a hologram this way. In particular, if we examine the color blur produced in white light (as seen well away from the top and bottom of the view zone), the blur of a point image produced by a master hologram is the same as the blur of a point image produced by an actual object.

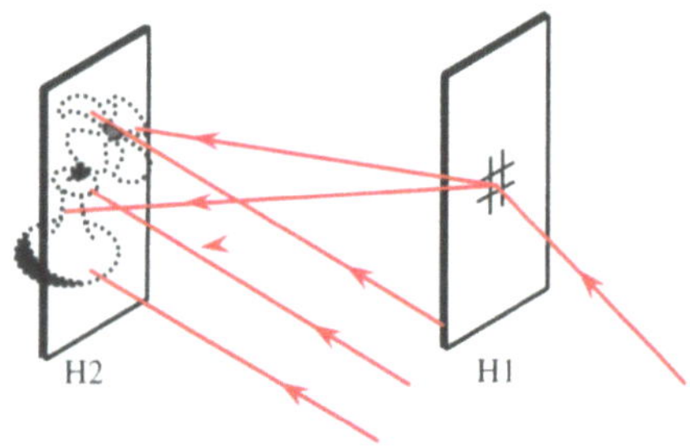

However, at the same time, the H2 is also making a hologram of the H1, and later projecting an image of the H1 into space, where it defines a *viewing zone* or *view window* for the final 3-D image. The H2 will record images of everything it sees, of course, but it is helpful to distinguish between the imaging of the object and imaging of the view-zone window as separate events.

Another useful "mental model" of the hologram's behavior is to consider the H1 to be acting like an array of small imaging systems: first as cameras recording perspective views, and then as projectors beaming those perspective images back into space. That is, every small area or patch of the hologram (perhaps three millimeters on a side, or somewhere between one and six) records a single perspective view of the object scene, as seen from the location of that particular patch. When the hologram is illuminated in phase conjugation, each patch projects its perspective view back in the direction it came from. One way of observing this in a real hologram is to probe small areas of the H1 with an undiverged laser beam, and note the perspective that is projected onto a white card at the intended location of the H2, and then to watch how this image changes as the probe beam is moved around the H1. The "mental model" of the H1 as an array of cameras and turned-back projectors becomes a powerful one when thinking about holograms transmitting images when phase conjugation is used, especially as things become more and more complicated. This is also an way of thinking about the behavior of holograms to which we will return when we consider how to synthesize them on computers and how to replace the developed plate with something electronic (and thus changeable).

Where the many projected perspective views overlap, a three-dimensional real image is formed from their sum; but let's consider the H2 instead to be making a recording of the sets of beams from only one of the H1 patches (as in the sketch—normally all of the patches would be exposed at the same time). Now, when the exposed and processed H2 is illuminated in turn by the phase conjugate of its reference beam, it sends back to the real image of each H1 patch the set of beams carrying the perspective view originally recorded from that location. An eye placed at that location sees that perspective view, and only that perspective view. As soon as the eye moves away from that patch, the image goes dark. When the eye moves to the location of the next patch, or rather the next real image of a patch, it sees a different perspective view of the scene (of course we will fill in the patches so that the view never goes dark).

Anything that disturbs the location of the real image of the H1 will change the location of the patches as a group, and thus will change (in a systematic way) the view that the eye sees at any particular location. This will look like a rotation of the object if it happens while the eye is fixed (more accurately, a shearing motion around the central plane of the hologram). That is, changing the illumination wavelength, divergence, or angle won't blur the image, it will just change its orientation and perhaps distort it a bit. We have enough math tools to be able to compute these effects analytically, but if you have access to a transfer hologram and a light source whose divergence and position can be changed, the authors recommend some hands-on playing around as the best way to develop an appreciation of what is really going on here.

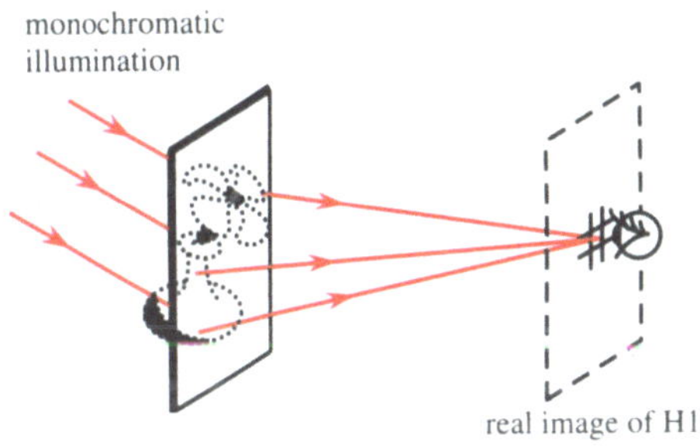

Color blur

One way of thinking about color blur, in terms of this new model, is to think instead of how the image of the H1 is affected by changing the wavelength of the H2's overhead illumination. If the wavelength is changed from red to green to blue, the image of the H1 (which we will usually call the "viewing zone") moves outward and, more importantly, downward (being rotated less radically). Thus, if the H2 is illuminated in blue light, the eye will see through the top of the image of the H1 hologram, and see the "high" perspective view of the object scene. More of the top of the object will be seen then, and only in blue light. Or, if the H2 is illuminated with red light, the eye will see through the bottom of the image of the H1 hologram, and see the "low" perspective of the object scene. Thus more of the bottom of the scene will be seen in red light.

If the hologram is now illuminated with "white" light, which presents red + green + blue simultaneously, all three differently-colored perspectives will be seen simultaneously. The various colored images will be the same and in register only where the 3-D image lies in the H2 plane. For image components out of that plane, the eye will see different perspective views in different colors, and where those views don't overlap in perfect registration, the eye will see "color blur." Because the differences between the perspectives are mainly in their vertical rotations, the color fringes appear mainly at the top and bottom of the image. Thinking of color blur as caused by a mixing of vertically-differing perspectives of various colors is often a more fruitful approach than our previous "spectral-blur of the scene image" model. It also makes it clear why the image seen from the center of the view zone can be "achromatic" or neutral-toned, which is attractive to many artists.

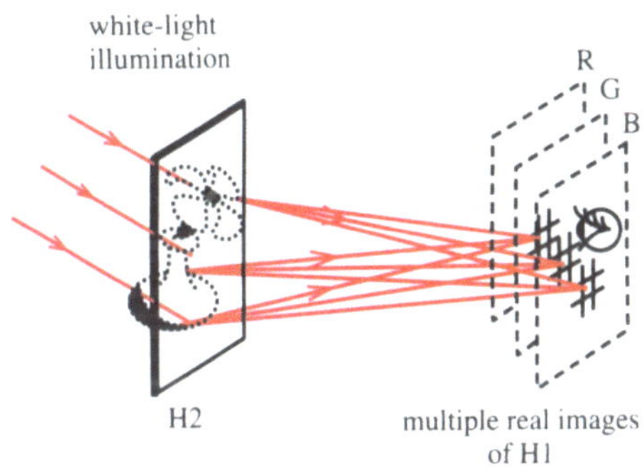

View-Zone Edge Effects

A characteristic of two-step H1-H2 holograms is that they present a viewing window that appears to hang in front of the hologram (assuming monochromatic illumination, or a reflection-mode hologram). The viewer's eyes must be inside the window area in order to see anything. If the view is at the same distance as the window, or viewing zone, the image "snaps" off as she or he moves across the edge of the zone, either horizontally or vertically. If the viewer stands back considerably further, she or he can see the edge of the view zone move across the image in a direction opposite to the viewer's motion, and perhaps perceive the edge as literally hanging in space like an open window. The wider the master plate, the H1, the wider the view zone will be, and the closer the H1 is to the H2, the wider the viewing angle the view zone width will allow. For full-aperture transfers, it is common to place the H1 as close to the object, and to the H2 afterwards, as is possible. However, for rainbow holograms it is necessary for the H1 to be at a carefully specified distance, usually much further away. Deliberately limiting the viewing angle increases the brightness of the image, as the same amount of light from the hologram (the incident power multiplied by the diffraction efficiency) is concentrated into a narrower beam.

If a full-aperture transfer hologram is illuminated with white light, some interesting things happen as the viewer's eyes move vertically across the top or bottom of the view zone (assuming vertically-inclined illumination). Moving upwards, for example, we see that the eyes move out of the blue-light view zone first, and then the green, so that an image in only red light remains, and is fairly sharp. That is, the width of the visible light spectrum is limited by the hologram geometry (the H1 edge) on the blue end, and by the end of the eye's spectral sensitivity (the visibility curve) on the infra-red end. Conversely, if the eyes move out of the bottom of the view zone, the image becomes deep blue in color, and again fairly sharp (although things never look quite as sharp in blue light as in red); the visible spectrum width is limited by red-end spectral cutoff by the H1 edge and by the ultraviolet end of spectral sensitivity of the eye.

Conclusions

Full-aperture transfer holograms, or image-plane holograms, have played an important part in the history of display holography, and are still important for reflection holograms. They provide vertical and horizontal parallax, and their images can project far into the viewer's space for dramatic effects. Although they are much less vulnerable to spectral and source-size blur than deep virtual-image holograms, their depth of field is limited to something less than 25 mm (one inch) with white-light illumination. The model of a master hologram, or H1, as an array of cameras and turned-around projectors is a valuable tool for thinking about these and other display holograms, as we move along to rainbow holograms and holographic stereograms.

References

i. Rotz, F. B. and A. A. Friesem, (1966). "Holograms with Non-Pseudoscopic Real Images," *Applied Physics Letters*, **27**, pp. 967-972.

ii. Brandt, G. B. (1969). "Image Plane Holography," *Applied Optics*, **8,** pp. 1421-1429.

iii. Bazargan, K. and M. R. B. Forshaw, (1980). "An Image-Plane Hologram with Non-Image-Plane Motion Parallax," *Optics Communications*, **32**, pp. 45-47.

iv. Benton disagrees with these common beliefs.

v. Phillips, N. J., A. A. Ward, R. Cullen, and D. Porter, (1980). "Advances in Holographic Bleaches," *Photo. Sci. Eng.*, **24**, pp. 120-124.

CHAPTER 14

White-Light Transmission "Rainbow" Holography

A Revolution in Holography

During the 1970s, two things happened that caused a revolution in display holography: the development of white-light viewable transmission holograms, and the development of very inexpensive processes to manufacture and distribute them. Both of these had their roots in the late sixties, and reached their full flower in the eighties, but the seventies were a time when everyone realized that important pieces of "the holography puzzle" were coming together to make display holography an industry at last.

Although holographic imaging had provoked a storm of popular interest in the middle sixties, following the announcements by Leith and Upatnieks of off-axis transmission holography, holograms continued to be things that you had to go to basements and museums to see—they were simply not bright enough to survive the glare of daylight. By 1972, McDonnell-Douglas Electronics had closed its pulsed-laser holography laboratory (which it had acquired with its purchase of Conductron, the University of Michigan spin-off company that had created so many impressive holograms for artists and industrial displays), and the rate of scientific publication in holography had fallen to almost nothing. There was a major economic recession going on at the time, and people's attentions turned to more immediately and economically promising technological challenges.

At Polaroid Corporation, a small laboratory had been established to study the applications of lasers to photographic problems, which also devoted a fraction of its efforts to display holography between manufacturing crises. In the course of some studies of full-aperture-transfer imaging and of bandwidth reduction concepts for electronic holography, a combination of the two ideas was found to hold promise for holographic television, and with a few changes it could instead produce *transmission* holograms that could be viewed with white light from ordinary sources, such as spotlights and the sun.[i,ii] The key was the elimination of vertical parallax from the image, so that only side-to-side variations of the image's perspective were presented—this was found to be sufficient for producing strong dimensionality in the image. White-light viewable *reflection* holograms had been known for several years, but the images they produced were dim, single-colored, and of low contrast (they will be described in a subsequent chapter). The new white-light transmission holograms, or "rainbow holograms" as they came to be known, produced very bright and multi-colored images that could be shown in rooms filled with light. They were quickly adopted for artistic and commercial displays because of their vivid imagery.

However, individual glass-plate and film holograms were still expensive to produce, usually costing thousands of dollars each. But at RCA Corp. a scientist had proposed that the technique that they had been using to manufacture LP records might be good enough to produce holograms cheaply, since LP record grooves were capable of diffracting light over fairly large angles if the music had high-frequency components.[iii] The new process involved producing a surface-relief or undulating-surface grating, electroforming a hard-metal copy of the relief pattern, and using it to emboss or cast a replica surface on a sheet of transparent plastic, which was subsequently mirrorized so that it could be attached to a surface with adhesive. This brought the cost of display holograms down to under a penny per square inch, cheap enough to be given away on magazine covers as attention getters, and eventually on credit cards and software packaging as counterfeiting deterrents. Over the years, these "silvery blob" embossed holograms have become a standard part of many printers' high-tech repertoire, and new variations are being developed constantly. To the extent that the general public are familiar with holograms, these embossed rainbow holograms are the ones they are most likely to have seen.

This chapter will describe the basic concepts of white-light transmission "rainbow" holography, and the next will pick up on some of the topics relevant to the state of the art in multi-color and embossed holograms. As we will see, the simplification in the viewing of rainbow holograms is won at the cost of some mathematical complexity in planning and making them. In particular, the details of *astigmatic* imaging will have to be taken into some account. We will first look at the process in the "forward" direction, from mastering to transferring to viewing. However, because of limitations in the viewer's distance that have to be anticipated, we will find that it is more often necessary to work "backwards," starting with the viewer's intended location, which specifies the transfer geometry, and then the mastering geometry.

Overview of the Process

Making a rainbow hologram involves two steps: first creating a master ("H1") hologram, and then using it as the object to create a transfer ("H2") hologram.

Mastering

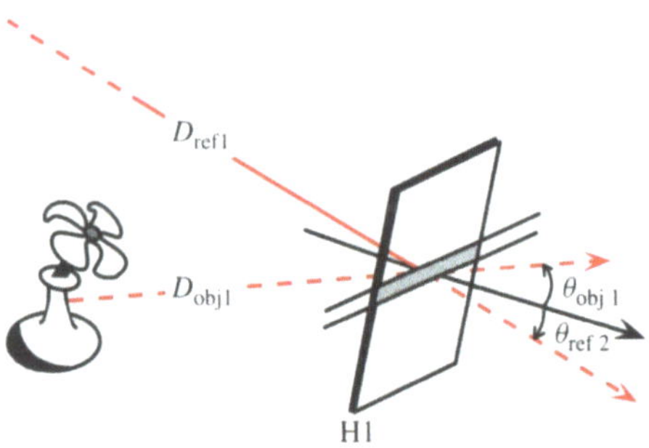

The process starts by recording a master hologram, or H1, although at a distance from the object that is usually quite a bit larger than that used for full-aperture transfers. We will see that the object-to-H1 spacing, D_{obj1}, will eventually determine the optimum viewing distance, D_{view}, along with all the reference and projection beam distances, and will have to be carefully reckoned. For now, let's assume that D_{obj1} is something handy, such as 300 mm (about a foot). As before, we can imagine that each small area of the plate, perhaps a half-millimeter on a side, records a unique perspective view of the scene corresponding to its location, from up to down and side to side.

Transferring

Again, the H1 is illuminated in phase conjugation (or at least approximate phase conjugation) by bringing an illumination beam (sometimes called the "projection beam") through its back surface in a direction opposite to that of the reference beam. The convergence of the projection can also match the divergence of the reference beam—they are both typically collimated—or not. The transfer hologram, or H2, is now placed so as to straddle the projected real image (which is pseudoscopic), making the maximum image depth as small as is practical (as a rule). A reference beam is introduced at an angle, θ_{ref2}, usually from below, and from a distance, D_{ref2}, that is as large as the table permits (if a second collimator is not available).

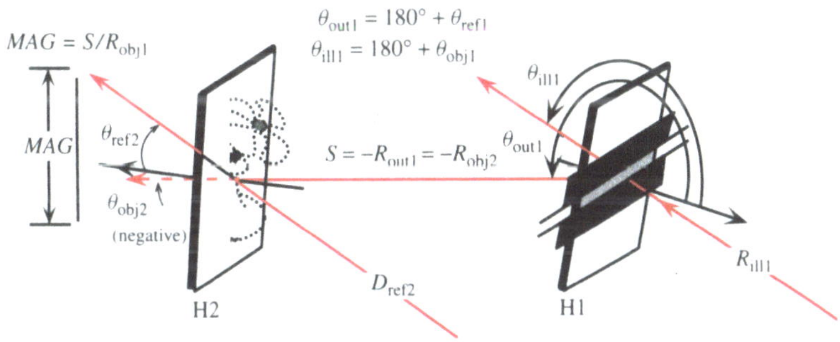

H1 as a line array of projectors

We wish to project a continuum of images differing only in right-to-left perspective content. Recall that each area of the hologram can project only the perspective view that it recorded. Thus, by illuminating only a narrow horizontal stripe of the H1, we can eliminate the up-to-down variations of perspective within the projected real image. The choice of the slit's width is determined by practical considerations mentioned below, and the choice of optics to "feed" the slit without wasting light are also topics for later discussion. But for now, we can imagine that we simply mask off most of the plate, perhaps with black photo tape applied directly to the back of the H1 (the front surface is typically index-matched to a clear-glass plateholder). Since we will get only one vertical perspective, it's of course necessary to set the slit's position to one that gives a nice-looking view of the object.

Viewing of the H2

The H2 may now be illuminated from above and behind with a monochromatic point source at the same wavelength with which it was recorded. The illumination is in the direction opposite to the reference beam, and the source distance is as large as possible so as to come as close to phase-conjugate illumination as possible. We can consider the H2's output in either of two ways: as an image of the real image projected by the H1, or as an image of the slit on the H1. Each point of view yields its own insights into the imaging process.

The H2 produces a pseudoscopic image of whatever its object exposure had been, which was itself a pseudoscopic image of the original object. "Two pseudos make an ortho," as we have seen before, so that a right-reading image is the final viewing result. It is visible from the direction of the image of the H1 slit, as before, but now we have to consider that slit image in more detail.

The real or aerial image of the H1 slit is formed at a fairly large distance from the H2, and its location is sensitive to the exactness of the phase conjugation of the illumination relative to the reference beam. Typically, for the longest beam lengths available on practical tables, the slit image is about 1.5× as far from the H2 as the H1 was during the exposure. This departure from perfection also means that the slit image will suffer from astigmatism, with consequences that we will explore shortly.

Ideal case: perfect conjugation

If the illumination is the perfect conjugate of the H2's reference beam (which usually means that the reference beam was converged with a large lens), the output slit will be located exactly as far from the H2 as the H1 had been. If the hologram is illuminated with a monochromatic point source of the same wavelength as the exposing laser, then the real image of the slit will be found exactly where the slit had been. All of the light diffracted by the H2 will focus through this slit image, and the viewer's eyes will have to be positioned accurately in that location. If the viewer moves up or down from there, the H2 abruptly turns dark! Moving side to side captures the light projected by various areas of the H1, which present images differing in horizontal perspective. This provides the difference between the right and left eyes' views, and motion parallax as the viewer moves from side to side.

If the wavelength/color of the light source is changed, the location of the slit real image changes in both angle and distance. It moves upwards for redder light, and downwards for bluer light, as was true for diffraction gratings. The redder image is also focused closer than the bluer image, as was true for Fresnel zone plates. Thus if the viewer moves up and down, instead of seeing different perspective views, *i.e.* different amounts of "look over" and "look under," she/he sees the same image but in different monochromatic hues.

Because of the limited change of wavelength in going from deep red to deep blue viewing, the range of output angle is only about 15°, which means that the "window" for seeing anything is somewhat limited in height. The viewer must also be at roughly the intended distance to see the entire image in a single color, such as "green." Moving too far backwards produces an image that is red at the top and blue at the bottom, while moving too near the hologram produces blue at the top and red at the bottom.

Let's go through the numbers for a simple "ideal" phase conjugation case. Assume that we locate the object 600 mm in front of the H1 "master" plate. The reference beam will be a collimated beam, and the laser wavelength will be 532 nm (a doubled-YAG laser).

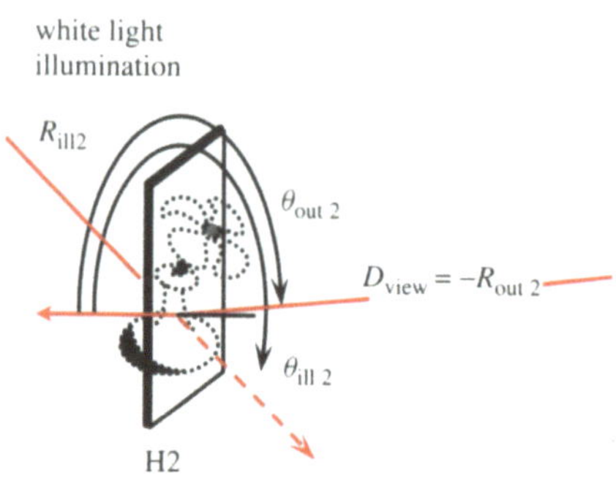

Upon back-illumination of the H1 with a collimated projection beam, the pseudoscopic real image of the object will be formed at unity magnification exactly 600 mm in front of the H1. This is where the H2 will be placed, so that it cuts the depth of the image roughly in two (the hologram plane "straddles" the image space). A slit is placed horizontally across the H1, blocking projection of the up-to-down variations of the views (the key step of the "rainbow" process). A collimated reference beam is used now for the H2, also of $\lambda = 532$ nm, to expose the final "transfer" hologram, and the beam is arranged so that it comes up from below the H1 at 45°, anticipating the eventual illumination direction.

After careful processing, the hologram is held vertically, and illuminated from 45° above and behind with a collimated white light beam (such as sunlight). Considering only the 532 nm green light component of the sunlight spectrum for the moment, we find that the image of the H1 slit is formed directly in front of the hologram, at a distance of 600 mm. An eye placed there, anywhere along the width of the slit image, sees an undistorted unity-magnification image of the object floating within the H2 as if the hologram were a window frame. Now, considering (for example) the 633 nm red component of the white light, the redder light is rotated more radically, and forms a slit image above the green-light slit image, and somewhat closer to the hologram. An eye placed at that new image location will see the same image as before, but in bright red light instead of green. The tonality and perspective will be the same—only the overall color will have changed. If the eye moves between these two locations, it will see the same image in a continuously changing spectral color, from green to yellow to orange to red. Contrariwise, if the eye moves downward, it will see the image in colors from green to cyan (blue-green) to blue to violet. It is the purity of these spectral colors that gave "rainbow" holograms their name.

As the eye moves from side to side within a single color zone, it picks up the images first captured by the corresponding regions of the master hologram, the H1. Eventually, the viewing zone "runs out of H1" and the image goes dark on the extreme right and left sides. Thus a viewing window is established that has its width determined by the width of the H1, and its height determined by the amount of spectral dispersion (typically about 15°).

Nonideal (typical) case

Most holographers have either no collimators or just one, because they are so expensive, and therefore perfect conjugation isn't available in either the transfer or the viewing stage, or both. This brings us to the *practical* side of rainbow holography, where our shop-math formulas help us place the image where we want it to be, regardless of the limitations of our equipment.

Let's assume that we have no collimators at all, so that we have to use diverging beams for reference #1, projection, reference #2, and illumination. The price we will have to pay is that the object will have to be smaller than the image we want to produce, and it will be closer to the H1 than the viewing distance. In addition, there will be some distortions of the image that we will have to live with, or partially compensate by pre-distorting the object in a complementary way. Let's assume for simplicity that all our beam-throws (wavefront radii) are the same, being 3 meters. The viewing distance will be 0.5 meter.

Other complications will also arise. First, the wavelength we shoot in will be different than our eventual viewing goal. The He-Ne laser wavelength is 633 nm (red), whereas our "target" viewing wavelength will be 550 nm (yellowish-green). Shrinkage of the emulsion layers will also be a concern, but for now we will assume that we "split the angle" between the object and reference beams for the H1, so that the fringes are vertical and thus unaffected in angle by shrinkage, and that we already know that to get a perpendicularly-exiting green beam from the H2, the red object beam has to have an angle of 7° to the perpendicular (see below).

Backwards Analysis

With all these points in mind, we are ready to start designing the exposure setup for creating a rainbow hologram that fills a certain prescription. We start by considering the setup for the H2. From here on, we will consider the radii of the relevant wavefronts, rather than the distances to their sources:

H2 "transfer hologram" optics

If we're going to solve this problem in reverse (from the light source and the viewer's eye to the exposure of H1), we need to specify something about the viewing conditions for H2. Let's suppose that we have a diverging white light source 3000 mm away and at 45° above the horizontal. The prescription above translates into illumination conditions of $\theta_{\text{illum2}} = -45°$, $R_{\text{illum2}} = +3000$ mm, and viewing conditions of $\theta_{\text{out}} = -180°$, $R_{\text{out,vertical}} = -500$ mm, $\lambda_3 = 550$ nm.

These conditions require a little more explanation. The output angle is 180°, or perpendicular to the hologram, which is the usual case for holograms that are meant to hang vertically, whether by wires or in a frame. It is easy to check, because the viewer will see his/her eyes reflected in the center of the hologram if the angle is right. The viewing distance is chosen arbitrarily, but generally depends on the size of the hologram—the bigger the hologram, the larger the viewing distance. People are used to looking a television at a distance such that the screen subtends an angle that is about one fist wide, held at arm's length (Try it! You aren't getting your money's worth at a movie if it isn't at least *three* fists wide!). So, a 4"×5" hologram is typically viewed at about half a meter (a bent arm's length) and an 8"×10" at a full meter, and so on. The viewing distance is represented in terms of a negative curvature of the output wavefront, because the H2 reconstructs an image of the H1 out in space, converging light into a "view zone" at the intended location of the viewer's eyes. What is somewhat subtle is that it is the vertical convergence that matters, so that the same color reaches the eye from the top and bottom of the hologram. Because of imperfect conjugation, the horizontal and vertical foci will be at noticeably different distances (the horizontal will be farther away), and we have to make sure we use the appropriate equations to calculate the two distances. Finally, the choice of wavelength is also arbitrary—green (550 nm) simply defines the center of the viewing window for convenience. If we are making multi-color holograms, we will choose two or three other wavelengths for our calculations instead.

Getting the angles right

The first equation we need to deal with is the $\sin\theta$ equation (3) from Chapter 8. We present it in symmetrical form to emphasize that the calculations proceed in both directions:

$$\frac{\sin\theta_{\text{out2},m} - \sin\theta_{\text{ill2}}}{\lambda_3} = -m\frac{\sin\theta_{\text{obj2}} - \sin\theta_{\text{ref2}}}{\lambda_2}, \qquad m = -1 \tag{1}$$

where the minus sign on the right side is introduced because of the convention for measuring the H2 angles that we discussed in the previous chapter. Inserting the values for the variables from the example, we find that we have an arbitrary choice of pairs of reference and object beam angles that will give the same spatial frequency at the center of the hologram. For example, 125.5° and –180°, 133.4° and –175°, 140.2° and –170°, and so on:

$$\frac{\sin(-180°)-\sin(-45°)}{550\text{ nm}}=\frac{\sin\theta_{\text{obj2}}-\sin\theta_{\text{ref2}}}{633\text{ nm}} \tag{2}$$

The preferred choice is determined by a factor that we haven't considered so far: that the hologram fringes form a "venetian-blind-like" structure in the emulsion that behaves like an array of tiny mirrors. Their angle has to be correct for the hologram to give maximum brightness when it is vertical, a phenomenon we will call "Bragg selection effects." The result depends on how much the emulsion shrinks during processing, and by how much its refractive index changes. Typically, for silver halide holograms, the emulsion shrinks by 7% and the refractive index drops from 1.64 to 1.59.

Solving the problem here requires two pieces of knowledge that we already have: first, the shrinking of the emulsion's thickness isn't going to change the spacing of the fringes on the surface (which is why we can still use the $\sin\theta$ equation in the way we just did above, without worrying about shrinkage effects); second, the tip angle of the fringes in the emulsion is exactly half-way between the angles of the beams forming them—and when we illuminate the hologram we want the beams again to be symmetrically disposed on either side of the fringes. So if we call the "before" tip angle of the fringes ϕ_{tip1} and the "after" ϕ_{tip2}, and the ratio of the "after" thickness to the "before" thickness s, we get the following trigonometric relationship:

$$\tan\phi_{\text{tip1}}=s\tan\phi_{\text{tip2}} \tag{3}$$

There's another number we have to take into account here: the index of refraction of the emulsion. The fringe angles are really happening inside the emulsion, so we need to use Snell's law to convert the outside beam angles into the inside beam angles (and we need to use the "before" index for exposure and the "after" one for reconstruction, too):

$$\theta_{\text{xxx,int}}=\sin^{-1}\left(\frac{1}{n_{\text{int}}}\sin\theta_{\text{xxx,ext}}\right) \tag{4}$$

Now, the main relationship we have to work with is going to be an expansion of Eq. (3) above:

$$\tan\left(\frac{\theta_{\text{obj,int}}+\theta_{\text{ref,int}}}{2}\right)=s\tan\left(\frac{\theta_{\text{out,int}}+\theta_{\text{ill,int}}}{2}\right) \tag{5}$$

and what we need is the solution to (2) above that also satisfies this relationship. Because of all the sines and tangents and so forth, there isn't an obvious way to solve that problem in closed form, and we will have to resort to an iterative solver program (or to a long evening with a lot of scrap paper). The basic procedure is to take the known output and illumination beam angles above, convert them to internal angles, and solve (5) for a pair of internal object and reference beam angles that when converted back to external angles also solve (2).

The result is a particular pair of object and reference beam angles:

$$\theta_{obj2} = -176.1°, \quad \theta_{ref2} = 131.8° \tag{6}$$

Note that the reference beam should come up to the H2 plate from "below" so that it can be illuminated from behind and above. This angle is usually difficult to arrange (unless you have a hole in the table, or some mirrors cleverly arranged), so the H2 is usually turned on its side so that the reference beam can travel horizontally across the table.

Since the mathematics of shrinkage are somewhat imperfect (imprecisely depending on chemistry and not closed-form) and thus resist our desire for an exact solution we can do quickly before building a laboratory setup, it's handy to keep this last result in mind and use it as a rough guide. Without shrinkage, the reference beam would have been 135° and the object beam –180°. Because of the change in fringe angle caused by shrinkage, we compensate by increasing the angle between the two beams, and because the fringes tilt more in the direction they're already tilted (if they were perpendicular to the surface, shrinkage wouldn't affect them) we compensate by rotating both beams down a little. So, in a pinch, you could do a rough shrinkage compensation for a common rainbow hologram by tilting the hologram just before the transfer exposure so that the object beam (otherwise on-axis) is a couple of degrees below the axis and the reference beam decreases by the same amount. The reference beam angle (before the plate tilt) should be calculated as usual based on the $\sin\theta$ equation—which you'll recall doesn't care about shrinkage.

We will revisit shrinkage in a later chapter when we look at reflection holograms, a type of display where recorded fringes are usually more parallel to the surface of the emulsion than in transmission holograms. We will think about the mathematics of shrinkage slightly differently, but the phenomena and the results are equivalent for both transmission and reflection holograms. In fact, every hologram has both transmission and reflection hologram behavior, to at least a limited extent!

Getting the distances right

The key distance to consider here is the viewing distance, D_{view}, at the intended wavelength, λ_3 and viewing angle (typically 0°, but the dependence on angle is quite small). And the key realization is that it is the *vertical* or color focus that is relevant—this is the peculiar astigmatic focus that we discovered is an effect in off-axis holograms, recall. The reason that this is the focus that matters is that we wish to see the same color, green in this case, coming from the top and bottom of the hologram. That is, we want to find the point where the green rays from the top, center, and bottom all cross. An eye placed there will see the entire hologram surface light up in bright green light! The other focus, where the green rays from the right, center, and left all cross determines where one can view the exact perspective captured by a region of the slit on the H1 master hologram (which doesn't matter at this point).

The vertical focus is determined by the $\cos^2\theta/R$ equation (7) from Chapter 10, which again we show in symmetrical form:

$$\frac{1}{\lambda_3}\left(\frac{\cos^2\theta_{\text{out2},m}}{R_{\text{out2},m,\text{V}}}-\frac{\cos^2\theta_{\text{ill2}}}{R_{\text{ill2}}}\right)=m\frac{1}{\lambda_2}\left(\frac{\cos^2\theta_{\text{obj2}}}{R_{\text{obj2}}}-\frac{\cos^2\theta_{\text{ref2}}}{R_{\text{ref2}}}\right) \tag{7}$$

Inserting the values for the variables that were discussed above, we find that the object beam must be a diverging beam with a positive radius of curvature of 397 mm. This means that the slit of the H1 must be 397 mm away from the H2, so we have determined the H1–H2 separation, usually called "S."

$$\begin{aligned}&\frac{1}{550}\left(\frac{\cos^2(-180°)}{-500\text{ mm}}-\frac{\cos^2(-45°)}{5000\text{ mm}}\right)\\&=-1\frac{1}{633}\left(\frac{\cos^2(-176.1°)}{R_{\text{obj2}}}-\frac{\cos^2 131.8°}{5000\text{ mm}}\right)\\&S=R_{\text{obj2}}=397\text{ mm}\end{aligned} \tag{8}$$

H1 "master hologram" optics

Now we are faced with the challenge of creating a master hologram, or H1, that will project a real image at the proper angle and distance (S) so as to straddle the H2 plane, or at least to put the image where we want it in front of or behind the transfer hologram surface.

The angles are fairly straightforward, given a couple of practical considerations. First, the projection beam for the H1 should be parallel to the reference beam for the H2, just for convenience in getting them both as long as possible within the constraints of the table. This gives the relation:

$$\left(\theta_{\text{obj1}}-\theta_{\text{ref1}}\right)=360°-\left(\theta_{\text{obj2}}-\theta_{\text{ref2}}\right) \tag{9}$$

Second, the object and reference beams should come in at equal but opposite angles to the perpendicular to the H1. This makes the resulting interference fringes perpendicular to the surface of the emulsion, so that there is no astigmatism in the focus of the image, and the fringe tip angle is insensitive to shrinkage of the emulsion—this widens our choices of processing chemistry considerably. It is also easy to check when the hologram perpendicular is at the right angle—the hologram will reflect the reference beam onto the object!

Assuming that the H1 and H2 are exposed at the same wavelength, there is no adjustment for wavelength change effects, and the exposing angles are simply:

$$\begin{aligned}2\left(\theta_{\text{obj1}}=-\theta_{\text{ref1}}\right)&=360°-\left(-176.1°-131.8°\right)\\\theta_{\text{obj1}}&=+26.05°\\\theta_{\text{ref1}}&=-26.05°\end{aligned} \tag{10}$$

The exposure distance, object-to-H1, is only slightly more difficult to find. The output wavefront must have a radius of "negative S" in order to converge at the required H2 location. The relevant axis of focus is now the *horizontal* or parallax focus, because the image focus is determined by the distance at which rays from the right, center and left areas of the H1 slit overlap. This we need the simpler "one-over-R" equation.

$$\frac{1}{\lambda_2}\left(\frac{1}{R_{\mathrm{out1},m,\mathrm{H}}}-\frac{1}{R_{\mathrm{ill1}}}\right)=m\frac{1}{\lambda_1}\left(\frac{1}{R_{\mathrm{obj1}}}-\frac{1}{R_{\mathrm{ref1}}}\right),\quad m=-1 \tag{11}$$

Substituting the values of the variables involved gives:

$$\frac{1}{633}\left(\frac{1}{-397}-\frac{1}{5000}\right)=-1\frac{1}{633}\left(\frac{1}{R_{\mathrm{obj1}}}-\frac{1}{5000}\right) \tag{12}$$

$$R_{\mathrm{obj1}} = 343\,\mathrm{mm}$$

That S is greater than D_{obj} means that the image will be magnified side-to-side by the same ratio, and magnified in depth by the square of that (which can become a lot!). Clever holographers often pre-distort their objects to compensate, so that intended spheres become small, shallow, dish-shaped objects. Previsualization of a hologram in the face of all these distortions becomes quite a challenge. Some folks use wire frames to help compose their scenes, or distorted checkerboards.

Other effects of imperfect conjugates

Not having enough collimators causes other problems, too. These are primarily apparent in the image projected by the H1, where the distances from the hologram are large, but can be seen in the way the H2 plays back too. The generic name for the effects of imperfect conjugates is "optical aberrations." Seidel identified and named these for conventional optical systems back in 1856, and we can adapt them for holographic discussions too.[iv] The five Seidel or primary aberrations are:

Spherical aberration: a lens with spherical surfaces doesn't usually produce a perfectly spherical wavefront—instead, it curves inward more sharply when measured further from the center.

Astigmatism: light passing at an angle through a lens generally has different curvatures in the direction toward the central axis (defining the *sagittal* focus) and perpendicular to it (defining the *tangential* focus).

Coma: even if astigmatism is cured, as the lens diameter increases the light will focus at different angles and distances, producing a diffuse comet-light tail around the sharp centrally-formed point.

Curvature of field: the image of a flat surface (or a constellation of stars) formed by a lens is only approximately flat—the surface of best focus is usually cupped slightly toward the lens.

Distortion: the image of a checkerboard is usually bowed inward or outward at the edges, termed "pincushion" and "pillow" distortion respectively. This arises from the output angle of the lens being non-linearly related to the input angle—its effects in holography are not discussed here.

An additional aberration that isn't one of Seidel's original five but is of importance to lensmakers (and holographers) is *chromatic aberration:* because the refractive index of glass is always higher in the blue than in the red, the focal length of a single lens is shorter for blue light, so the image is out of focus in other colors. This can be corrected by matching lenses made of very different glasses, but more complex forms of chromatic aberration also arise. The wavelength dependence of holographic lenses is much stronger than that of glass lenses.

When an image is projected by an H1, aberrations arise when an imperfect conjugate wave is used for illumination. The biggest problems are caused by spherical aberration, which causes the hologram image of a flat surface to curve away from the hologram plane, being closest directly in front of the viewer, and to "roll" as the viewer moves from side to side. Coma also arises, which causes the image of a point to move up and down as the viewer moves from side to side. The "trail" of a point in the hologram plane can trace out some strange shapes, instead of the straight horizontal line predicted by simple theory, which can cause eyestrain in extreme cases. The particular mix of aberrations found reflects the holographer's choices of equipment, and sometimes it's possible to identify a particular holographer's work just from the shape of the "trails" of bright points in the image!

Slit Width Questions

One of the perennial questions for rainbow holographers is, "How wide a slit should I use?" The best answer can vary between 0.5 mm and 25 mm, and depends very much on the nature of the image. A thin (0.5–2 mm) slit gives very sharp images over great depths (perhaps 150 mm in front of and behind the hologram), but with high speckle contrast. As the slit is widened, the speckle slowly decreases in contrast until it becomes nearly invisible (8–25 mm), but the image starts to be blurred at shallower depths. Only a few experiments will provide a useful answer, and will typically require a compromise between depth and speckle.[v]

As a practical matter, as much of the H1 illumination as possible is fed to the slit area by using cylindrical lenses to spread the beam upstream of the H1. If more beam width control is needed, crossed cylindrical lenses of very different focal lengths are used, often with a collimating lens to control the spreading of the beam. Cylindrical lenses can be expensive, but a test tube full of mineral oil, or a carefully-chosen section of polished glass rod, can usually suffice.

Limitations Due to Horizontal-Parallax-Only Imaging

Rainbow holograms are white-light viewable because they sacrifice one axis of parallax—they produce "horizontal parallax only" images (HPO images). Conceptually, we can say that the entropy of the hologram (its information content) has been reduced to match the reduced entropy of the light source (its temporal coherence). However, there are other techniques for producing HPO images, such as the use of lenticules (small vertical cylinders embossed/cast into the surface of a plastic sheet), as seen on 3-D postcards. All HPO images share certain limitations or optical effects that should not be attributed to holograms in particular:

Inherent astigmatism: In a horizontal plane, the rays from an image point fan out from the point's apparent location behind the hologram surface, a central principle of stereoscopy. But the rays fanning out in a vertical plane always have their common center on the hologram surface—a point on the horizontal "track" of the image point. The result is an astigmatic ray bundle, or a wavefront that has different curvatures in the horizontal and vertical directions—with a difference that increases as the image location moves further from the hologram surface.

Depth of field: The human eye can tolerate only a limited amount of astigmatism before eyestrain results (the eye continually refocuses to try to sharpen the image). Optometrists usually allow "one quarter diopter" of astigmatism before changing a prescription to correct it. In our terms, that translates to

$$\begin{aligned} \Delta_{\text{astig1}} &= \frac{1}{D_{\text{near}}} - \frac{1}{D_{\text{holo}}} \leq 0.25 \text{ m}^{-1} \\ \Delta_{\text{astig2}} &= \frac{1}{D_{\text{holo}}} - \frac{1}{D_{\text{far}}} \leq 0.25 \text{ m}^{-1} \end{aligned} \qquad (13)$$

Thus for a hologram viewed from 500 mm away, the image point can be 56 mm in front of the hologram, or 71 mm behind the hologram, before viewing becomes stressful. Art holographers deliberately violate this limit as a matter of course, assuming that nobody will be looking at any one image for very long. But we should also recall that someone with 1/4 diopter of uncorrected astigmatism will be able to tolerate more depth on one side of the hologram, and less on the other.

Viewer distance limitations: The same astigmatism effect produces a distortion of the image when the viewer is not at the correct distance (defined now as the distance to the horizontal focus of the H1 image). The image of a spherical object, or ball, floating in front of the hologram will appear squashed, or shrunken up-to-down, as the viewer moves further than the intended distance, and stretches up-to-down as the viewer moves closer. Fortunately, the human eye is quite tolerant of height-to-width distortions, so that a useful range of viewing distances can be accepted.

Spectrum tip: For most people, the correct viewing distance is the one at which the image appears in a single color from top to bottom. That is, the one formed by the vertical focus of the H1 image). This is usually calculated for the middle-green wavelength, 550 nm. If the eye moves upward, a yellower, then redder image is seen. However, we should note that the optimum viewing distance also shrinks considerably, so that the surface of optimum viewing turns out to be a plane that is tipped forward. The angle of tip is what we identified earlier as the "achromatic angle," or α, and is somewhat greater than the angle of the illumination beam. Recall from Chapter 11 that

$$\tan\alpha = \sin\theta_{\text{ill}} \qquad (14)$$

Conclusions

The principal advantages of rainbow holograms are that their images are sharp and deep when viewed with commonly available light sources, and that they can be very bright. No light is wasted by narrow-band filtering to make the source monochromatic, and the light that is diffracted (the image light) is sent into a beam that is quite narrow, vertically (only about 15° high). Thus relatively weak unfiltered spotlights can be used to illuminate them—even flashlights and candles work well! Unfortunately, they are a little difficult to produce, especially if the highest levels of quality are desired. The exposure system must be carefully designed in order to produce the desired effect under the specified illumination and viewing conditions.

Fortunately, the tolerance for error is high enough to allow a wide range of viewing conditions to produce acceptable images, and mass-produced rainbow holograms have become very popular. The next chapter will consider some of the issues of making practical rainbow holograms, including multi-color holograms and embossed holograms, all based on the same principles that are outlined here.

References

i. Benton, S. A. (1969). "Hologram Reconstructions with Extended Incoherent Sources," *J. Opt. Soc. Amer.*, **59**, 10, pp. 1545-1546.

ii. Benton, S. A. (1977). "White-Light Transmission/Reflection Holographic Imaging," in Marom, E. and A. A. Friesem, eds., *Applications of Holography and Optical Data Processing*, Pergamon Press, Oxford, UK, pp. 401-409.

iii. Gerritsen, H. J. private communication.

iv. Born, M. and E. Wolf (1980) *Principles of Optics*, Pergamon Press, Oxford, UK, pp. 211-218.

v. J. C. Wyant (1977), "Image Blur for Rainbow Holograms," *Optics Letters*, **1**, pp. 130-132.

CHAPTER 15

Practical Issues in Rainbow Holography

Stephen A. Benton, Michael Halle, and V. Michael Bove, Jr.

Introduction

In this chapter, we will investigate several topics that extend white-light transmission rainbow holograms from single-color, limited production displays into the multiple-color and mass-produced holograms that proliferate the industry today. First, we'll look at multi-color rainbow holograms, building on our previous tools for analysis and intuition to understand two major methods for multi-color recording. Since these holograms require the exposure of multiple master holograms into a single H2 transfer, we'll discuss some of the consequences of multiple images and multiple exposures. Finally, we'll touch on some topics related to embossing, the most common method for mass-producing holograms.

Multi-Color Rainbow Holograms

White-light transmission rainbow holograms represented a huge leap toward making holography accessible to the public. Artists and designers were able to use rainbow holography to create bright, clear, dimensional images illuminated with a single point light source instead of a laser. For artists experienced in using color as an element of their creative work, though, the single-color look of the rainbow hologram represented a significant limitation. Several techniques were developed to provide additional color options using white-light transmission holography.

Before we begin a discussion of multi-color holographic techniques, we must first come to some agreement about what we mean by "color" in the context of holography. There is much room for ambiguity here; after all, we use a single-color laser to make our exposures, and a standard rainbow hologram is very colorful when viewed. These terms are not highly rigorous; the definitions we give them here are common if not universal. Also note that in general we are talking about displayed color only—unless we explicitly state otherwise, readers should assume that the color of an object's holographic image is not correlated with its color in the original scene.

The terms *single-color* or *monochrome* mean that an image appears in a single wavelength or a small range of wavelengths narrow enough to give the impression of a single saturated color or spectral hue. The color itself may be constant, or it may change with the position of the viewer (as is the case with a "classic" rainbow hologram).

Achromatic, or *black-and-white*, means that an image is composed using multiple wavelengths in such a way that it presents as unsaturated or neutral in hue, much like a black-and-white photograph or television image. Achromatic images can be produced in a number of ways, including a set of perceptually equal and complementary primaries, or a transmission hologram illuminated with a vertical line source.

Multi-color is a broad and inclusive term for images where different parts of the scene appear in different hues simultaneously, in a controlled way. ("Controlled" means that the color effect is intentional. Ordinary rainbow holograms can be viewed so that the image appears in a range of hues from top to bottom, but are not considered multi-color holograms.) This definition puts no restriction on number of colors, whether the images are posterized or continuous tone, or whether the images of different colors are registered or aligned together.

Full-color holograms are a subset of multi-color holograms where the image can span a wide gamut of colors, commonly including white. Full-color emissive displays are usually based on a mixture of three primary colors that the human visual system interprets as a wide spectral range. The images of the different primary colors should be in registration when viewed from a variety of angles. However, full-color displays may suffer from significant color shift dependent on the location from which they are viewed.

Natural-color or *true-color* displays are full-color images where the appearance of the original scene and its image are perceptually similar and relatively independent of any changes in viewing location. Matching recorded and displayed color is a challenge in all media; holography is more difficult than most. For example, one full-color recording technique uses multiple lasers in different wavelengths to illuminate the scene. While the laser primaries may span a wide color gamut, they sample the object's reflected spectrum at only a limited number of wavelengths. As a result, these "true-color" replicas of the scene may appear very different from how the scene itself would look when viewed under a continuous spectrum white light.

In this chapter, we will concentrate on multi-color holograms without the goal of achieving natural or true color: most holographers don't have three different colors of lasers, and the apparatus required to record a scene in perfect alignment during three different color exposures is very complicated and demanding.

We can think of creating a multi-color hologram as making several overlapping "component" holograms, each chosen to bring a different wavelength to a focus at the viewer's position. In accord with television practice, we will discuss only three wavelengths, intended to represent "red," "green," and "blue" light. However, we hasten to point out that making a strong link between the wavelengths of an image and its perceived color is very risky in view of perceptual research originated by Edwin Land.[i] Optimal selection of primary-color wavelengths remains the subject of research (and some controversy) in the field of color holography as well as in electronic imaging. For the following discussion, we will arbitrarily choose 633 nm red, 532 nm, green and 470 nm blue as our primaries.

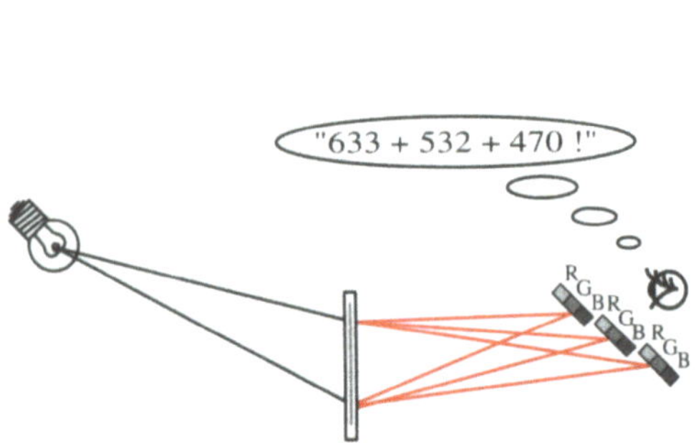

To understand how a multi-color white light transmission hologram works, consider an ordinary white light transmission hologram. Recall from Chapter 14 that an H2 hologram of an H1 slit master that is centered in front of it, recorded in 633 nm light, will under phase conjugate illumination replay an image of that slit master so that 633 nm light is centered in front of the display. Other wavelengths present in the illumination beam will be diffracted at different angles: a green image of the slit can be seen just below and behind the red image, and a blue image further below and behind.

We've also seen that it's possible to center the spectrum of the image of the slit seen under white light illumination by compensating for the wavelength difference between 633 nm recording and the central wavelength of the visible spectrum (say, green at 532 nm). Adjustment of the H2 object or reference beam angle will center this spectrum.

Now, imagine a superposition of the two holograms described above. A viewer viewing the new display "straight on" will see the first image in red, and the second one in green. In the parts of the display where both images are bright, the scene will appear as a blend of red and green, which we perceive as yellow. (Note that the stimulus here is completely different in a physics sense from spectral yellow light, but our eyes and brains make the two appear to be the same hue.)

If the observer moves down a little, he or she will see the first image in green, and the second image in blue, yielding a "cooler" image of the object. Moving down still further, the observer will no longer see the second image at all, in any wavelength, and will see only the first image in blue. Moving above the center viewing position, the observer can see a similar effect: in this region, the first holographic image contributes nothing, and the second image appears red. At the risk of redundancy, we repeat that the only change is in the displayed image color, not somehow in the recorded color of the scene.

Ignoring for a moment how we might make such a composite display, we bravely forge ahead and add another component H2 hologram to the mix. This hologram is exposed so that a red image of its H1 slit master plays out even higher above the midline of the display. A viewer looking "straight on" to only this component hologram would see an image in blue. When this hologram is composited with the other two we have just discussed, a vertically-centered viewer can see parts of the scene in any mixture of red, green, and blue light.

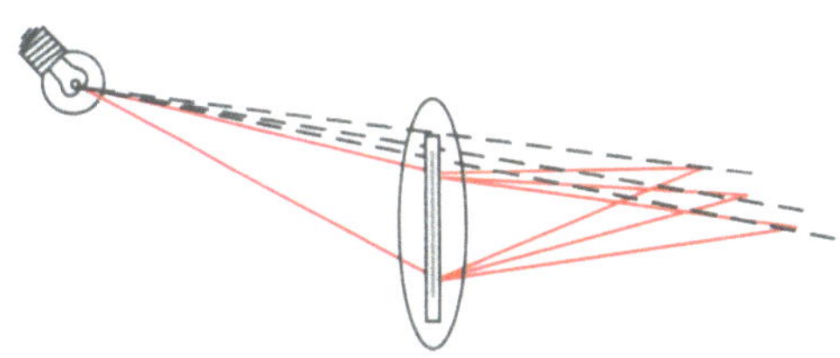

We have, during this thought experiment, created a multi-color (yes, even a full-color!) hologram by carefully overlapping the output spectra of three different H1 holograms. If our goal is to produce an achromatic hologram, these three H1s could be identical. Ironically, however, producing a high quality achromatic hologram that displays as a nice white tone with minimal color shift is just as demanding (if not more) as producing a good color image. We will not be discussing achromatic holograms as a separate topic for the remainder of this chapter.

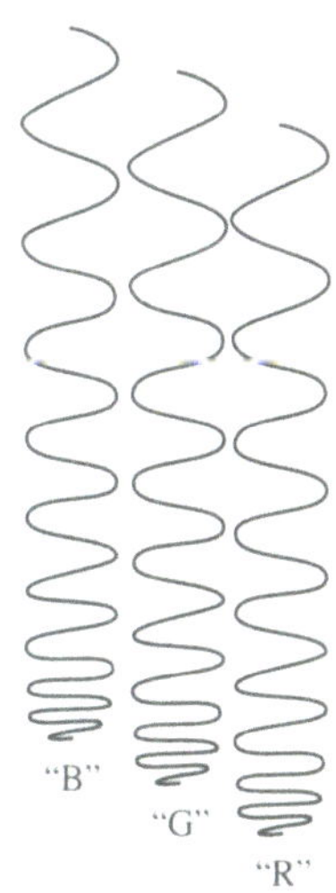

Fortunately for us, creating a composite hologram containing the information from three H1s is straightforward, if somewhat involved and subject to several caveats. A holographic emulsion is capable of storing fringe patterns from several different exposures, and playing the composite image back when illuminated. Instead of recording three different H2 holograms, each with its own H1, we will record multiple H1s onto a single H2. This H2 will be our composite hologram.

Our first step in producing a composite H2 hologram is to record the set of H1 holograms that will store the scene information for each color to be displayed. We need to be able to specify the output angle and distance for reconstruction of these three H1 images so that each image plays out on-axis in its wavelength at the intended view distance, taking account of the wavelength shift from the exposing laser wavelength. Since the illumination distance and angle will be the same for each H1 image (the final H2 hologram will be illuminated by a single point light source), we have only the reference angle and distance and the object angle and distance as possible parameters to adjust. Either set of parameters can be used in practice; each approach has different tradeoffs.

For either the multiple-object-beam or multiple-reference-beam geometries, the H2 hologram is exposed, processed, and illuminated in the same way as any single-master rainbow hologram.

Multiple-Reference-Beam Holograms

In this technique, the position of the master H1s (or single H1, if we're making an achromatic H2) is held constant with respect to the exposed H2 for three holographic exposures during holographic transfer, while the position and angle of the H2 reference source are changed for each H1. It would seem as if we could do all the exposures at the same time, but in practice each H1 exposure must be done separately for two reasons: first, the three H1 slits occupy the same position; and second, multiple reference beams would interfere with each other and produce diffraction patterns that reduce the efficiency of the image-bearing part of interference pattern.

In the transfer apparatus, the three reference beams are set up at the same time. When a particular exposure is to be made, the appropriate H1 hologram is placed into position and the appropriate reference beam leg is unblocked. This technique, then, requires three different sets of reference beam optics and usually results in a very crowded holographic table.

To calculate the required reference angles and distances for the three exposures, we use our now-standard holography equations. The $\sin\theta$ equation provides the angles we need. For distances, we need to choose between the horizontal focus equation ($1/R$) or the vertical focus equation ($\cos^2\theta/R$). To make this choice, we consider the slit as a horizontal feature during viewing. Horizontal lines are vertical detail, and vertical detail is sharpest at the vertical focus, so the $\cos^2\theta/R$ equation is the appropriate choice for the distance calculation.

To find the angles, we first note that the reference angle for the blue master will be steeper than for the other two H1s. We'd like to keep all exposure angles under 60° from the normal to the plate (which means between 120° and 180° in the transfer geometry) to avoid loss from reflection. For convenience, we decide to illuminate the hologram from 45° above the plate. So, using the $\sin\theta$ equation for the blue transfer, we put in the following constraint:

$$\text{for } \lambda_2 = 470\,\text{nm},\ R_{\text{obj}} = 300\,\text{mm},\ \theta_{\text{ref}} = 120° \tag{1}$$

and we calculate that

$$\theta_{\text{obj}} = -175° \tag{2}$$

We can use that same object angle for each H1. Now we need to find out what the reference beam angles will be for the other two exposures. For green,

$$\lambda_1 = 633\,\text{nm}, \lambda_2 = 532\,\text{nm}, \theta_{\text{obj}} = -175°, \theta_{\text{ill}} = -45°, \theta_{\text{out}} = 0°, m = -1 : \theta_{\text{ref}} = 131° \tag{3}$$

For red,

$$\lambda_1 = 633\,\text{nm}, \lambda_2 = 633\,\text{nm}, \theta_{\text{obj}} = -175°, \theta_{\text{ill}} = -45°, \theta_{\text{out}} = 0°, m = -1 : \theta_{\text{ref}} = 142° \tag{4}$$

Now, let's solve for distances in a similar way. We shall constrain the illumination distance to be 1500 mm, the object distance to be 300 mm, and the projected slit distance at illumination (the viewing distance) to be 500 mm (*e.g.* $R_{\text{out}} = -500$ mm). Since we know that the blue exposure will have the longest reference distance, we start with blue calculations to make sure we won't fall off our optics table. (If it does, one choice is to use the longest possible distance the table allows and accept any slight registration error that may occur, or else use a collimator if one is available and doesn't interfere with the red or green H1 reference beams.) Using the $\cos^2\theta/R$ equation, and the angle results from above,

$$\lambda_1 = 633\,\text{nm}, \lambda_2 = 470\,\text{nm}, R_{\text{obj}} = 300\,\text{mm}, R_{\text{ill}} = 1500\,\text{mm}, R_{\text{out}} = -500\,\text{mm}, \\ m = -1 : \\ R_{\text{ref}} = 1520\,\text{mm} \tag{5}$$

For green,

$$\lambda_1 = 633\,\text{nm}, \lambda_2 = 532\,\text{nm}, R_{\text{obj}} = 300\,\text{mm}, R_{\text{ill}} = 1500\,\text{mm}, R_{\text{out}} = -500\,\text{mm}, \\ m = -1 : \\ R_{\text{ref}} = 810\,\text{mm} \tag{6}$$

And for red,

$$\lambda_1 = 633\,\text{nm}, \lambda_2 = 633\,\text{nm}, R_{\text{obj}} = 300\,\text{mm}, R_{\text{ill}} = 1500\,\text{mm}, R_{\text{out}} = -500\,\text{mm}, \\ m = -1 : \\ R_{\text{ref}} = 632\,\text{mm} \tag{7}$$

You might imagine that optimizing these equations and following them exactly is a time-consuming process when done by hand. When possible, we highly recommend using a computer program or spreadsheet to simplify the process. As a further hint, getting the distance at which the different slit images are reconstructed to be exactly the same may not be essential in all applications. If the different masters reconstruct at slightly different distances, the viewer may see some banding or color inconsistencies; such artifacts might be acceptable for particular holograms.

The geometry we've just calculated is sketched below.

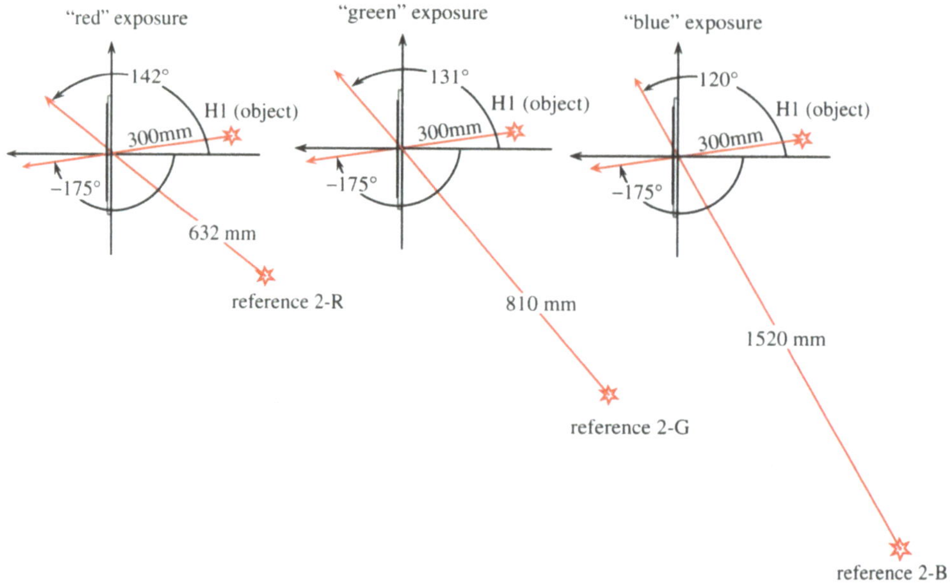

Multiple-Object-Beam Holograms

The alternative to the multiple reference beam approach is to use a single reference beam and expose all three slit H1 master holograms at once, with each H1 in an appropriate position so that the spectra of the three slit images will overlap correctly during illumination.

This "multiple-object-beam" technique has one chief advantage over multiple reference beams: the three slits can be recorded by the H2 simultaneously. Besides the fact that the three H1s are spatially separated, the light coming from each H1 is relatively weak compared to the reference beam. So while the three H1 images will interfere with each other, the magnitude of the resulting unwanted fringe pattern is small compared to the image-bearing interference between the reference beam and each slit.

The ability to expose the slits simultaneously means the exposure geometry can be simplified somewhat. The position of the H1 holograms at exposure lie approximately along a line similar to the "achromatic angle" introduced in Chapter 11. (Remember, though, that the red and blue slits are reversed from the position where they would be in a spectrum emerging from a hologram precisely because we're compensating for the spectral characteristics of the diffractive display.) The H1 slit masters can be individual holograms, or they can be individual exposures on a single large holographic plate. In either case, the three slits can be simultaneously illuminated using a single collimated beam, eliminating the need for three sets of projection optics.

Let's compute the numbers for a multiple-object-beam example analogous to the one we did in the preceding section. Again, first we calculate the angles:

$$\begin{aligned}
&\lambda_1 = 633\,\text{nm}, \theta_{\text{ill}} = -45°, \theta_{\text{out}} = 0°, \theta_{\text{ref}} = 135°, m = -1,\\
&\text{for } \lambda_2 = 633\,\text{nm}, \theta_{\text{obj}} = -180°\\
&\text{for } \lambda_2 = 532\,\text{nm}, \theta_{\text{obj}} = -172°\\
&\text{for } \lambda_2 = 470\,\text{nm}, \theta_{\text{obj}} = -166°
\end{aligned} \tag{8}$$

Next we determine the distances for the H1s. Recall our constraints: R_{ill} = 1500 mm R_{out} = –500 mm. This time, we solve for R_{obj} (rather than R_{ref}) as a function of λ_2. We set R_{ref} to 1000 mm so the reference beam fits on a moderately large holographic table. Using the $\cos^2\theta/R$ equation:

$$\begin{aligned} &\text{for } \lambda_2 = 633\,\text{nm}, R_{obj} = 353\,\text{mm} \\ &\text{for } \lambda_2 = 532\,\text{nm}, R_{obj} = 300\,\text{mm} \\ &\text{for } \lambda_2 = 470\,\text{nm}, R_{obj} = 258\,\text{mm} \end{aligned} \tag{9}$$

The resulting geometry for the transfer setup is illustrated below.

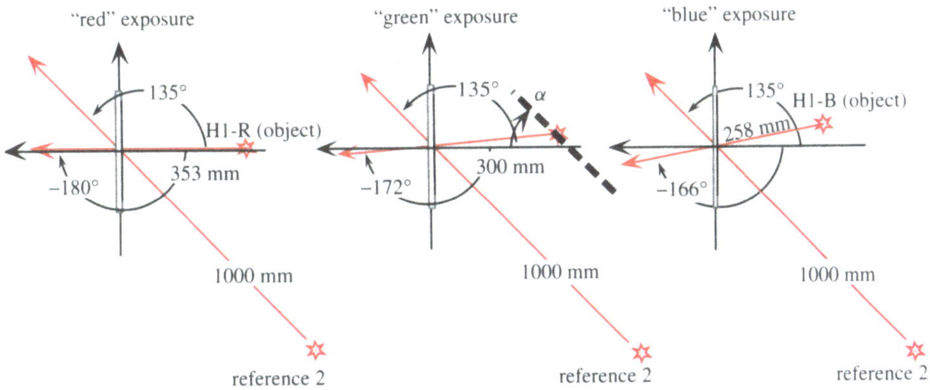

Note that the locations of the slits are on an approximately-straight line tilted at an angle that is the same as the "achromatic angle" derived previously, roughly given by:

$$\tan\alpha = \sin\theta_{ref} \tag{10}$$

Because they are on a line, the H1 holograms can all be projected with a single slit-beam of illumination that passes across all of the H1s, or the H1s can be made upon a single large plate or film hologram and illuminated with a single wide collimated beam.

Comparison of the Multi-Color Methods

Let's look in more detail at the advantages and disadvantages of the two multi-color rainbow hologram methods just described:

	advantages	disadvantages
multiple-reference-beam	exact registration possible simpler mastering layout good for photoresist	low diffraction efficiency long "blue" reference beam more complex transfer layout
multiple-object-beam	single exposure high diffraction efficiency simpler transfer layout	approximate registration more complex mastering layout

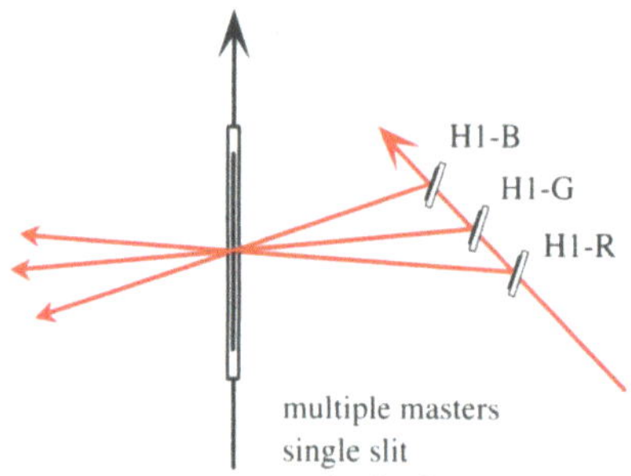

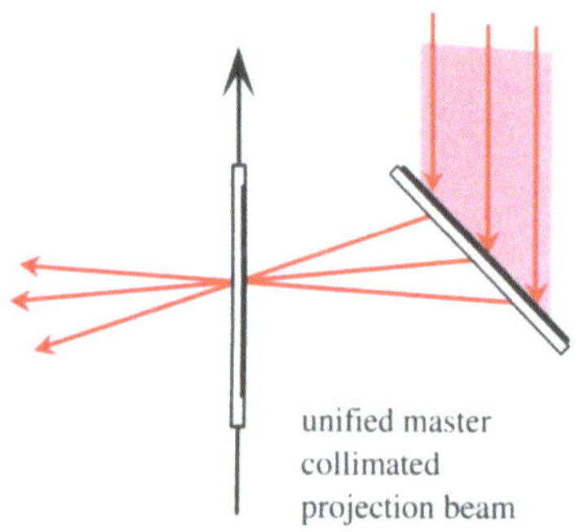

Complexity

In both cases, changes to the scene that produce different combinations of output colors (*e.g.*, painting an object black or uncovering an object for one or more exposures) must be made between H1 master exposures. Depending on the exact exposure technique, the multiple-reference-beam is usually simpler to master (the three H1 exposures use the same geometry), but that simplicity is negated by the more complex transfer setup. Conversely, the multiple-object-beam technique requires three H1s mastered in different positions, but can take advantage of a simpler transfer geometry.

Registration

The difference in registration between the two methods results from the different processes of recording the component H1 holograms. In the multiple-reference hologram case, the H1 masters for the three exposures are in the same location, and capture the same visual perspective of the scene. In the multiple-object-beam case, the three H1 masters capture different perspectives of the object: the master that will appear blue looks down on the scene from above, the "green" master slightly less so, and the "red" master closer to straight-on to the object. Since rainbow holography eliminates vertical parallax, the recordings of these perspectives will be forever different. Registration of these different perspectives can be correct for only a single depth: a deep image will suffer from increasing misregistration as the distance from this registered "sweet-spot" increases. Thus, for images with more than modest depth that require exact registration, the multiple-reference-beam approach may be the only practical option.

For completeness, we should note that it is possible to use the multiple-object beam approach using masters originally exposed at the same location, thus avoiding the problem of multiple vertical perspectives described above. Aligning the different masters and the respective projection beams during transfer is a rather difficult process that negates at least some of the advantages of the multiple-object-beam technique.

*Efficiency and the "one-over-*N*" law*

We have seen that the multiple-reference-beam technique requires three separate transfer exposures, while the multiple-object-beam method requires only one. This difference can have an influence on diffraction efficiency and thus on the brightness of the resulting display hologram. Multiple incoherent exposures onto a single piece of holographic recording material are less efficient than simultaneous coherent exposures of the same total energy: as a general rule, the efficiency of each image's exposure falls off as $1/N^2$ (where N is the number of exposures), and the efficiency of the composite hologram of N exposures is thus $1/N$ compared to a single exposure. Accordingly, multiple-reference-beam holograms are less efficient than multiple-object-beam holograms.

This "1-over-N" law holds because of the peculiar non-linear relationship between intensity of holographic exposure and the modulation of fringes in the holographic recording material. Recall that in our discussion about diffraction (Chapter 6), we found that diffraction efficiency varies with the square of modulation. It's necessary to split the total available dynamic range of the material among the number of exposures made—to record N different exposures into a single hologram without overexposing any part, we must reduce the exposure (and thus the modulation) of each by N, which in turn reduces diffraction efficiency by N^2.

Another reason for the "1-over-N" nature of holographic exposure is a problem known as "bias buildup". If we make a simple hologram of multiple points, the fringe patterns that represent each point combine coherently to form the resulting holographic fringe pattern, usually combined with some base exposure due to the brightness of the reference beam. Coherent exposure means that the patterns from individual points can add to and cancel each other out in some places, reducing as well as adding to the base exposure amount and achieving more efficiency. In contrast, incoherent exposures can only add to previously made exposures; there's no way to reduce any exposure that's already been made. With each exposure, this base exposure builds, "biasing" the meaningful part of the fringe pattern and minimizing the dynamic range it can have.

*The "order effect" amendment to the "one-over-*N*" law*

Even beyond the diffraction-efficiency cost of multiple exposures, there is more bad news. Peculiarities of the physics of the silver halide process cause the first exposure to dominate in holographic effect.[ii] If three equal sub-hologram exposures are given to a composite hologram, the first-exposed sub-hologram will be brighter than the second, which will be brighter than the third. An approximate compensation can be made by giving them unequal exposures in the ratios of

$$t_1 : t_2 : t_3 = \frac{2}{9} : \frac{3}{9} : \frac{4}{9} \tag{6}$$

Different materials, and even newer silver halide materials, will display different "order effects" or perhaps no order effect at all. For example, it is often the case in photoresists that only the last substantial exposure matters, and for photopolymers it is only the very first exposure that matters.

This section wouldn't be complete without clarification of what the "incoherent" in "multiple incoherent exposures" means. Basically, if it is impossible for the object beams for the various sub-holograms to interfere with each other, they are effectively "incoherent." In the examples we've been discussing in this chapter, they are separated by time, but they could also be separated by polarization, by wavelength, and by other effects.

Slit-Illumination Beam Forming

Rainbow holograms are created by limiting the amount of the H1 "master" hologram used to a narrow horizontal slit, usually accomplished by a combination of masking of the H1 and concentrating the light illuminating the slit thus formed. However, it is important to control the radii of the illumination in both the horizontal and (less important) vertical directions by a suitable choice of optics. This is

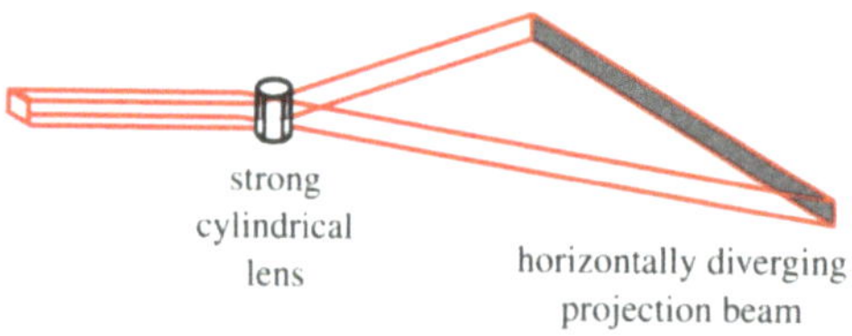

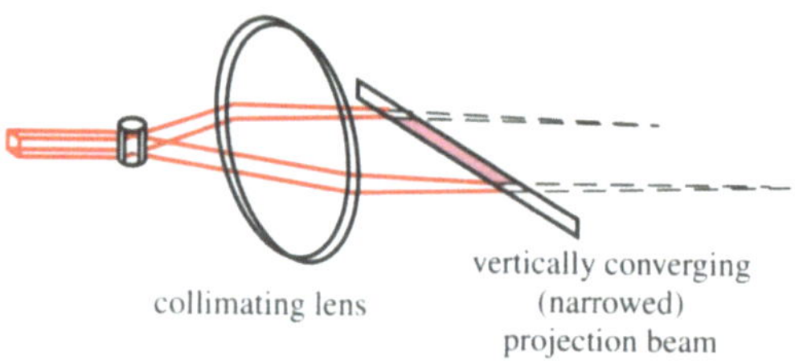

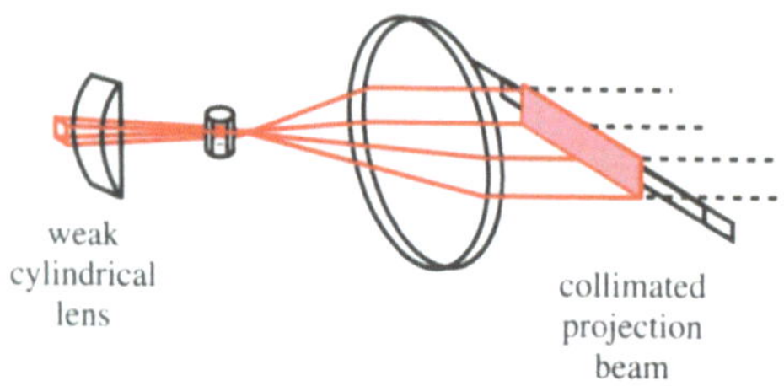

usually easy if the slit width is roughly the diameter of the raw laser beam, but more careful shaping of the beam requires more elaborate optics.

Simplest case: diverging the raw beam
The usual starting point is simply the horizontal spreading of the raw laser beam by the use of a vertical cylindrical lens (all directions are with respect to the hologram frame; typically the slit is vertical on the table, and the cylindrical lens' axis is horizontal). Although very good short-focus (around 10 mm focal length) cylindrical lenses are available in the optics catalogs, a small glass test tube filled with mineral oil, or a polished glass rod, will often do just as well. The only caution is that a strong laser beam can cause convection currents in the mineral oil, degrading the holographic recording.

Collimating the slit illumination beam
Whenever possible, we would prefer to illuminate the H1 with a collimated beam so as to minimize the distortion in the resulting image. However, simply putting a collimator one focal length from the diverging cylindrical lens is not usually adequate, because the beam will start to converge in the vertical direction downstream of the collimator. The result will be a narrower slit than before, and increased speckle in the image. To keep the beam of constant width, which allows as much distance between the collimator and the H1 as needed, and to increase the beam width when desired, we can use a long-focus cylindrical lens upstream of the diverging lens, spaced so that the foci of the two lenses coincide. The ratio of long to short dimension of the slit beam will then be the ratio of the focal lengths of the cylindrical lenses, which can be varied widely.

Embossed Holograms

We have spoken of diffraction gratings and holograms mainly as repetitive variations of absorptance or transmissivity, but recall that repetitive variations of light delay or phase modulation will also cause diffraction. In bleached holograms, this is mainly due to variations of the refractive index of the emulsion, but the same effects can be produced by variations in the thickness of the emulsion. These two effects usually accompany each other, but it is possible to make holograms that have only thickness variations, which are usually called "surface relief" holograms.

One nice feature of surface relief holograms is that they can be simply and cheaply replicated by transferring the thickness variations to a piece of transparent plastic through some combination of heat, pressure and perhaps softening agent. However, surface relief holograms are vulnerable to physical damage, such as by scratching, so it is also useful to mirrorize the surface-relief side (by vacuum evaporation of aluminum, for example), and then view the hologram in reflection mode. This causes endless confusion between true "volume reflection holograms" (the Denisyuk sort of thing) and "reflective rainbow holograms" (the Benton sort of thing—a transmission rainbow hologram with a mirror behind it), which you will sometimes have to sort out by context.

Making a stamping master

Silver-halide materials may be processed so as to produce a prominent surface relief pattern, usually by using very strongly hardening chemicals, and by rapid drying. However, the depth of the pattern is very dependent on the spatial frequency of the pattern, and is not usually prominent beyond a few hundred cycles per millimeter.

For commercial hologram production, special materials have been developed that produce only surface relief, and they are called *photoresists*. These are either photo-polymers, which are cross-linked by exposure and thus made less soluble in a developer bath (called negative-working photoresists), or long molecules whose bonds are broken by exposure to deep-blue light—a process called *scission*—and become more soluble (positive-working photoresists). The latter are the type usually used for microelectronic fabrication, and the same materials have been adapted for holographic use. Because these materials are sensitive mainly to deep-blue light, only krypton (413 nm), helium-cadmium (442 nm) and argon (458 nm) gas laser lines are useful. The large shift between exposing and viewing wavelengths makes careful compensation for wavelength effects essential. And because these materials have low sensitivity (20 mJ/cm^2 is typical), exposures can run to almost half an hour and thus high-quality setups and very careful technique are required.

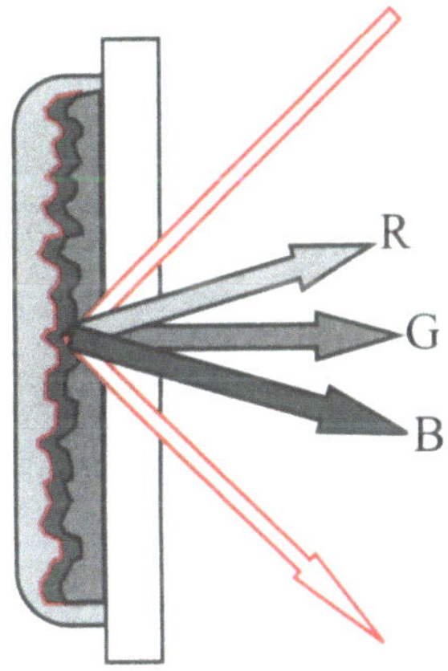

"Development" is usually accomplished by agitating in an alkaline bath for a few seconds, lengthy washing, and then careful drying so as to leave no spots or dust. The resin is itself too soft to serve as a mold or stamper, and so is replicated by coating with nickel metal that is peeled away to be the stamper or shim. The nickel is electrically deposited to a thickness of a millimeter or two, but there must first be a "starting electrode" that is either vacuum deposited gold or aluminum, or "electrode-less nickel" that is formed by a chemical reaction. The first nickel shim is often used as a "mother" to replicate several "daughter" shims, which in turn may give birth to "granddaughters," so that a single photo-resist exposure may produce millions of eventual embossed holograms.

Early evaluation of an embossed hologram also represents a considerable challenge, but is essential if exposure, beam ratio, and development are to be properly chosen. Transmission viewing of the dry photoresist master will give a certain RMS hologram phase modulation (luckily, the photoresist development can be resumed after drying). Viewing of the dry nickel master will give about four times as much modulation, and viewing of the final embossed hologram will give about six times as much modulation. The trick is to wind up with a high modulation, and thus high image brightness, without *overmodulating*, which causes a milky white blur to appear. Only an experienced embossing holographer can accurately judge the outcome when looking at the photoresist plate while still in the lab!

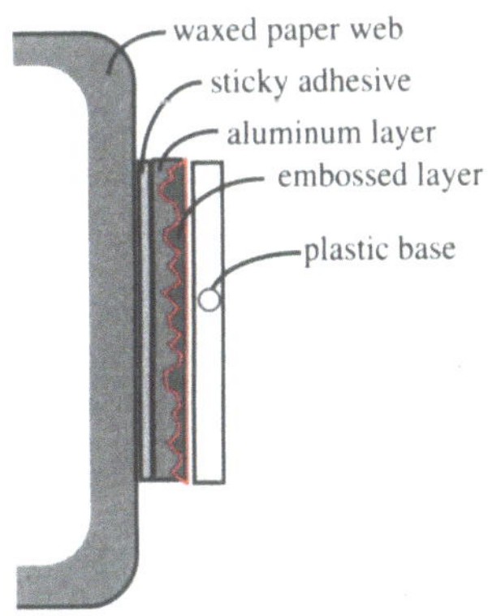

There are two types of embossed holograms in common production. The first, historically, is the thick "sticker" hologram that is stuck to a waxed paper carrier, and transferred (often by hand) to a product surface. However, this is too slow and expensive a process for very large product run, so a newer process has been evolved from the traditional hot-stamping foil process. The foil has a very thin sur-

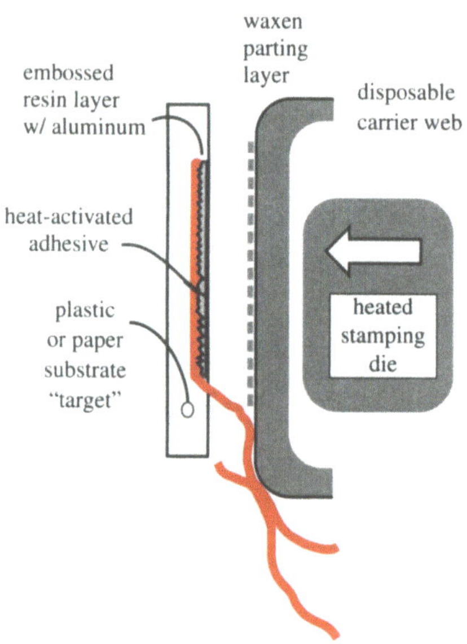

face relief layer on it, which is applied to the product surface by a combination of heat and pressure. If conditions are right, the hologram can be pressed below the surface of a credit card, which makes it almost impossible to remove without destroying both it and the card. Because hot-stamp holograms are so thin, they are especially sensitive to the texture of the product surface, such that coarse paper cannot be used because its texture overwhelms the surface relief of the hologram.

A newer process impresses surface relief in the coating that is often applied to fine paper while it is being manufactured, and then the whole roll is aluminized and varnished to protect the hologram layer. The resulting diffraction paper can be printed upon, so that instead of adding a hologram to a page, it is "removed" by being printed over. The results are so cheap that they are often used for wrapping paper and other wide-roll applications. However, it is fair to say that the surface quality is not high enough to allow deep three-dimensional images to be reproduced. The old-fashioned "sticker" holograms still provide the best image quality for that purpose.

Shrinkage Compensation

When holographic materials are exposed and processed, they typically undergo a change of average thickness and refractive index. For silver halide materials, both changes are due to the fact that some material is removed from the emulsion. About 17% of the volume of a typical holographic emulsion is silver bromide microcrystals, of refractive index 2.25, and 83% is gelatin, of refractive index 1.54. Depending on how the emulsion is processed, up to half of the silver halide may be gone at the end, and the layer mechanically collapses (depending on how it was hardened during the processing) and drops in refractive index. The following diagrams suggest how this might change as a function of exposure for three common process types (assuming no hardening occurs):

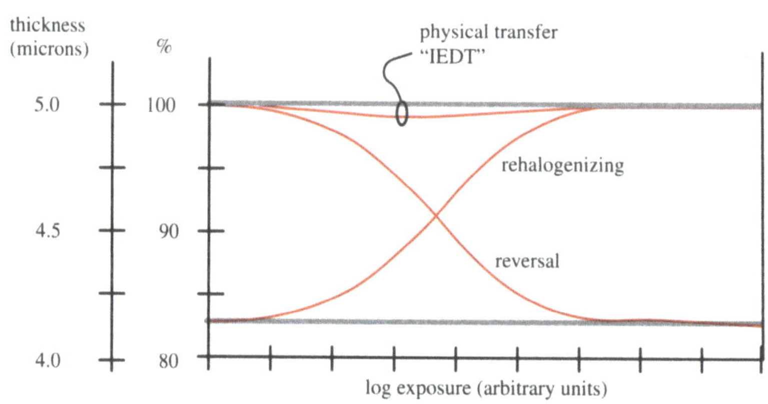

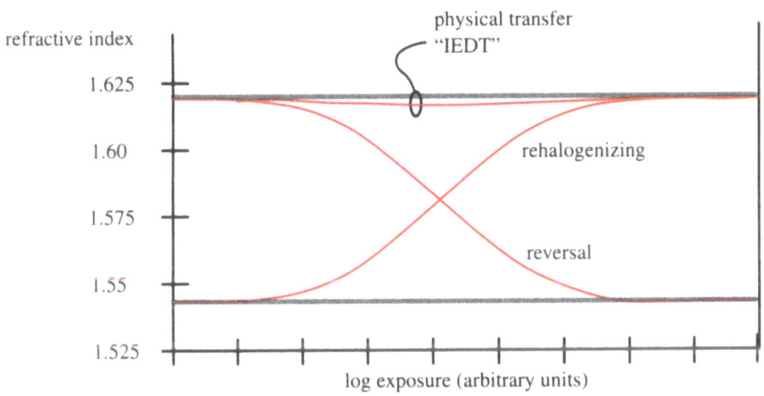

The results of applying the model we discussed in connection with Eqs. (3)-(5) of the preceding chapter to these conditions yields the following recommendations for object and reference beam angles for producing a 532 nm "green" image on axis.

shrinkage	θ_{obj}	θ_{ref}
minimum: $t_2 = t_1$ $n_2 = n_1 = 1.62$	−175.83°	129.48°
50%: $t_2 = 0.92\ t_1$ $n_2 = 1.58$	−174.80°	131.07°
maximum: $t_2 = 0.84\ t_1$ $n_2 = 1.54$	−173.70°	132.71°

If we specify a "50%-shrinkage" process (there are several options for such) then the appropriate exposure geometry will be:

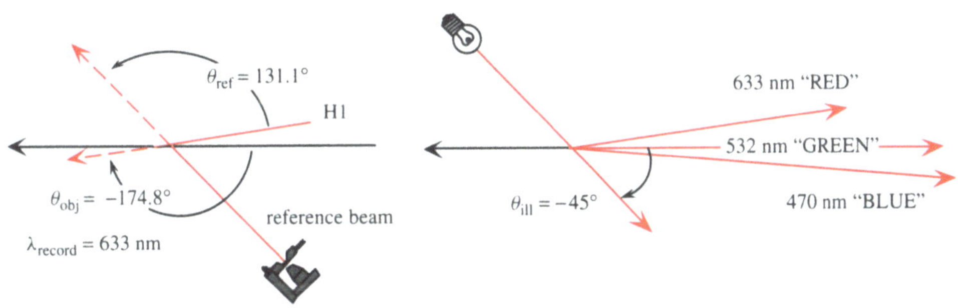

In this case, only the 532 nm green light will be maximally diffracted. The tip angle of the fringes is not quite right for red and blue rays, and even for 5 micron thick emulsions some falloff due to "Bragg angle mismatching" will be apparent—but not so much so as to detract from the beauty of the hologram.

Conclusions

We've devoted extra discussion to practical issues relating to rainbow holograms because of the commercial importance of this variety of hologram. There are many special features of these display holograms that are constantly being developed by ingenious artists and designers. Some of them are designed to change the "look" of a hologram and to give it some visual distinctiveness, and some are designed to lower the costs or speed up the process. Holography is still an exotic and expensive process in the world of graphics, and there is much more progress left to be made in the world between optical science and commercial innovation.

References

i. Land, E. H. (1959). "Experiments in Color Vision," *Scientific American*, **200**, 5, pp. 84-99. See also Land, E. H. (1977). "The Retinex Theory of Color Vision," *Scientific American*, **237**, 6, pp. 108-128, and McCann, J. J. and J. L. Benton, (1969). "Interaction of the Long-Wave Cones and the Rods to Produce Color Sensations," *Journal of the Optical Society of America*, **59**, 1, pp. 103-107.

ii. Johnson, K. M., L. Hesselink, and J. W. Goodman (1984), "Holographic Reciprocity Law Failure," *Applied Optics*, **23**, pp. 218-227.

CHAPTER 16

In-Line "Denisyuk" Reflection Holography

Introduction

We have been thinking of transmission holograms as producing images by means of an array of many overlapping negative lenses of various locations and focal length. Each elemental lens forms a virtual image behind the hologram plane. With each negative lens there is an associated positive lens that forms an image in front of the hologram too, and unlike glass lenses these diffraction lenses form images in different locations for different colors. But there is another way of forming a virtual image of a point illumination source: by reflection from a convex (outwardly curving) mirror surface.

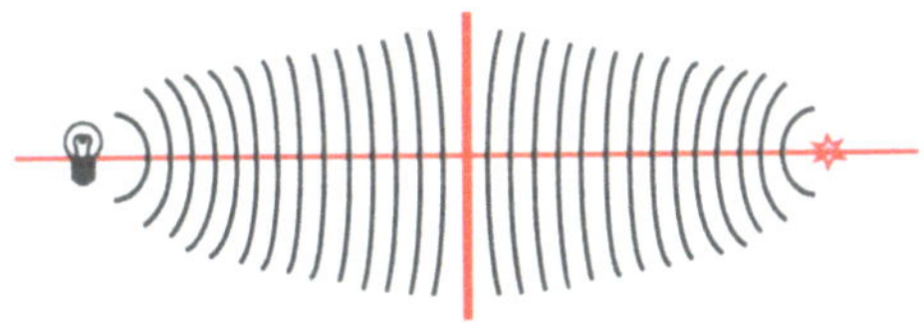

Consider a point source of light, and a location at which we would like to form an image. There are an infinite number of combinations of curvatures of mirrors (positive = convex, none = planar, negative = concave) and corresponding locations that will do the job. For example, a flat mirror forms a virtual image at a distance behind the mirror equal to the distance to the source in front (recall that your image in a flat mirror is as far behind the mirror as you are in front). Any of these mirrors at its appropriate location will produce the image we seek, as would a stack of barely-reflecting mirrors if the variation of curvature of each one with location were correct. We are going to think of a reflection hologram as a slice or sample though such a stack (illustrated here).

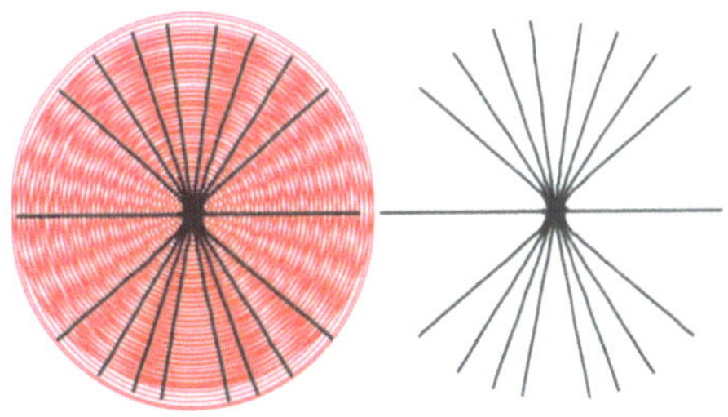

But how to fabricate such a stack of mirrors? The solution to this puzzle was provided by the Russian physicist Yurii Nikolaevich Denisyuk in 1958 (published in Russian in 1962[i] and in English in 1963[ii]). Recall the interference pattern formed between two point sources, such as we considered in Chapter 4. One of the surfaces of constructive interference, or reinforcement, is a flat plane halfway between the two sources. To the right of that mid-point, the surfaces curve toward the right-hand source, as spheres near the axis but stretching out to become hyperboloids of revolution. To the left, they curve in the opposite direction.

Denisyuk's insight was to use these interference patterns to expose a very-high-resolution emulsion, and to process the patterns to produce reflecting "fringes" that would be nested with exactly the proper shape to serve as the mirrors mentioned above. Note that Denisyuk's work preceded the invention of the laser by about four years! He used a special high-pressure mercury lamp which produced light that was weak but highly coherent—it was probably a superradiant source, in today's terms. Denisyuk's ideas about capturing the shape of optical wavefronts by interference, and reconstructing them by diffraction, were met with deep suspicion by the Russian Academy of Sciences, and his work was suppressed until the later work by Leith and Upatnieks drew international attention. Then Peter Kaptiza required that his critics write letters of support for the Lenin Prize (their highest scientific recognition), which Denisyuk received in 1969. From then onwards, holography was a prominent feature of the Soviet Union's scientific profile, along with space technology, nuclear power, and high-power lasers. We refer to such holograms as "Denisyuk," or "volume reflection" or "volume dielectric" holograms (especially to distinguish them from the "reflective rainbow" holograms mentioned previously).

The source of Denisyuk's remarkable idea was a boyhood reading of the story "Star Ships" by the noted Russian science fiction author J. A. Efremov (also translated into English).[iii] In it he described travelers who found a multilayered metal disk. When the sun shone on it, 3-D images of humanoid faces appeared. Denisyuk took on this scientific challenge, and soon realized that it was similar to the "interference color" images of Gabriel Lippmann. It took a while, but Denisyuk eventually realized his boyhood dream.

Making a Denisyuk Hologram

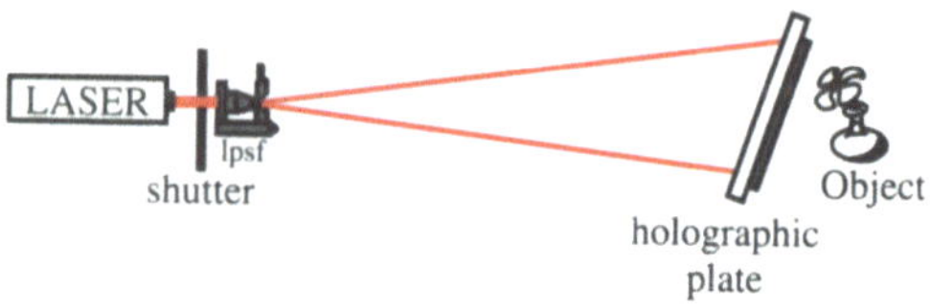

The "classical" or "single-beam" Denisyuk technique simply shines a diverging laser beam through a holographic plate, which is so finely grained that it absorbs very little of the light, and onto the subject of the hologram. The light reflects back from the subject to the plate, where it overlaps the incoming light to produce the desired interference pattern. Of course the subject must be closer to the plate than one-half the laser's coherence length, but otherwise the technique is very simple and direct, and can produce results of very high quality. It is well suited to very large holograms, because no supplemental optics are required and the system is readily engineered to be resistant to vibration.

But a more important property of reflection holograms is that they can be viewed with white light from concentrated sources. That is because they reflect only a narrow spectrum, usually centered at the same wavelength as that of the exposing laser. This is because the stacked mirrors are uniformly spaced by half the wavelength of the reflected light, just as the interference fringes that produced them were. Light first reflects from the first curved mirror surface, but most of it passes on to the deeper layers. The reflection from the second layer comes back out after a delay of one full wavelength. All the following reflections are delayed by one more wavelength each. If there are enough such reflections of roughly equal strength, then only one wavelength is strongly reflected (the one for which all the reflections emerge in phase), and the strength drops to one-half its maximum if the wavelength varies by one-over-M of its central value (where M is the number of reflections that come back).

Optics of In-Line Reflection Holograms: Distances

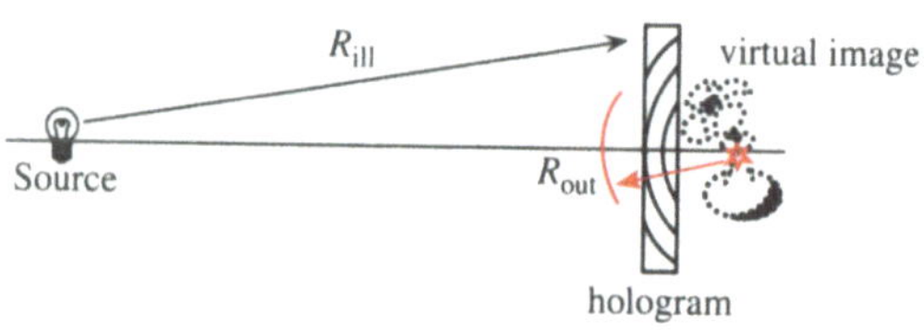

The distance-determining optics of the reflecting diffractive mirrors are the same as for transmissive diffractive lenses, and we will cite the relevant equations without proof. Unlike normal mirrors, they have a strong wavelength dependence, but otherwise they have many of the same properties as conventional mirrors.

$$\frac{1}{R_{out}} - \frac{1}{R_{ill}} = m\frac{\lambda_2}{\lambda_1}\left(\frac{1}{R_{obj}} - \frac{1}{R_{ref}}\right) = \frac{1}{R_{effective\ mirror}} \tag{1}$$

Note that all optical radii (and the mirror radius) are positive in the example diagrammed alongside. The value of m depends on which side of the hologram is being illuminated; $m = +1$ if the illumination is coming from the same side as the reference beam—there is no other diffracted order if the hologram is reasonably thick. However, the physical mirrors in the hologram layer are indeed curved, and illuminating the hologram from the opposite side produces the effect of a concave mirror instead of convex, which is usually a real image focused in space.

The Optics of In-Line Reflection Holograms: Angles

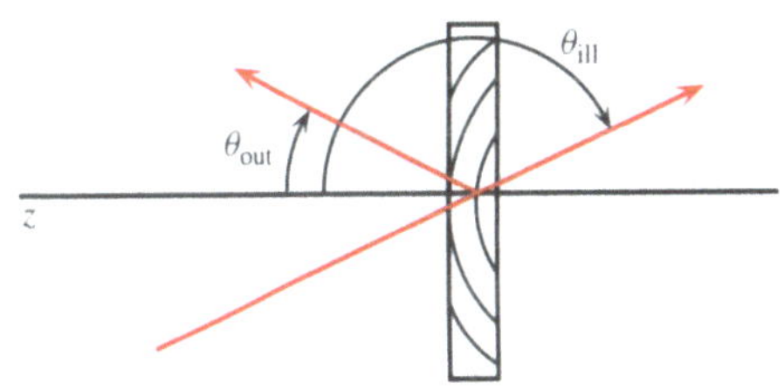

$\theta_{out} = 180° - \theta_{ill}$

The reflecting fringes in the holograms we are considering here are parallel to the emulsion surface (ignoring their curvature for the moment), because the rays that create them are incident at equal but opposite angles (zero degrees, in this case). These are called "conformal fringes" because their shape conforms to that of the emulsion surface, which allows them to act like mirrors in their geometrical properties too. That is, the diffracted light leaves the hologram as though it were reflected by a flat mirror parallel to the surface, independent of the wavelength that is reflected.

$$\theta_{out} = 180° - \theta_{ill} \tag{2}$$

Emulsion Swelling Effects

One of the fun and interesting properties of reflection holograms is the effect of swelling or shrinkage of the emulsion during viewing. Recall that when we discussed emulsion shrinkage in transmission holograms we noted that if the fringes are perpendicular to the surface, a change in the emulsion's thickness won't affect the behavior of the hologram; here we have precisely the opposite situation! If you breathe upon the processed hologram (so as to condense water upon the emulsion, which is quickly absorbed by the sponge-like gelatin layer to swell it slightly), the reflected color red-shifts to a longer wavelength. Conversely, heating the hologram (*e.g.* with a hair dryer) to dry it out and shrink it a bit will cause a blue-shift to a shorter wavelength. Depending on how the hologram was processed (especially on whether it was cross-linked or hardened), it will show these effects more or less strongly. For exposure and illumination perpendicular to the plate, the reflected wavelength will vary according to

$$\lambda_{2,0} = \frac{n_2 t_2}{n_1 t_1} \lambda_1 \text{ (times angle effects, to be discussed)} \tag{3}$$

where t_1 and t_2 are the physical thicknesses of the emulsion during exposure and viewing, and n_1 and n_2 are the refractive indices of the emulsion during exposure and viewing. All of these can be controlled to some extent—one popular method of creating colors different from that of the exposing laser is the pre-swelling of the emulsion by imbibition of solutions of sugar and water, which wash out during processing and produce a controllable blue-shift effect. These are called "pseudo-color" processes, because the colors are not actually those of the objects portrayed.

Viewing Angle Effects: the "Blue Shift"

Another effect that is easily observed is the variation of image color as the hologram is tipped away from the illuminating beam, so that the angle of illumination increases from zero to some value. The reflected image blue-shifts in color, although the color shift may be too small to notice in a deep-red image. The reason is that the time delay between the light reflected by adjacent hologram layers actually *decreases* when the light comes in at an angle, which is the opposite of what you might expect.

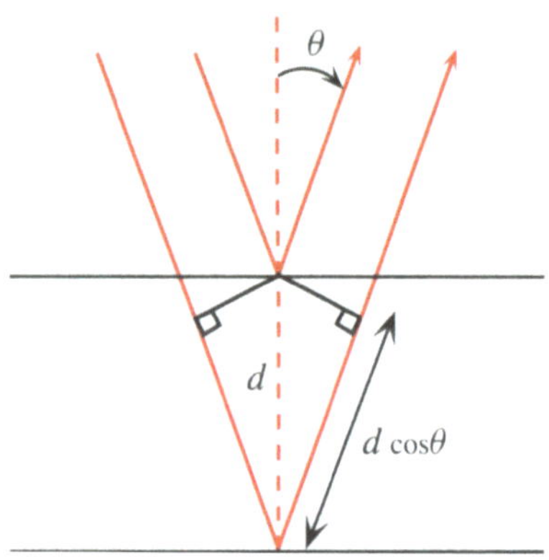

To examine this unusual behavior, consider the sketch in the margin. The key to understanding the effect is to note that it is the extra distance that the second reflection must travel that matters, but that the "race" between the two reflections begins at the point where they are last "abreast" and ends when they cross another line that is perpendicular to both rays. There are two effects: the second ray does spend more "time" between the reflecting layers as the illumination angle increases, but it also "cuts a corner" with increasing angle, and this effect dominates to produce the net decrease of delay between the reflections. The delay produces a wavelength shift given by

$$\lambda_{2,\text{external}} = 2n_2 d \cos\theta_{\text{int}} \tag{4}$$

The result is that, contrary to some people's intuitive expectation, a reflection hologram image "blue shifts" as it is illuminated at steeper angles.

Conversely, if the hologram is exposed at an angle and viewed perpendicularly, the reflected wavelength is longer than the exposing wavelength: a "red shift" occurs. Assuming that the object and reference beams are coming in at opposite angles, so as to produce fringes that are parallel with, or conformal to, the hologram surface, and that the illuminated beam reflects at the mirror angle. That is:

$$\theta_{\text{ref, ext}} = 180° - \theta_{\text{obj, ext}}, \quad \theta_{\text{ill, ext}} = 180° - \theta_{\text{out, ext}} \tag{5}$$

Now the fuller version of Eq. (3) becomes:

$$\frac{\lambda_{2,\text{ ext}}}{\lambda_{1,\text{ ext}}} = \frac{n_2 t_2 \cos\theta_{\text{ill, int}}}{n_1 t_1 \cos\theta_{\text{ref, int}}} \tag{6}$$

Diffraction Efficiency of Reflection Holograms

It should be clear that we seem to have arrived at the opposite end of the scale from thin gratings (where mostly what concerns us is the spacing between the fringes on the surface and the profile of the variation in intensity or index; these gratings exhibit what is called *Raman-Nath diffraction*). *Volume gratings* like the ones we're discussing here exhibit what is called *Bragg diffraction*; we've already seen some of the rules of this regime when we considered how to match the beam angles with the fringe angles in making transfer holograms. The distinction between thin and thick gratings is often made by looking at a parameter Q.[iv] If we call the thickness of the grating t and the fringe spacing d

$$Q = \frac{2\pi\lambda_2 t}{n_2 d^2} \tag{7}$$

then a thin grating is one for which Q is small (less than one) and a thick grating is one for which Q is large.

The amount of light that an "in-line" reflection hologram will reflect depends on more factors than for transmission holograms, especially on the wavelength of the light involved. A typical reflection hologram will have a reflectance or diffraction efficiency spectrum having roughly the shape plotted in the margin, where $\lambda_{2,0}$ is the central wavelength of the reflection spectrum, which we have seen above is determined by the exposing wavelength, shrinkage effects, and the angle of incidence. The maximum height of the curve, or DE_{max}, and the width of the reflection spectrum at its half-height points, $\Delta\lambda_2$, are what we are about to discuss. We have to be content with a low-efficiency-only model here, although it can be extended to more practical levels with more advanced mathematics.

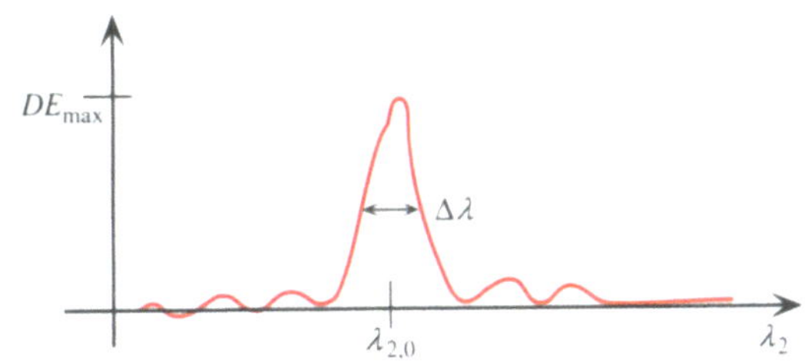

First we will consider the amount of reflection at the spectrum's peak. We will model the reflection hologram as a stack of alternating layers of high and low refractive index, n_{low} and n_{high}, each of equivalent thickness of one-quarter wavelength. Each high-to-low or low-to-high boundary will reflect a small amount of light amplitude, as given by the venerable equations of Fresnel (this simple form is for perpendicular incidence):

$$R_{ampl} = \frac{n_{high} - n_{low}}{n_{high} + n_{low}} \tag{8}$$

Note that a transition from high- to low index gives a positive reflection, while a transition from low to high gives a negative reflection. However, because there is a round-trip delay of one-half wavelength between them, they arrive in-phase and add together.

Now the wavelength inside the emulsion is λ_0/n_2, and if the emulsion is of thickness t_2 there will be a total of

$$M = \frac{t_2}{(\lambda_{2,0}/2n_2)} = \frac{2n_2 t_2}{\lambda_{2,0}} \tag{9}$$

each of the high- and low-refractive-index layers. Assuming that the reflectance of each layer is low enough that we can ignore the effects of double reflections (the Born single-scattering approximation), the amount of light reflected by each transition will be equal and the total reflected amplitude will be:

$$R_{ampl,\ total} = 2MR_{ampl} = M\frac{n_{high} - n_{low}}{\left(\frac{n_{high} - n_{low}}{2} = n_2\right)} \tag{10}$$

It remains only to adjust slightly the value of the refractive index modulation to take into account that the actual variation is nearly sinusoidal instead of a step-like variation, and has a peak-to-peak modulation of Δn_2. The ratio is the same as that between an electrical square wave and its fundamental frequency:

$$\left(n_{high} - n_{low}\right)_{square\ wave} = \frac{\pi}{4}\Delta n_{sinusoid} \tag{11}$$

Thus the total intensity reflectance, or diffraction efficiency, can be approximated by:

$$
\begin{aligned}
DE_{+1} &= \left|R_{\text{ampl, total}}\right|^2 = 0.62M^2\left(\frac{\Delta n}{n_2}\right)^2 \\
&= \left(\frac{\pi}{4}\frac{t_2\Delta n}{\lambda_2}\right)^2
\end{aligned} \tag{12}
$$

As Δn and thus DE_{+1} increase, this simple model fails to account for all the important phenomena, especially the dropoff of the illumination as light propagates deeper into the hologram, and the multiple reflections of the diffracted light. The next level of complexity was provided by the analysis of the "coupled wave model" by Herwig Kogelnik in 1969[v]. The result shows that as the modulation, Δn, further increases, the diffraction efficiency stops growing so quickly, and eventually rolls over to approach asymptotically its maximum value of unity. The analytical expression for the diffraction efficiency includes the hyperbolic tangent function, tanh(x):

$$
DE_{+1} = \tanh^2\left(\frac{\pi t_2 \Delta n}{\lambda_2 \cos\theta}\right) \tag{13}
$$

where θ is the angle of the illumination within the emulsion (zero for the examples so far). Readers interested in a comparison of analytical models and measured results may wish to consult Liu *et al.* (1995).[vi]

Spectrum Width

For many applications of reflection holograms, it is important to keep the width of the reflected spectrum to a minimum. The simple low-diffraction-efficiency model gives a FWHM (full-width at half-maximum) bandwidth of roughly

$$
\Delta\lambda = \frac{\lambda_{2,0}}{M} \tag{14}
$$

One way to understand this number is to recall that, at the central maximum, the reflections from all the M layers arrive in-phase. In particular, the reflection from the back layer arrives $(M-1)\cdot 2\pi$ delayed with respect to the first layer. But at, for example, a shorter wavelength (say $\lambda_{2,L} = \lambda_{2,0} - \Delta\lambda$), there will be an added phase delay of

$$
\begin{aligned}
\Delta\phi &= 2\pi\frac{2n_2t_2}{\lambda_{2,0}-\Delta\lambda} - 2\pi\frac{2n_2t_2}{\lambda_{2,0}} \\
&\approx 2\pi\cdot 2n_2t_2\cdot\frac{\Delta\lambda}{\lambda_{2,0}^2}
\end{aligned} \tag{15}
$$

If the phase delay from the back of the hologram is 2π, then there will be a layer near the center of the hologram for which it will be π, and the reflection from the front layer and the middle layer will cancel out. For the next layer into the emulsion, there will be another reflection just beyond the middle layer that will cancel it out, and thus there will be cancellation by pairs throughout the depth of the emulsion, and the total reflection (that is, the diffraction efficiency) will drop to zero.

Substituting the key variables back into the equations yields the expression for the required wavelength shift as:

$$\Delta\lambda = \frac{\lambda_{2,0}^2}{2n_2t_2} = \frac{\lambda_{2,0}}{M} . \tag{16}$$

Some analyses will produce an $M+1$ or $M-1$ in the denominator instead of M, but because M is typically more than ten, we will ignore this detail. We also assert that the distance from the spectral peak to the first zero will be the same as the FWHM spectral width, to within the accuracy needed for our purposes.

Note that increasing the thickness of a hologram both decreases the spectral blur in white-light illumination and increases the peak diffraction efficiency. This would suggest that making reflection holograms very thick indeed would be a good idea, and yet this is usually not the case, for reasons that we will explore only briefly.

Anomalous Spectral Shapes

It often happens that the reflection spectra of real holograms are very different from the idealized "sinc-squared" spectrum that a simple Fourier analysis predicts. There are three reasons that we ought at least be aware of:

1. The reflectance of the first few layers may be much higher than the simple theory can accommodate, or
2. The reflectance of the various layers may be very different due to the differing chemistries that occur as solutions diffuse through the emulsion during processing, or
3. Similar depth-dependent chemistry changes may cause an uneven swelling or shrinkage of the hologram through its depth, which is often described as a "chirping" of the hologram's fringe spacing or spatial frequency (one indicator of #2 and #3 is that the hologram may look very different when viewed from the emulsion side and the support side).

For some applications (hologram jewelry made with dichromated gelatin is one example), a deliberately wide spectrum with high diffraction efficiency is desired, in order to give a silvery or golden image effect. Very dramatic processing is used, involving baths of hot alcohol and vigorous washing, to give the combination of high refractive index change and exaggerated chirping that is needed.

Conclusions

While the Denisyuk type of reflection hologram works a little differently from what we'd been considering up to now, readers should have enough mathematical and intuitional tools at this point to be able to understand the principles. In coming chapters we're going to make the geometry a bit more extreme and see what happens.

References

i. Denisyuk, Y. N. (1962). "Photographic Reconstruction of the Optical Properties of an Object in Its Own Scattered Radiation Field," *Soviet Physiks Doklady*, **7**, pp. 543-545.

ii. Denisyuk, Y. N. (1963). "On the Reproduction of the Optical Properties of an Object by the Wave Field of Its Own Scattered Radiation," *Optics & Spectroscopy*, **18**, pp. 365-368.

iii. Denisyuk, Y. N. (1992). "My Way in Holography," *Leonardo*, **25**, 5, p. 425.

iv. Klein, W. R. and B. D. Cook (1967). "Unified Approach to Ultrasonic Light Diffraction," *IEEE Trans. on Sonics and Ultrasonics*, **SU-14**, pp. 123-134.

v. Kogelnik, H. (1969). "Coupled Wave Theory for Thick Hologram Gratings," *Bell Systems Technical Journal*, **48**, pp. 2909-2947.

vi. Liu, D., G. Manivannan, H. H. Arsenault, and R. A. Lessard (1995). "Asymmetry in the Diffraction Spectrum of a Reflection Hologram Grating," *J. Modern Optics*, **42**, 3, pp. 639-653.

CHAPTER 17

Off-Axis Reflection Holography

Michael Halle

Introduction

In the last chapter, we considered the properties of the on-axis reflection hologram. In this chapter, we will combine these properties with what we've learned about transmission and rainbow holography to show how reflection holograms can be used for practical holographic display. Fortunately, much of the intuition we've developed will be useful for modeling the behavior of reflection holograms (with a few tweaks and extensions to the equations).

Qualitative Comparison of Transmission and Reflection Holograms

Let's begin by recapping the interesting properties of reflection holograms, compared to our more familiar transmission ones.

Both transmission and reflection holograms are recorded using an informationless reference beam interfering with light from possibly many points on an object, as illustrated here. During transmission hologram recording, the object and reference beams are on the same side of the plate; in reflection holography, the object and reference beams are on opposite sides of the recording medium.

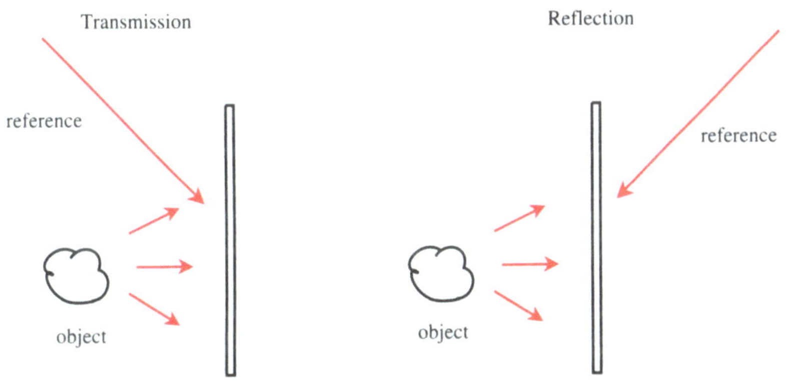

The orientation of a reflection hologram's fringes in an emulsion of sufficient thickness produces a volume grating composed of many semi-transparent mirrors. At illumination, this volume grating filters light passing through it based on wavelength and angle, passing only a limited range of wavelengths.

A transmission hologram, in contrast, passes a broad spectrum of light, splitting white light into a rainbow-like fan of diffracted color. Since the peak wavelength passed by a reflection hologram depends on the internal spacing of the fringes in the emulsion, any change of thickness (or index of refraction) of the emulsion will change the holographic image's primary color.

The "zero-th" order approximation of a transmission hologram, when illuminated, is a clear piece of glass; the base behavior of a reflection hologram acts like a mirror. This fact means that the illumination source is always on the opposite side of the hologram from the viewer in a transmission hologram (barring the use of any

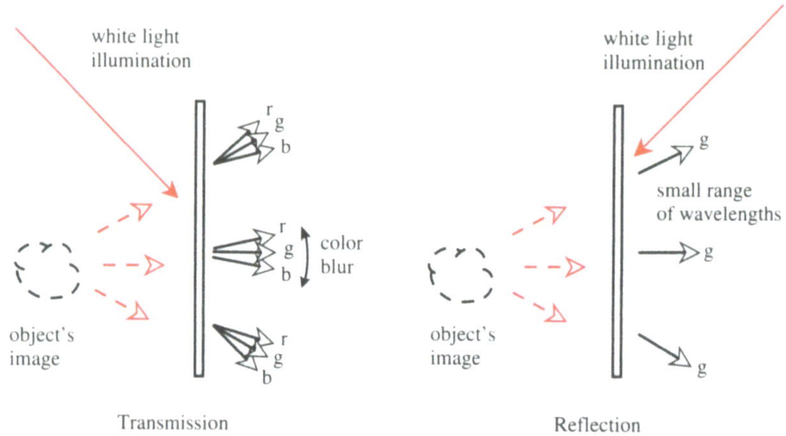

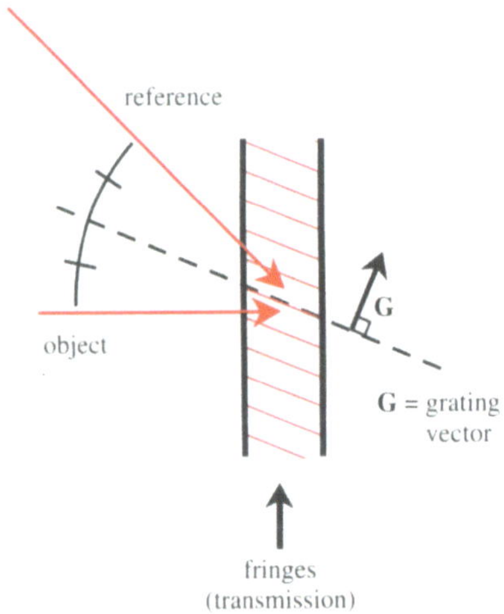

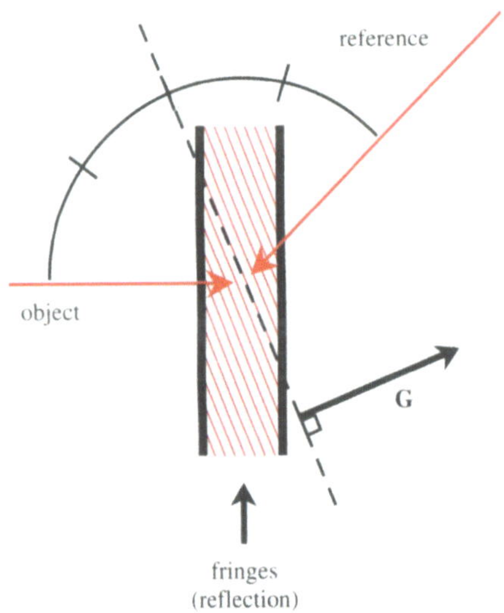

mirrors), where the illumination for a reflection hologram is always on the same side as the viewer. More interesting transmission holograms act as multiple overlapping focusing lenses; reflection holograms act like multiple focusing mirrors.

The standing wave pattern in space caused by the interference of the two beams is similar in both transmission and reflection holography; however, transmission and reflection holograms record a plane of this pattern from two different locations in "fringe space" with quite different characteristics. In a transmission hologram, the fringes are usually close to perpendicular to the emulsion (generally no more than about 40 degrees from perpendicular). In reflection holograms, the fringes are almost parallel to the emulsion surface (generally no more than 40 degrees off parallel).

Interestingly, none of these properties directly depends on a hologram's on- or off-axisness: on or off axis. Our intuition is that we should be able to get the same benefits of convenient overhead illumination by using an off-axis illumination (and corresponding exposure) geometry for reflection holograms in the same way we did for transmission holograms. This intuition proves to be accurate.

It turns out that the off-axis geometry isn't required to get rid of overlapping multiple orders in reflection holography (as we needed to do in off-axis transmission holography): diffraction orders other than $m = 1$ and $m = -1$ aren't in general propagated through the emulsion. On the other hand, overhead illumination is the only practical way to light holograms in most gallery or museum contexts and minimizes annoying reflections off the glass holographic plate.

Deconstructing Reflection Holograms

While the previous comparison might make it seem that transmission and reflection holograms are irreconcilably different, in fact it's relatively straightforward to approximate at least some of the reflection hologram's behavior using transmission holography. Reflection holography has three major parts: the "imaging" part, the "reflection" part, and the "filter" or wavelength selectivity part. The imaging part can be modeled using a transmission hologram, and considering only the first order beams ($m = \pm 1$). The reflection part changes the hologram from being "windowlike" to being "mirrorlike", moving the illumination to the observer's side of the hologram. The filter part limits the output wavelengths that emerge from the hologram:

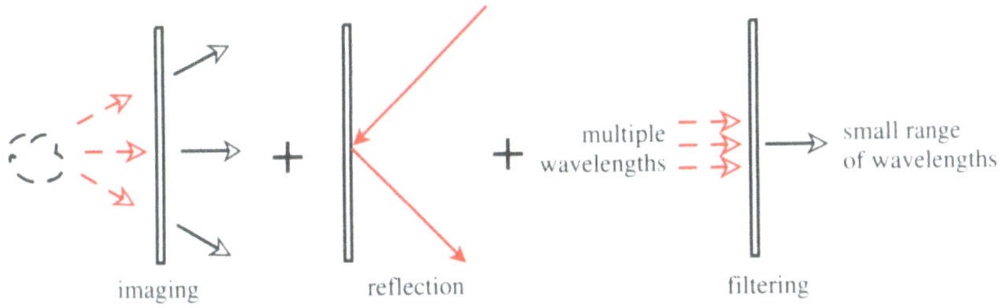

for a bright image (or any image at all!), light must be diffracted by the hologram and pass through the filter. We will look into the characteristics of this filter in more detail a little later.

Each of these conceptual components contributes to the reflection holograms's appeal. The imaging part brings the pedigree of holography as a high-fidelity 3-D display medium. The reflection part has practical benefit for galleries and other locations where wall mounting and ceiling illumination is necessary.

But perhaps most uniquely, the ability of reflection holograms to selectively filter light provides enormous flexibility in designing white light illuminated holographic displays. This filter greatly reduces the spectral blur that is the bane of full-parallax transmission holography. It also can produce a much purer, more consistent, and more stable primary color than is possible with white light transmission hologram (either full aperture or rainbow types). Finally, several filters can be embedded in a single emulsion using multiple holographic exposures. This process results in a color display that can accurately render the appearance of real-world or synthetic scenes, in full-parallax holographic 3-D.

Mathematical Modeling of Reflection Holograms

The reflection part of the reflection hologram model can be simulated by mirror-flipping the reference and illumination beams along the axis of the hologram when we compute angles. Here's how this flip works. Let's say we have transmission hologram exposure geometry we'd like to turn into a reflection one. Draw the hologram plane and the reference and object beams on a sheet of paper. Now, imagine that the reference and object beam lines are actually on two pages of thin newspaper with the hologram line as the seam. "Turn the page" of this newspaper so that the reference beam moves over to the other side of the hologram line from object beam.

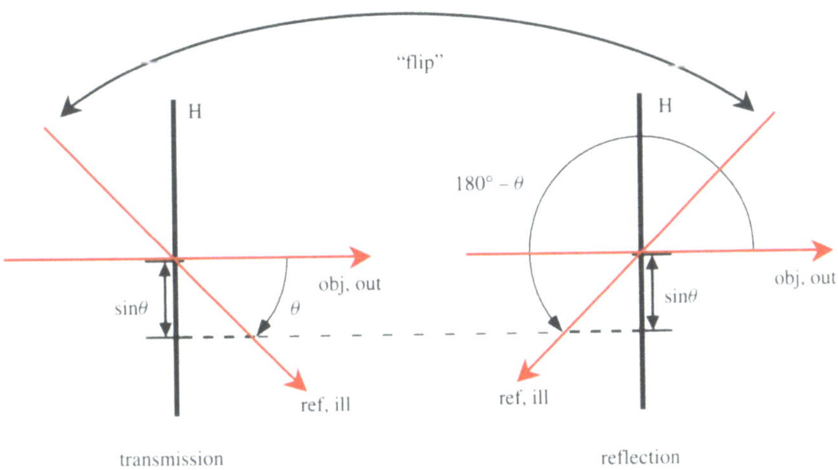

The new diagram, with the object and reference on different sides of the plate, is the equivalent reflection hologram exposure geometry. Of course, we could also have done the same process in reverse, starting with a reflection hologram and working our way to the corresponding transmission hologram exposure geometry.

Continuing this example, let's figure out where light from a reflection hologram goes. For now, ignore the wavelength-selective effect of the volume grating. Beginning with the reflection hologram exposure geometry, we use the above technique to find the "dual" transmission hologram by flipping the reference beam across the plate. ("Duals" are often used in science to help relate less understood problems to ones we already understand.) We then simulate or calculate the output beams for the +1 (or possibly –1, if the hologram is illuminated in phase conjugate mode) output orders, using our knowledge of transmission holography to choose an appropriate illumination angle and position. Once we calculate the illumination and output information, we convert to reflection mode by flipping the illumination back across the plane of the hologram. At this point, the output rays, and the viewer, will be on the same side of the hologram as the illumination source. After you've performed this exercise a few times, you'll gain enough intuition about the relationship between transmission and reflection geometry that you'll be able to think in reflection terms "natively".

An interesting property of this "flipping" of the refererence and illumination rays is that it is completely compatible with our transmission holography angle calculator, the sin theta equation, at least if we're careful. Recall from trigonometry that

$$\sin\theta = \sin(180° - \theta) \tag{1}$$

That's our flip (refer to the preceding figure), but it also means that $\sin\theta_{\text{transmission}} = \sin\theta_{\text{reflection}}$ for object, reference, illumination, and output beams.

In practice, then, this mathematical equivalence means we can use the $\sin\theta$ equation as we always have in transmission holography, but apply it as well for predicting the behavior of reflection holograms. We do need to keep our wits about us, though. When it comes time to take the inverse sine function to solve for our unknown variable, the equation provides us no guidance about which solution (on which side of the plate) to choose. We must handle this inherent mathematical ambiguity by using our knowledge of the type of hologram we're making. If it's a transmission hologram, light from the illumination will propagate through the hologram to form the output orders that (hopefully!) will be visible to an observer on the other side of the plate from the light source. If it is a reflection hologram, the illumination rays will bounce off the hologram to form the output orders visible to a viewer on the source's side of the plate. Using this information, it's straightforward to figure out which value of inverse sine to use.

This result is quite exciting, because it means that we don't need to learn a new equation to figure out where the interference pattern recorded in a reflection hologram tries to bend light of a particular wavelength. Since the sin theta equation was derived from the diffraction equation, and the diffraction equation gave rise to the horizontal focus equation (*1/R*) and vertical focus equation ($\cos^2\theta/R$), it follows that these tools in our "shop math" toolbox also hold for either transmission or reflection holograms. We just need to be careful to keep track of our angles throughout the entire unit circle and keep our wits about us and keep track of where light really can go.

Modeling Wavelength Selectivity in Reflection Holography

Unfortunately, though, reflection holography does introduce a major wrinkle in our mathematical simulation plan. Up until this point, we've ignored the effect of the wavelength selectivity of the reflection hologram. While the sine theta equation tells us where the volume grating of the hologram wants to direct light of a particular wavelength, it says nothing about how much of that light, if any, can actually get through the interference filter formed by the fringes. One of the major advantages of reflection holograms is this wavelength selectivity; it's very important to know how much light, and at what wavelengths, actually makes it to the viewer.

Modeling the behavior of the interference filter component of reflection holography is generally only approximate: differences in the material, thickness, and processing of a holographic material can control the range of angles through which any one wavelength can be seen, or the range of wavelengths that can be seen from any one output angle. The previous chapter provides some guidance into the spectral bandwidth of reflection holograms. For those needing more exact answers to these questions, we refer you to more definitive and detailed texts on the subject. For the rest of you, what we can do is offer two methods for beginning to understand wavelength selectivity in off-axis reflection holography. The first method is a general vector-based analysis. The second, the $\cos\theta$ equation, is a simpler method that is similar in many ways to our previous shop math approach.

Understanding Fringe Geometry

To broaden our holographic understanding to include off-axis reflection holography, we take a closer look at the orientation of fringes in a thick emulsion. We return to our vector model of interference based on wave and grating vectors. The grating vector **G** (the vector perpendicular to the fringes) is the vector difference of the object and reference wave vectors, where each wave vector has a direction that corresponds to the internal angle of the corresponding beam and with length equal to the reciprocal of the wavelength of the source in the emulsion:

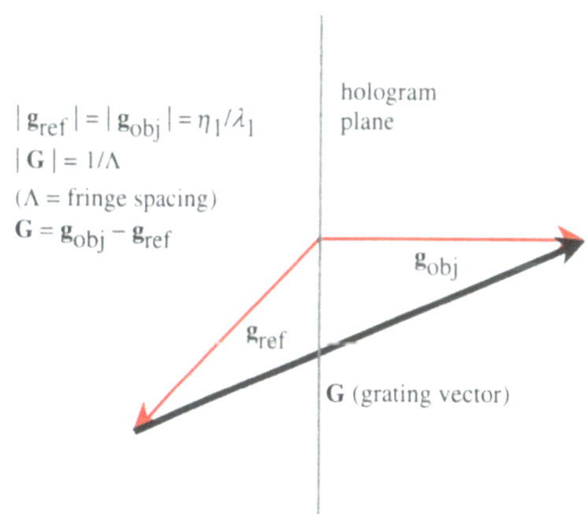

$$\frac{\eta_1}{\lambda_1} = |\mathbf{g}_{obj}| = |\mathbf{g}_{ref}|$$

$$\mathbf{G} = \mathbf{g}_{obj} - \mathbf{g}_{ref} \tag{2}$$

$$|\mathbf{G}| = \frac{1}{\Lambda}$$

The figure in the margin shows a graphical version of this relationship. **G** has length equal to the spatial frequency of the grating, where Λ is the fringe spacing. **G**'s direction is perpendicular to the surface of the fringes in the emulsion. The longer the grating vector, the higher the spatial frequency of the grating. Some texts name the grating vector as **K**, where

$$|\mathbf{K}| = \frac{2\pi}{\Lambda} = 2\pi|\mathbf{G}| \tag{3}$$

For our purposes, the 2π term drops out of the equations, and this needlessly complicates them. This formulation is similar to the Ewald spheres approach used in crystallography.

As a sanity check, we can confirm that if the reference and object beams are opposite from each other in an inline configuration, the difference between them is the sum of their magnitudes:

$$\begin{aligned} |\mathbf{G}_{\text{inline}}| &= 2\frac{\eta_1}{\lambda_1} = \frac{1}{\Lambda_{\text{inline}}} \\ \Lambda_{\text{inline}} &= \frac{\lambda_1}{2\eta_1} \end{aligned} \tag{4}$$

This result is the same one we found during our on-axis reflection analysis in the previous chapter. (Also recall that this extremely fine fringe spacing, on the order of half a wavelength of light, requires a high-resolution emulsion and extreme motion stability to record effectively!)

To review, the vector **G** represents the grating created using interference of the reference and object sources. It also represents the diffractive power of the grating to "transform" the illumination beam into the output beam. For now, assume that the *emulsion material* goes through no changes between exposure and illumination. Represent the illumination and output beams as vectors in exactly the same way as we did the reference and object beams above: directions corresponding to the beam directions, magnitude corresponding to 1 divided by the wavelength in the emulsion:

$$\frac{\eta_2}{\lambda_2} = |\mathbf{g}_{out}| = |\mathbf{g}_{ill}| \tag{5}$$

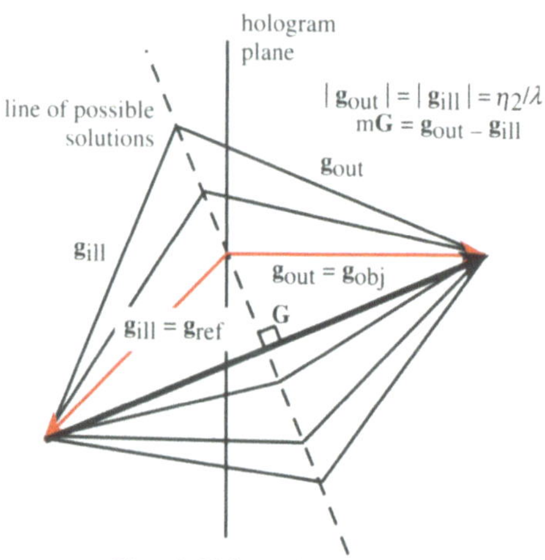

Platonic Holography: $m = 1$, $g_{ill} = g_{ref}$, $g_{out} = g_{obj}$ and no changes in the emulsion. Many other solutions exist!

For the grating to propagate light from illumination to output, the grating vector **G** must form the difference between these illumination and output wave vectors, as shown in the illustration here, with the addition of *m*, an order term (just as we used in transmission holography):

$$m\mathbf{G} = \mathbf{g}_{out} - \mathbf{g}_{ill} \tag{6}$$

Let's set the order term *m* to 1 for now to simplify the discussion. With no change in the emulsion, the equation and thus both diffraction and Bragg condition are satisfied if the illumination matches the reference beam, and the output beam matches the object. This is our "platonic" holography condition, which shows the correspondence between exposure and illumination that makes holography "magic".

Looking at the equation for **G** again, and at the graphical representation in the figure, we see that there are other solutions that exist that satisfy this condition: in fact, any two equal-length vectors that touch the tips of **G** will do. The shortest pair of vectors, corresponding to the longest wavelength of light that can pass through the hologram at any angle, occur when $\mathbf{g}_{ill}$ and $\mathbf{g}_{out}$ are equal and opposite and thus parallel to **G**. This result corresponds to the unintuitive answer we found in the preceding chapter (Eq. (4)) that off-axis illumination of an on-axis hologram always shifts towards the blue end of the spectrum. From the vector analysis, we can see that this result holds true no matter what the angle of the reference or object beams: the diffracted and Bragg-matched output vector is always shortest (and thus the wavelength is always longest) when the illumination is in the direction of the fringe vector (and perpendicular to the fringes).

The wave vector analysis explains the cosine terms in Eq. (6) of the last chapter as well: the case of off-axis reference and object beams creating a hologram that shifts red when illuminated and viewed perpendicularly. When the reference and object beams are both off axis, the grating vector becomes shorter by a function of the cosine of the beams' angle to the emulsion. The length of the illumination and output vectors that satisfy the Bragg condition also vary as a function of the cosine of the angle to the emulsion, with the shortest vector, and reddest reflected color, coming on axis when $\cos\theta$ is greatest.

Unlike the previous chapter's results, though, this vector analysis model holds true for any angle of exposing or illuminating beams; we can match illumination and output vectors to the resulting grating vector independent of whether the fringes are parallel to the emulsion.

As you can see, this vector analysis is extremely powerful for understanding how holograms—any holograms—behave.

Changes to the Emulsion

Up until this point, we've assumed the emulsion at exposure and illumination is identical. From transmission holography, we know that the emulsion can change its index of refraction as well as its thickness. The vector model can accommodate those kinds of changes, at the cost of some complexity.

Changes in index

The index of refraction change is the easiest to model. We simply need to plug in a different "*n*" when considering the illumination and output vector lengths and directions. For instance, if the index of refraction increases after the emulsion is processed, the wavelength of the illumination and output vectors required to meet the Bragg condition will have to be proportionally shorter so that the fraction η_2 / λ_2 remains a constant.

Changes in thickness

Just as in transmission holography, the emulsion of a hologram can shrink or swell based on processing, changes in temperature or humidity, or due to the addition of swelling agents. For our purposes, changes in thickness occur only in the direction perpendicular to the emulsion (by our convention, the z axis): we continue to assume no change in geometry across the face of the hologram. This means that the grating vector **G** is modified by shrinkage or swelling by scaling its z component by the factor t_1/t_2, where t_1 is the thickness at exposure, and t_2 is the thickness at illumination. The new vector, which we call $\mathbf{G}'$ here, is given by:

$$\mathbf{G}'_z = \mathbf{G}_z \frac{t_1}{t_2} \tag{7}$$

For example, should the emulsion shrink, the magnitude of the grating vector will increase in magnitude as the fringes get closer together. The combined effects of index of refraction and thickness changes on the length of the grating vector corresponds to the scalar result we saw in Chapter 16, equations (3) and (6).

Controlling color

Control of the thickness and refractive index of the emulsion allows us to manipulate the color of a reflection hologram. With no change in thickness or refractive index from exposure to illumination, and a direct correspondence between the exposure and illumination geometries, a reflection hologram lit with a white light source will pass a narrow range of wavelengths centered around the laser's wavelength. A full color hologram can be made this way, but only at the expense of multiple exposure lasers.

Alternatively, we can change the emulsion before or after the exposure. If the emulsion shrinks, for example, the final hologram's appearance will shift towards the blue end of the spectrum because the spatial frequency of the fringe pattern will get higher. Shifting the output color of a hologram may be important for aesthetic reasons, or it may allow a designer to choose a color to which the human eye is more sensitive than the typical He-Ne 633 nm red (say, orange or green).

The key to effective color control, though, is to understand, predict, and limit changes in the emulsion to achieve the desired color changes. The process of color control is complex and beyond this text to discuss in-depth. Several techniques used for emulsion control and manipulation include the following:

- The use of developers that change the emulsion thickness (by removing silver halide material) or rigidly maintain it by "tanning" the emulsion's proteins into a more rigid structure by cross-linking (just like leather tanning!);
- The use of a solution containing TEA, or triethanolamine, to pre-swell the emulsion to a larger thickness before exposure,
- The use of "in situ", or "in solution" holographic exposure, where the holographic plate is exposed while inside of a tank filled with alcohol and water in precisely controlled proportions.

TEA and in situ processing can both be used for multiple exposure holograms, where several different color primaries are exposed into the hologram in sequence, with some change in thickness in between. This process results in multiple fringe patterns of different spatial frequencies being recorded in the holographic emulsion, producing a multi-color display.

Modeling Filter Bandwidth

Real-world reflection holograms (or transmission ones, for that matter!) allow more solutions than the ones modeled by this exact wave vector model. That's because the emulsion isn't infinitely deep, and thus isn't a perfect filter. While we won't model the exact bandwidth of any hologram, we can use the wave vector model to describe how to think about bandwidth.

The wave vector diagram lives in a frequency world, which is an inverse of the spatial domain. As such, "thick" things in the spatial domain (like the thickness of an emulsion) are very small in the frequency domain, and vice versa. A thick hologram's extent in the frequency domain, then, will be very small: the very tip of the grating vector, for instance. A thin hologram is represented by a relatively wide extent in the frequency domain. The frequency response of any hologram (its shape and extent in the frequency domain) is the Fourier transform of is spatial characteristics (including thickness and a variety of other properties).

While you will have to consult Goodman (2005)[i], Kogelnik (1969)[ii] or other authorities to find the exact bandwidth of a particular hologram type, the orientation of the frequency response is something interesting that we will consider. This frequency response region of frequency space means tells us which range of combinations of wavelengths and angles of illumination and output beams will satisfy Bragg condition *besides* the ones indicated by the tip of the grating vector (as illustrated here). For planar holograms, the range of solutions lies along a line centered on the tip of the grating vector and perpendicular to the emulsion surface (again, by our convention, the z axis). The tip of the grating vector represents the hologram's highest diffraction efficiency; the frequency response may have local maxima and minima, but it will generally fall off with distance from the grating vector tip.

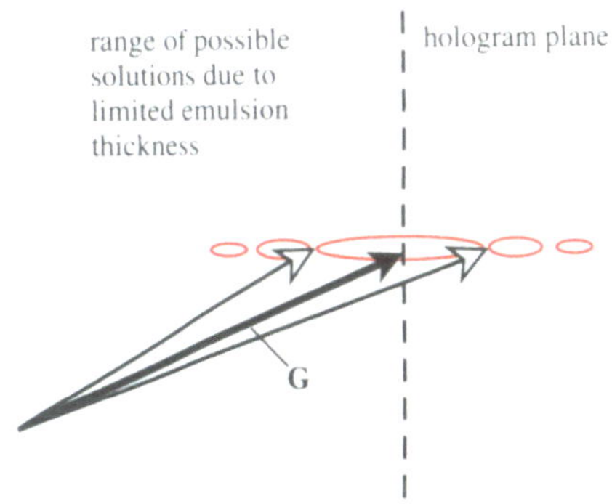

Let's take one further step and approximate the extent of this range of solutions; we follow Goodman's (2005)[iii] simplest approach here. The emulsion truncates the fringe pattern recorded in it. In signal processing terms, the fringe pattern is multiplied by a "rect" function of width t_2. (A rect function truncates a function outside a given range of values.) The transform of the rect function in frequency space is the normalized sinc function:

$$F(\text{rect}(x/t)) = C\,\text{sinc}(ft) \tag{8}$$

where C is a constant we'll ignore. What that means is that the rect function manifested by the limited emulsion thickness becomes a sinc-shaped phenomena in our frequency space (actually, because we're looking at power, the function is sinc squared, but that doesn't affect our calculations here).

The center lobe of the normalized sinc function extends from –1 to 1 in frequency space, so much of its energy is in the part of the lobe from –0.5 to 0.5. When we transform the rect of width t_2 into frequency space, the width of this part of the scaled sinc function has width $1/t_2$. Therefore, our grating vector **G** has uncertainty of approximately $1/t_2$ in the z direction in our standard coordinate system due solely to the finite thickness of the emulsion.

"Cos-Theta" Equation

Now let's turn this vector formulation into a more convenient 2-D form, one more similar to our standard holography equations. If we restrict ourselves to the x-z analysis plane, our wave vectors can be modeled as functions of angle to the z-axis:

$$\left(\frac{\eta\sin\theta}{\lambda},0,\frac{\eta\cos\theta}{\lambda}\right) \tag{9}$$

Start with this:

$$\mathbf{g}_{\text{out}}-\mathbf{g}_{\text{ill}}=m\left(\mathbf{g}_{\text{obj}}-\mathbf{g}_{\text{ref}}\right) \tag{10}$$

then pull out the lengths of the vectors (the "hat" notation indicates a vector of unit length):

$$\eta_2\frac{\left(\hat{\mathbf{g}}_{\text{out}}-\hat{\mathbf{g}}_{\text{ill}}\right)}{\lambda_2}=m\eta_1\frac{\left(\hat{\mathbf{g}}_{\text{obj}}-\hat{\mathbf{g}}_{\text{ref}}\right)}{\lambda_1} \tag{11}$$

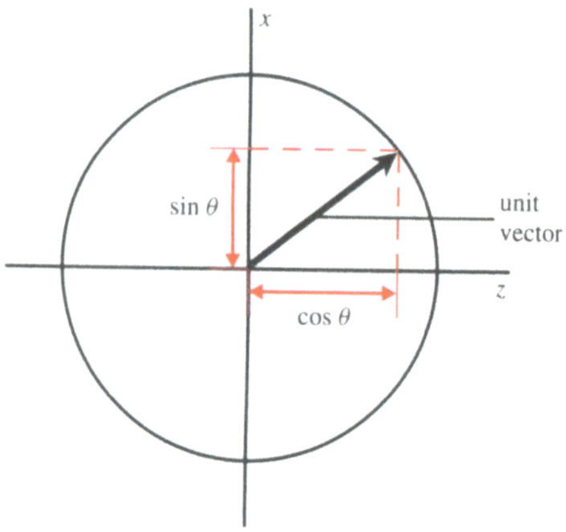

We can consider each component of the vector separately. The x component of the unit vector is just the sine of the angle of the vector with respect to the z-axis (as in the figure). So, using just the x component turns the vector equation into our old friend the $\sin\theta$ equation. Recall that Snell's law doesn't influence the x component of the vector, so we can drop the index of refraction term and use exterior angles to the emulsion:

$$\frac{\left(\sin\theta_{\text{out}}-\sin\theta_{\text{ill}}\right)}{\lambda_2}=m\frac{(\sin\theta_{\text{obj}}-\sin\theta_{\text{ref}})}{\lambda_1} \tag{12}$$

The $\sin\theta$ equation models the component of the holographic fringe vector that is perpendicular to surface normal the holographic emulsion, the so-called "thin hologram" behavior. For thick holograms, of which reflection holograms are a subset, the component of the fringe vector parallel to the hologram surface normal describes the emulsion's volume behavior. In our two-dimensional shop math analysis, this new vector component is quantified by the $\cos\theta$ equation:

$$\eta_2\frac{\left(\cos'\theta_{\text{out}}-\cos'\theta_{\text{ill}}\right)}{\lambda_2}=m\eta_1\frac{(\cos'\theta_{\text{obj}}-\cos'\theta_{\text{ref}})}{\lambda_1} \tag{13}$$

where $\cos'\theta$ is an internal angle. In the $\cos\theta$ equation, unfortunately, index of refraction and internal angles must be carried around. Since this equation is the z component of the wave vector, we can add thickness changes into the equation as well; the length of the z component of the grating vector, modified by shrinkage, is given by the following equation:

$$\eta_2\frac{\left(\cos'\theta_{\text{out}}-\cos'\theta_{\text{ill}}\right)}{\lambda_2}=m\eta_1\frac{t_1}{t_2}\frac{(\cos'\theta_{\text{obj}}-\cos'\theta_{\text{ref}})}{\lambda_1} \tag{14}$$

This relation gives us the following equation for the central wavelength satisfying the Bragg condition, which we call the $\cos\theta$ equation. Rearranging terms:

$$\cos'\theta_{out} = m\frac{\eta_1 t_1 \lambda_2}{\eta_2 t_2 \lambda_1}(\cos'\theta_{obj} - \cos'\theta_{ref}) + \cos'\theta_{ill} \quad (15)$$

This equation works hand in hand with the $\sin\theta$ equation; after all, now we know they're just two different components of the same vector formulation. To a large extent, the $\sin\theta$ equation models the diffractive part of the hologram, and the $\cos\theta$ equation the filter part. For reflection holograms, m is limited to +1 or −1.

We can augment the $\cos\theta$ equation to provide an estimate of the bandwidth of a hologram due to limited emulsion thickness. We do so by adding in the uncertainty term found in the previous vector analysis, which was wholly contained in the z component of the grating vector. This bandwidth envelope is about $1/t_2$ wide, centered around the end of the grating vector: $\mathbf{G}'_z \pm 1/(2t_2)$.

So, the combinations of output and illumination angles that will meet Bragg conditions must satisfy this equation:

$$m\mathbf{G}'_z = \eta_2 \frac{(\cos\theta_{out} - \cos\theta_{ill})}{\lambda_2} \pm \frac{1}{2}\frac{\lambda_2}{\eta_2 t_2} \quad (16)$$

solving for the output angle:

$$\cos'\theta_{out} = m\frac{\eta_1 t_1 \lambda_2}{\eta_2 t_2 \lambda 1}(\cos'\theta_{obj} - \cos'\theta_{ref}) + \cos'\theta_{ill} \pm \frac{1}{2}\frac{\lambda_2}{\eta_2 t_2} \quad (17)$$

To use this equation in practice, first set the bandwidth range term to zero and solve for the output angle, being careful to manage all of the internal angles and indices of refraction appropriately based on conditions at exposure and illumination. The m term should be chosen based on whether the hologram is being illuminated in direct or phase conjugate mode. This center output angle solution is the angle that best satisfies Bragg conditions.

Then, solve the equation twice, once with the bandwidth term positive, the second time with it negative. As expressed above, this pair of equations will reveal the angle to which a particular output wavelength will be propagated above some limit of efficiency. (This analysis assumes, of course, that the illumination source is sufficiently broadband to contain this range of wavelengths; the reflection hologram can only filter the wavelengths that enter it!)

Alternatively, the equation can be used to calculate the approximate range of wavelengths propagated to a particular angle. Rather than solving for the range of the output angle, solve instead for the illumination wavelength using a fixed output angle and the bandwidth term. The two solutions (the positive and negative bandwidth terms) represent the emulsion's bandwidth limits due to finite emulsion thickness. It is important to note that should the emulsion become very thin, or the geometry of the holographic exposure produce a transmission-like hologram, factors other than emulsion thickness may limit the bandwidth of the hologram. Similarly, unusual exposure and display geometries, such as those used in edgelit holograms, may have more complicated bandwidth characteristics.

$$\cos'\theta_{out} = m\frac{\eta_1 t_1 \lambda_2}{\eta_2 t_2 \lambda 1}(\cos'\theta_{obj} - \cos'\theta_{ref}) + \cos'\theta_{ill} \pm \frac{1}{2}\frac{\lambda_2}{\eta_2 t_2}$$

The "Cosine-Theta" equation

Conclusions

Reflection holography offers a wide range of creative and practical possibilities for holographers, far too many to discuss here. While we have only touched on the basics of off-axis reflection holography, you should now have a more complete grasp of the similarities and differences between these holograms and our more familiar transmission holograms. Many of the mathematical tools we have developed work with both reflection and transmission holograms, including the angle, horizontal focus, and vertical focus equations; it is important, however, to keep track of signs and angles and to have an intuition about where light is actually likely to go. The vector analysis presented here spans both transmission and reflection hologram types, as well more unusual kinds of holograms such as the edgelit displays described in a later chapter. Finally, the $\cos\theta$ equation models the filter part of a reflection hologram, and has a similar mathematical appearance to our other shop math holography equations.

References

i. Goodman, J. W. (2005). *Introduction to Fourier Optics,* Roberts & Co., Englewood, CO.

ii. Kogelnik, H., (1969). "Coupled Wave Theory for Thick Hologram Gratings," *Bell System Technical Journal,* **48**, pp. 2909-2947.

iii. Goodman, J. W. (2005). *Op. cit.*

CHAPTER 18

Edge-Lit Holography

William Farmer

Introduction

The simple and flexible edge-lit display configuration provides strong advantages which motivate the pursuit of this geometry. Using an edge-introduced reference beam for recording, and similar illumination beam for display allows a simplified display configuration for holograms. Eliminating the need for an external, distant illumination source, this display integrates the hologram, its supporting display structure, and the illumination source into a compact, integral device.

In this chapter, we will take a practical look at the recording of edge-lit holograms, give consideration to recording tools and techniques, and see how edge-lit holography offers insight into our perception of transmission versus reflection hologram types.

With this chapter, this book enters into topic areas where both the theory and the practice are incomplete and still in active development.

Traditionally, reflection and transmission holograms are considered as distinct types, each having unique optical properties. The distinction is established by the direction of the reference beam during recording, incident from the same side of the recording medium as the object beam for transmission holograms, or from the opposite side for reflection holograms. A key differentiator between the two hologram types, resulting directly from the different directions of their respective reference beams, is the geometric orientation of their fringes, expressed with simplified generalization as running perpendicular to the face of the hologram for transmission types, and parallel to the face of the hologram for reflection types.

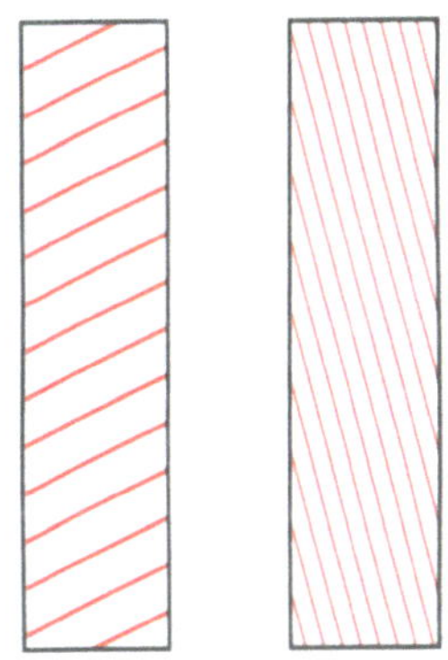

Transmission (left) and reflection (right) fringe structure

There is an optical space that lies between the transmission recording region and reflection recording region that reinforces this perception of uniqueness between these two types. This intermediary zone, dubbed "inaccessible" by renowned holographer Emmett Leith (1966),[i] precludes certain recording geometries by limiting reference beam angles within the recording layer for both transmission and reflection holograms. The theory of the inaccessible zone hinges on the large distinction between the indices of refraction between air ($n = 1.0$) and the holographic recording layer ($n = 1.63$ for silver halide); this refractive index gap severely limits the maximum angle of the reference beam within the emulsion. For reference beams impinging from air at very steep angles—near 90°—the maximum intra-emulsion reference beam angle after Snell refraction is approximately 38°. The 104° gap between the maximum reference beam angles achievable within the emulsion for transmission and reflection types is the inaccessible zone.

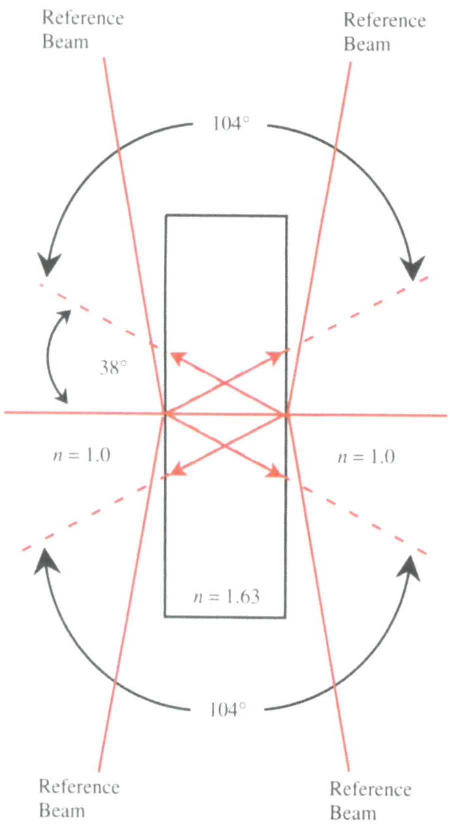

The small "accessible" zone for reference beam angles results in highly differentiated fringe structure and illumination beam angle of incidence on the fringes for transmission versus reflection holograms. This perception of these two hologram types as being highly distinct is captured in our characterizing model of transmission holograms as a window-imaging system, and reflection holograms as a mirror-filter-imaging system. We will see later in this chapter that edge-lit holography opens up the inaccessible zone and bridges the distinct gap between transmission and reflection holograms.

Early work in edge-lit holography focused on penetrating the inaccessible zone by introducing the reference beam through the edge of the substrate glass that supported the emulsion,[ii] through use of total internal reflection,[iii] or use of evanescent waves.[iv] Introducing the reference beam through a medium with an index of refraction closer in value to the emulsion's index, results in reduced refraction at the emulsion interface and steeper reference beam angles within the emulsion.

The theme of introducing the reference beam through the bottom edge of a glass layer supporting the emulsion was picked up by Juris Upatnieks (1992),[v] who applied this geometry to the recording of holographic optical elements for use in compact heads-up displays. Upatnieks introduced his reference and illumination beams through the bottom edge of a thicker glass plate to which the hologram's glass substrate was optically coupled, thus using the thicker glass block as a wave guide. This glass block as wave guide became the principal tool in the development of edge-lit display holograms.

We'll start our consideration of edge-lit holograms by looking at the recording geometries of several types of edge-lit holograms.

Recording Geometries

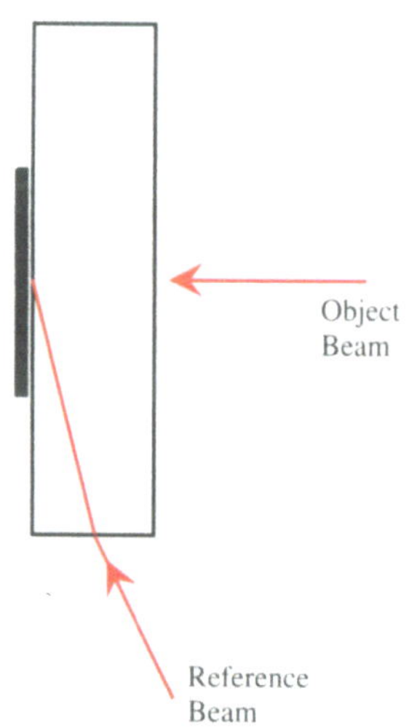

Edge-lit holography is also referred to as steep reference angle holography, this latter name giving better insight into the unique element of its recording technique. But beyond this steep reference beam angle, the mathematics and general recording geometries of edge-lit holograms are consistent with standard holographic recording techniques.

Steep reference angle constraint

Distortion free playback of a virtual or real image requires that the hologram be illuminated by a wavefront that either is identical to the reference beam, or is its complex conjugate. In standard recording practice, the reference beam is often collimated, allowing its real image to be reconstructed with a collimated beam. For white light viewable holograms, the subsequent illumination beam is often a point source at a distance sufficient to permit the spherical wavefront to flatten into an approximation of a collimated wavefront.

The compact display of edge-lit holography is characterized by a nearby, divergent illumination beam. Thinking backwards from our edge-lit display and its distinctive illumination beam gives insight into reference beam selection (and avoidance) suitable for edge-lit recording.

Our display configuration is simple—that is the whole point! We have an illumination source in the base of a display apparatus. The illumination beam enters the bottom of a clear block—commonly acrylic plastic—and illuminates a hologram with a sharply diverging wavefront.

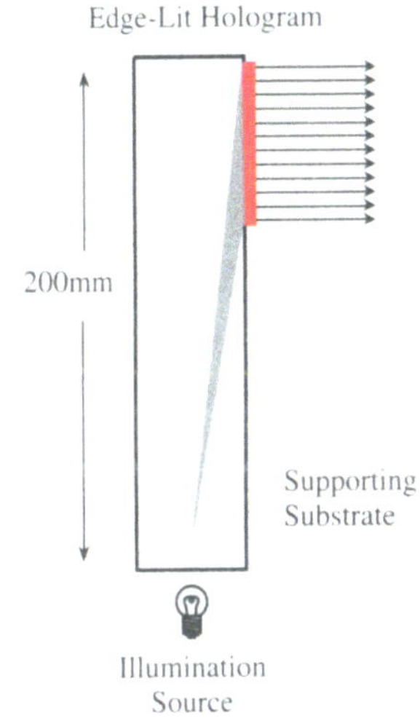

The implications of this diverging illumination are the primary drivers in considering our recording geometries. First, consider the case where we choose to use complex conjugate illumination to play back our final edge-lit image; our sharply diverging illumination beam is the complex conjugate of a sharply converging reference beam. To create a reference beam that fully irradiates the holographic plate and converges to a nearby point—the point from which the illumination beam would diverge—would require a very large, very optically fast, very expensive convex lens. This is impractical.

A second potential reference beam approach would be to follow traditional practice and collimate the reference beam, setting up playback with collimated, complex conjugate illumination. This, however, pushes the burden of collimated playback onto the illumination beam and into our display device design, whose fundamental design goal is simplicity. So while this is a practical solution, especially with the potential use of low cost gratings in the base to effect collimation, we prefer a different constraint for edge-lit reference beams.

The constraint that we prefer for edge-lit holography is to *limit the reference beam to be identical to the intended illumination beam*, and always to play out the virtual image from the displays with direct illumination. And, as you will see from the following examples, this is not a particularly harsh constraint.

Edge-lit reflection hologram—one-step

Let's start with a most basic recording example, identical to the recording of any reflection hologram, except for two noteworthy points. First, according to our constraint, the diverging reference beam replicates the intended diverging illumination beam, which will play out a virtual image from our simple display unit.

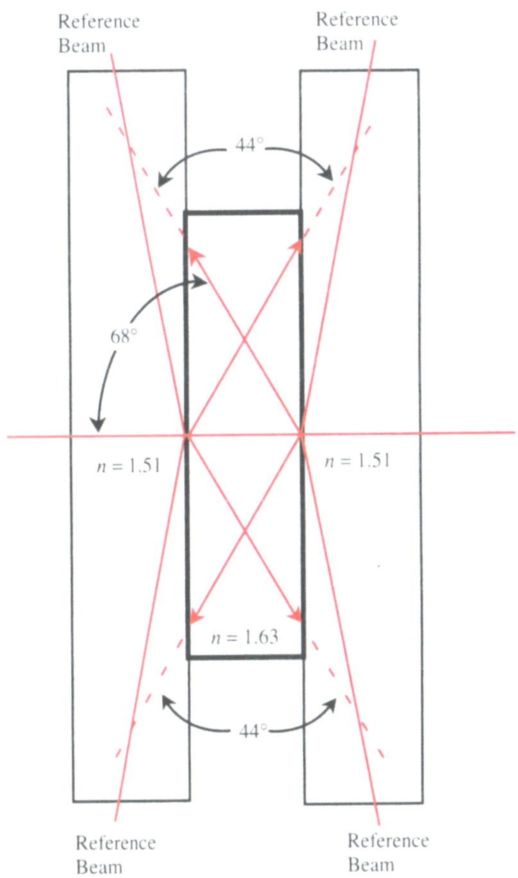

The second, more interesting point deals with the reference angle within the emulsion.

Illumination from the base of the glass block allows our reference angle to approach 90° within the glass block. For illustrative purposes, let's say it is at 80°. Using indices of refraction of 1.51 and 1.63 for the glass block and the silver halide emulsion respectively, from Snell's Law, the beam angle within the emulsion is 66°.

The maximum intra-emulsion reference beam angle for beams introduced through air is around 38° (and this is the boundary of the inaccessible zone). By introducing the beam through the edge of a glass block, with the glass block having an index of refraction much closer than air's to that of the emulsion, we are able to achieve intra-emulsion reference beam angles up to 68°. This reduces the extent of the inaccessible zone from 104° to 44°.

Edge-lit rainbow holograms—an extra step required

As you have studied earlier, traditional rainbow holograms are made with a two-step recording geometry. An H1 hologram is played back with complex conjugate illumination to create a real image of the original object. A slit filters the H1 projection, throwing away verti-

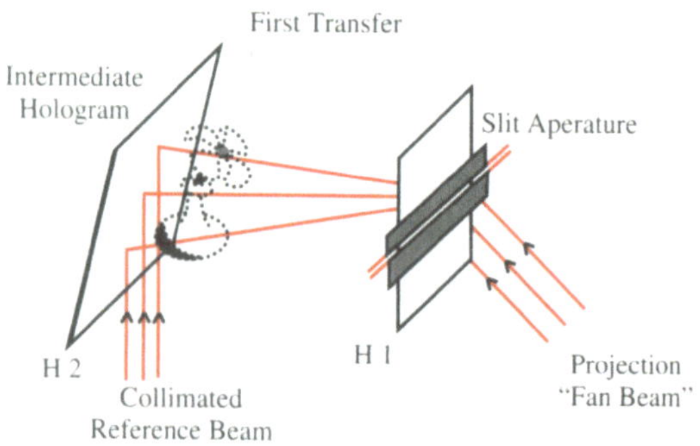

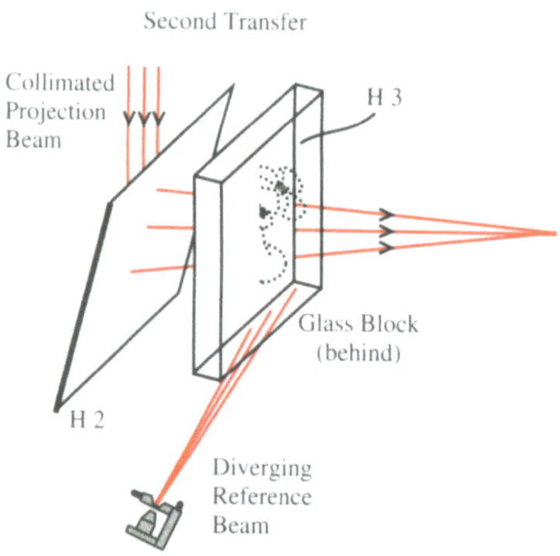

cal parallax, in compensation for the chromatic dispersion of white light viewing of transmission holograms. An H2 records the filtered image reconstructed from the H1. The H2 is recorded in the plane of the original object, with a collimated reference beam. For viewing, the H2 is illuminated with approximate complex conjugate white light—the distant point source cited earlier—and projects a chromatically dispersed rainbow of real images of the H1's slit as the viewing window.

The complex conjugate illumination beam of traditional rainbow holography violates our edge-lit constraint of direct illumination for viewing.

The workaround to adhere to our constraint for edge-lit rainbow holograms requires a third step be added to the recording process. The H1 is recorded identically to traditional rainbow holography—significant in that all existing H1 masters are therefore suitable candidates for edge-lit reproduction. The H2 recording step introduces our workaround. The slit-filtered real projection from the H1 propagates beyond the original object plane—or equivalently, beyond the H1's real image focal plane. Further downstream, an H2 is recorded on an oversized recording plate—oversized to fully capture all of the information now diverging from the real image focus. A collimated reference is used for recording this H2.

In the new, additional H3 recording step, collimated complex conjugate illumination is used to project the H2's real image back upstream to the original object plane. Here the real image is recorded in an H3, with a nearby, diverging reference beam, satisfying our constraint.

For viewing, this H3 is played out with direct illumination, and the wavefront that had projected from our H2 is now regenerated and the rainbow of view windows is created.

Edge-lit Ultragrams—two steps

In the production of classical stereograms, a series of 2-D perspective views are sequentially rear projected onto a diffusing projection screen and are recorded into a linear series of very narrow slit holograms, comprising the H1 slit master. The H1's real image is projected back to the projection plane and recorded in an H2. The H2 is then illuminated with complex conjugate illumination to project the real image of the slit master, which becomes the view window. The viewing experience of classical stereograms is often compared to looking through a picket fence—the slits being the space between the pickets—at two different perspective views, which form a stereo pair.

To record a classical stereogram in the edge-lit format would require a third step, like rainbow edge-lit holograms, to compensate for the H2 being played out with complex conjugate illumination. The development of the "Ultragram," which breaks the geometric constraints of the classical stereogram through computer image data manipulation, allows these stereograms to be recorded in a two-step process.[vi]

In Ultragram geometry, the slit plane and the view plane—coincident in classical stereograms—are separable. After a predetermined view plane is selected, the geometry of slit plane to projection plane to view plane is mapped. Data image for a specific view position is mapped through the projection plane into a specific slit location on the slit plane. By this method, the view for any specific viewpoint is sorted into projection plane positions and slit positions. The image date is then re-sorted to construct the appropriate projection data for any individual slit—which now serves a variety of view positions—and an Ultragram H1 slit master is recorded through sequential projection and recording of slit holograms.

The real image of the H1 Ultragram slit master is reconstructed with complex conjugate illumination, and then propagates to its focal plane - the original projection plane. Here an H2 transfer hologram is recorded. Because the image data will continue to travel in the same direction towards the view plane for viewing, this H2 can be played out with direct rather than complex conjugate illumination. This satisfies our constraint, and the H2 is recorded with a steep angle reference beam consistent with our geometry.

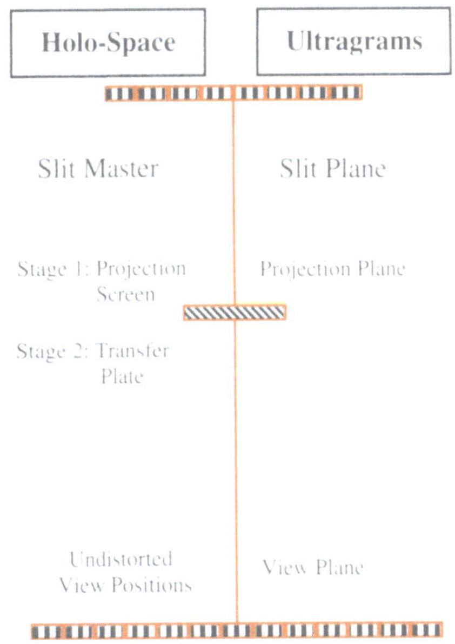

Recording summary

The recording of edge-lit holograms adheres to the same mathematics and closely aligns with the recording geometries and processes of traditional holograms. Unique in the recording geometry of edge-lit holograms is the introduction of a highly divergent, steeply angled reference beam through the edge of an optical element, originally a glass block. The close proximity of the index of refraction of the optical element to that of the recording layer enables uniquely steep reference angles within the emulsion layer. The constraint of a highly divergent reference beam requires a workaround for holograms whose white light image would traditionally be reproduced with complex conjugate illumination.

A Practical Issue with Steep Reference Angle Recording

While we have described the recording of edge-lit holograms as being a relatively straightforward application of standard hologram recording techniques, the use of a steep angle reference beam creates a severe practical problem.

Woodgrain

Holograms produced in the steep reference angle geometry have been marred by a defect resembling a woodgrain pattern in the emulsion layer. This pattern both creates an observable viewing defect in our recording, and also robs the emulsion of diffraction efficiency. This problem was sufficiently perplexing to cause Lin, the original experimenter with the edge-lit format, to abandon his study. It continues to be, along with stability during recording, a primary practical consideration for those producing edge-lit holograms.

The woodgrain pattern is a very low frequency fringe pattern, produced by Fresnel reflections at the various optical interfaces of the recording geometry.

Fresnel reflections

At interfaces of mismatched indices of refraction, an incident wave will be partially transmitted and partially reflected. For S polarized

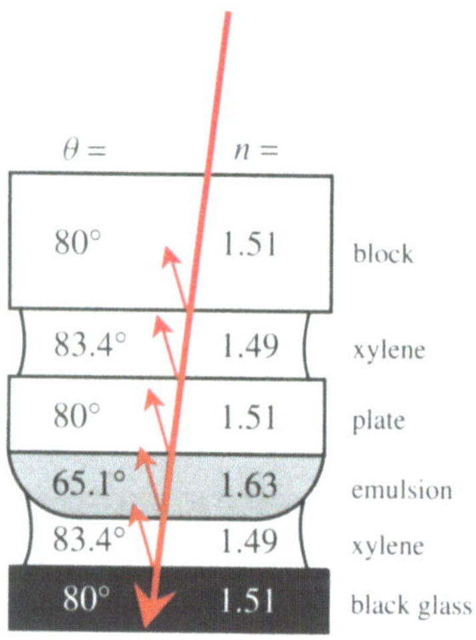

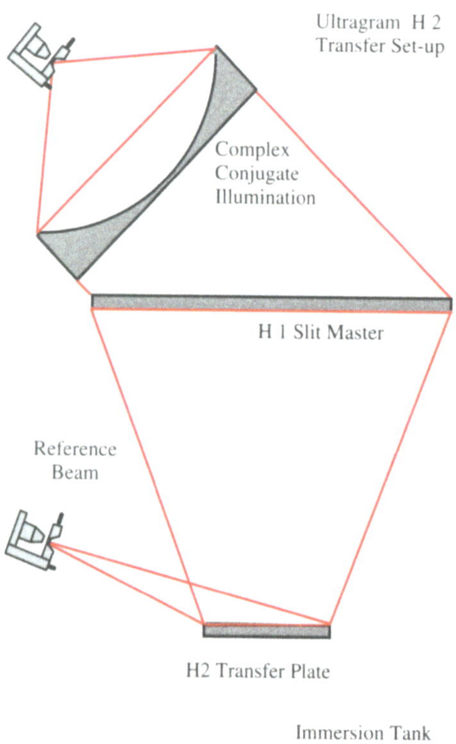

light—the likely choice for hologram recording—the amplitude of the reflected wave is given by:

$$R = \frac{n_1 \cos\theta_1 - n_2 \cos\theta_2}{n_1 \cos\theta_1 + n_2 \cos\theta_2} \tag{1}$$

The intensity of the reflected wave is equal to the amplitude squared.

The typical "glass block" recording geometry for edge-lit recording has a stack of several critical interfaces. The emulsion ($n = 1.63$) is carried by a glass substrate ($n = 1.51$) that is optically coupled to the glass block ($n = 1.51$) with an intermediary layer of xylene ($n = 1.49$). Xylene is also used to mate light absorbing black glass ($n = 1.51$) to the emulsion layer, to absorb, as much as possible, any transmitted illumination that, on reflection, could rob diffraction efficiency from the emulsion.

Two considerations quickly link the woodgrain defect to the steep angled reference beam used in edge-lit recording. First, because the spatial frequency of fringes increases with the angle subtended between the interfering beams, the very low spatial frequency of the woodgrain pattern must be created by beams traveling in the same direction with very small included angles. The woodgrain fringes are caused by portions of one of our two incident beams being doubly reflected from two separate interfaces and interfering with the incident beam itself, inside the emulsion layer.

Because reflection increases at an interface with its angle of incidence, of the two potential candidates—the image beam and the reference beam—the steeply angled reference beam rather than the normally incident image beam is easily understood to be the cause.

The woodgrain effect was clearly tied to Fresnel reflection of the steep angle reference beam. However, the relative contributions of the two major index mismatches—mismatch of the immediate emulsion to glass index, and mismatch of the less proximate, but very large glass to air index—were not as clearly understood.

Immersion tank recording

To isolate the two perceived contributors to the woodgrain effect, an immersion tank recording device was devised. With immersion tank recording, the final edge-lit hologram plate sits in a bath of xylene. The tank is designed to remove the strong glass-air index mismatch from the proximity of the emulsion layer, leaving only the smaller emulsion-glass mismatch to contribute to woodgrain.

After initial experimental success in mitigating the woodgrain defect, the immersion tank was redesigned with improvements to further remove spurious reflections and make use of the tank more efficient and safer.

Recording of the final hologram within the immersion tank does have an optical cost. First, the introduction of an optical element—half of the immersion tank—into the path of the projected image to be recorded in the edge-lit hologram, shifts the location of the focal plane of the projected image. This can be easily accommodated through visual adjustment of the focal plane by the careful holographer.

Secondly, because of the non-linear effect of Snell refraction on the various rays of the projected image as they leave the air and enter the immersion tank, rays are variously laterally shifted from their correct focal positions, and distortions are introduced into the image being recorded. Holographers have seen that the adjustment of the recording focal plane to compensate for the half-tank that is introduced into the image path greatly compensates for the Snell-refraction-induced distortion, leaving only a non-prominent distortion that has minimal detrimental effect on the viewed edge-lit image.

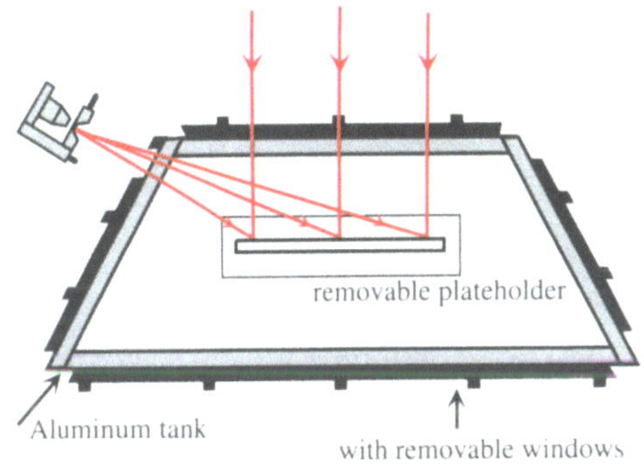

The glass block revisited
The emergence of photopolymer recording materials has regenerated interest in the use of the glass block recording apparatus.

Photopolymer recording materials, such as those produced by DuPont Corporation, have an index of refraction on the order of 1.49. The much closer proximity of this index (compared with silver halide emulsion at 1.63) to the glass index in our block recording setup potentially eliminates one of the culprit index mismatches that generated the woodgrain fringes.

Experimentation with photopolymer using the glass block recording device has demonstrated a significant reduction in woodgrain effect. The residual woodgrain when recording with photopolymer is partially attributed to the mylar substrate which carries the photopolymer material. This substrate has a refractive index of 1.66. Use of photopolymer materials that allow removal of the mylar substrate and allow direct application of the recording material to an alternative glass plate, mitigates this last significant index mismatch in the vicinity of the recording layer.

Woodgrain summary
The woodgrain defect that frustrated early experimentation with edge-lit holography has been successfully mitigated through use of the immersion tank to record silver-halide emulsion holograms, and through use of the glass block to record photopolymer recording materials that have had their mylar substrates removed and replaced by glass. By logical extension, glass plate photopolymer recording in-situ with the immersion tank will address both index mismatches, and produce the brightest, most efficient, woodgrain free edge-lit holograms.

Characteristics of Recording Within the Inaccessible Zone

The perception that reflection and transmission holograms are distinctly different is challenged by the recordability of the edge-lit hologram in both transmission and reflection geometries in the extremes of the inaccessible zone. The properties of reflection and transmission holograms recorded in this region are more similar than different. To see this, let's first reexamine the inaccessible zone, and then work with a simple and familiar fringe model.

Accessing the inaccessible zone
Earlier, when describing edge-lit reflection holograms, we noted that with edge-lit recording we had penetrated and thereby reduced the

extent of the inaccessible zone. We have seen that, recording with silver halide emulsions ($n = 1.63$), and accessing the hologram from air, limits the intra-emulsion reference beam angle to a maximum of around 38°. But, again with silver halide, when we introduce from the base of a glass block, with a steep reference angle within the glass ($n = 1.51$), we can achieve a reference beam angle of up to 68° within the emulsion.

To bridge the gap between transmission and reflection holograms, let's see if we can further reduce the inaccessible zone. The ultimate reduction of the inaccessible zone would be to achieve an intra-emulsion reference beam angle of 90°, in essence eliminating the inaccessible zone.

To achieve a 90° reference beam within the emulsion, we note from Snell's Law that we need for the emulsion to have a lower index of refraction than glass, and to have the reference beam incident on the emulsion at the critical angle. We can create that situation using photopolymer ($n = 1.49$) interfaced with glass. For illustrative purposes, using an 80° reference angle in the glass block produces an intra-photopolymer recording layer reference angle of 86°. For point of reference, the critical angle for the glass to emulsion interface, at which an intra-emulsion reference bean angle of 90° is realized, is 80.7°.

Now that all intra-emulsion reference beam angles are accessible, our range of potential fringe angles is also fully expanded.

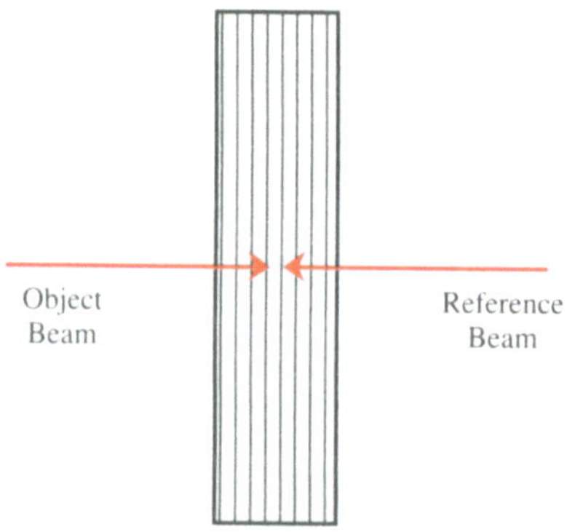

Full fringe model

We'll make use of the common model of fringes as a Venetian blind of partially reflective mirrors to examine fringe structure and its consequences in our now fully expanded edge-lit space.

If we start our model in reflection mode with co-axial and oppositely directed beams, the fringes will bisect this 180° angular separation, and will be recorded at an angle of 90°, parallel to the face of the emulsion.

As we rotate our intra-emulsion reference angle counter-clockwise, our fringes will also rotate counter-clockwise at half the angular rate of the reference beam (because the bisector of an angle will change as half the change in the angle itself).

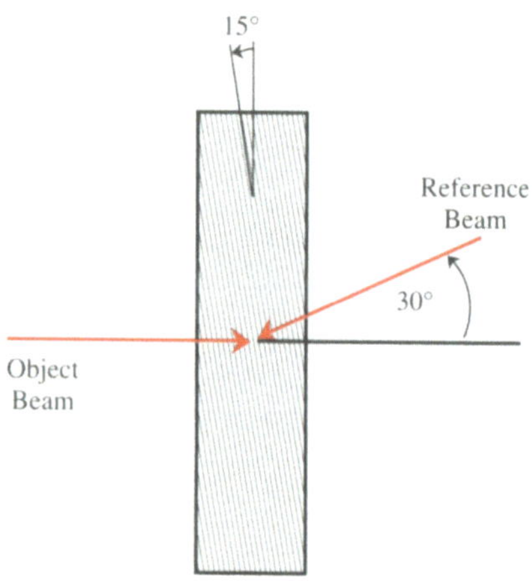

Thus as our reference beam circumnavigates the emulsion to return to its original co-axial reflection geometry, our fringes will rotate 180°, that is they will flip once and return to an orientation parallel to the face of the plate.

Of most interest in the examination of the edge-lit fringes is when the reference beam approaches and swings through 90° (and later through 270°). Here the recording mode changes from reflection to transmission (and later transmission to reflection). But these extreme reflection and transmission geometries share similar fringe characteristics. And the direct illumination will impinge on these extreme transmission and reflection fringes at very close relative angles. We rightly expect these extreme transmission and reflection holograms to share optical properties. But what will these shared optical properties be in this region between classical transmission and classical reflection hologram spaces?

Optical characteristics of the extreme edge-lit region
In earlier chapters, we broke the optical properties of classical transmission and reflection holograms into parts. We characterized transmission holograms as having a window part and an imaging part. Reflection holograms, in contrast, had a mirror part, an imaging part and a filter part.

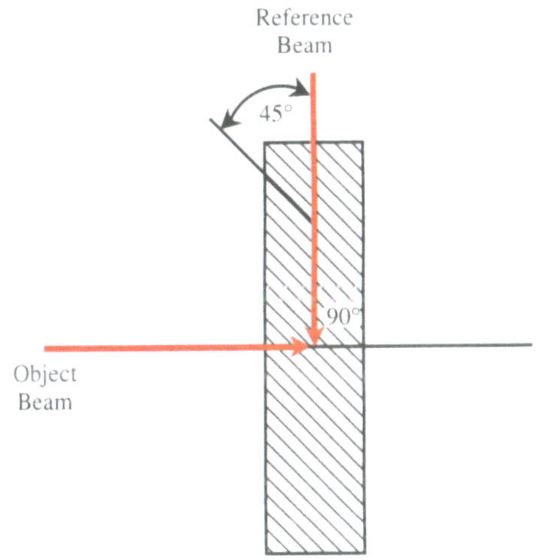

In resolving the shared properties of transmission and reflection holograms in the extreme region of edge-lit space, let's start with the imaging part. In previous chapters, we found that both transmission and reflection holograms can be modeled with the now familiar $\sin\theta$ equation. As intuition would suggest, this equation will also govern the imaging part of both transmission and reflection holograms in the extreme edge-lit region.

The window versus mirror part—the passive optical element part—can be evaluated by consideration of the fringe pattern in the extreme. To simplify consideration, an intra-emulsion reference beam at 90° will produce a fringe structure reclining at 45°. The passive optical element part of the shared characterization is a periscopic mirror, with its angle varying slightly as the reference beam swings from transmission side to reflection side. As you might have imagined, this periscopic mirror splits the difference between window and mirror.

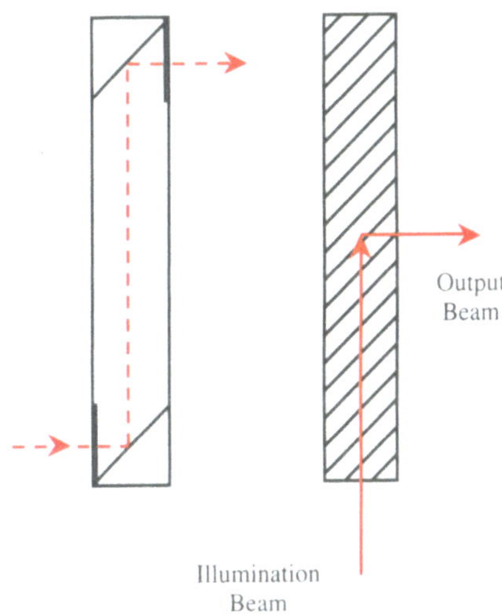

In consideration of the filter part—present in reflection holograms but not a characteristic of transmission holograms—we assert that the filter is not really tied to the type of hologram, but rather to the thickness of the hologram as seen by the illumination beam. While we generally view the inaccessible zone as having divided holograms into two clear types—transmission and reflection—we could take the perspective that the inaccessible zone divides holograms into virtually thin and virtually thick types.

The virtual thickness of a hologram is a measure of two attributes, the intra-emulsion path length of the illumination, and the spatial frequency of the grating structure, which together equate to the number of fringes encountered within the emulsion by a ray of the illumination beam.

Recalling the grating vector from our Chapter 17 discussion of off-axis reflection holograms, the classical transmission hologram has a short grating vector (low spatial frequency) with an extremely small horizontal component (suggesting minimal volume effects, of which filtering is a key effect). This attributes virtual thinness to classical transmission holograms. In contrast, the long, nearly horizontal grating vector of a classical reflection hologram asserts virtual thickness and volume effects.

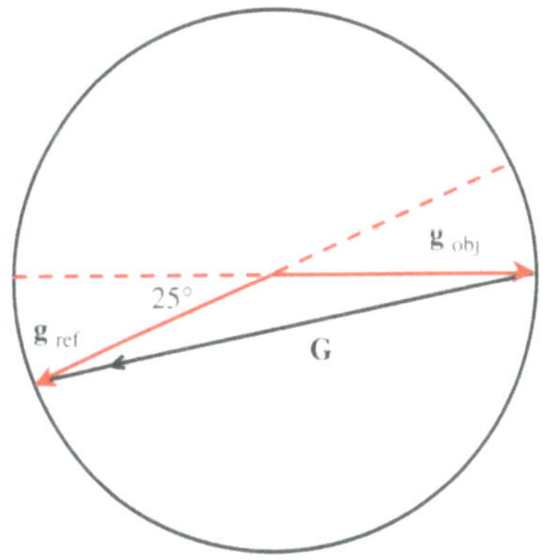

Reflection grating vector model for 25° intra-emulsion reference angle

Let's take an intuitive look at the virtual thickness of an extreme edge-lit hologram, by comparing it to classical holograms. In a classical reflection or transmission hologram with an intra-emulsion illumination angle of 25° (corresponding to 45° in air) and with emulsion thickness of 6 microns (typical of silver halide), the full transit across the emulsion of the illumination beam will be about 7 microns (6 microns / cos 25°). In the same emulsion, with a steep illumination beam of 60°, the intra-emulsion path length of the illumination will be about 12 microns (6 microns / cos 60°), almost double that of the classical holograms.

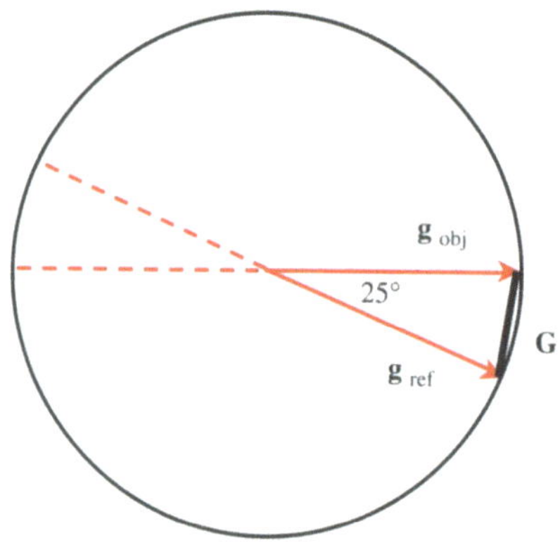

Transmission grating vector model for 25° intra-emulsion reference angle

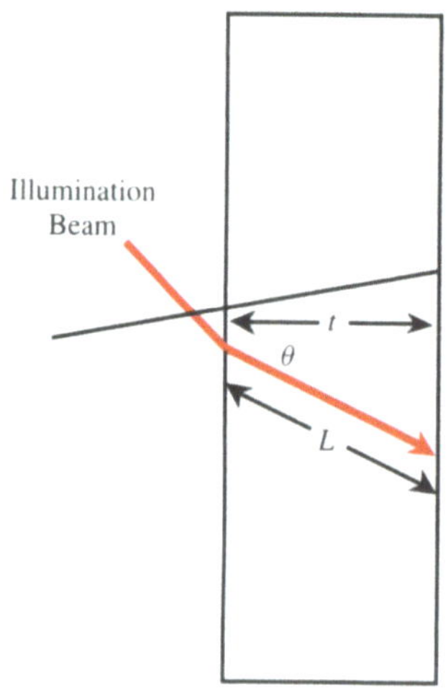

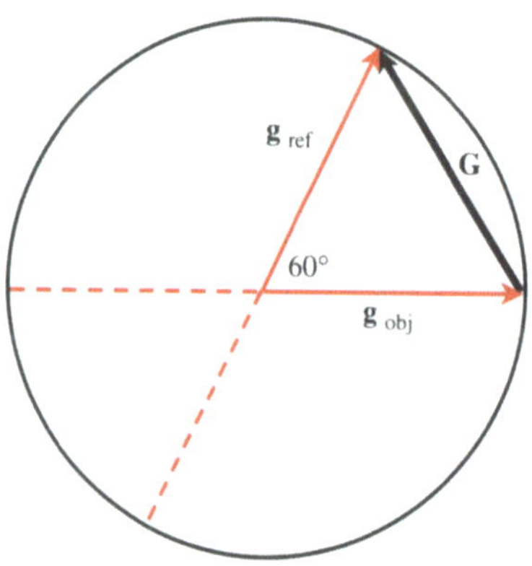

Edge-lit transmission grating vector model for 60° intra-emulsion reference angle

The second intuitive aspect of the virtual thickness of the filter part is the spatial frequency of the fringes. In Bragg filtering, the higher the spatial frequency, the more fringes encountered, the stronger the effect of the constructive and destructive interference of wavelengths that are the basis of the filter. Fringe spacing, the inverse of spatial frequency, is given by:

$$d = \frac{\lambda}{2\sin(\theta/2)} \tag{2}$$

where θ is the angle between the object and reference beams, and λ is the wavelength within the emulsion.

We use a λ of 633 nm in air, and an index of refraction of 1.63 for silver halide. For our reflection hologram, with a θ of 205°, we get about a 0.2 micron fringe spacing; for our edge-lit example with a θ of 60°, (this is then a transmission hologram; in a reflection hologram our θ would be 120°, and the fringe spacing would be slightly denser) we get a fringe spacing of around 0.4 micron. The edge-lit transmission hologram's fringe spacing is twice that of our classical reflection hologram, or equivalently, it has half the fringe density (spatial frequency).

To complete this consideration, we look at a classical transmission hologram with an intra-emulsion reference beam of 25°. Its illumination beam's path length across the emulsion will be the same as that for the 25° reference beam in the reflection case, around 7 microns. Its fringe spacing calculates to about 0.9 micron, about a quarter of the spatial frequency of the reflection hologram. The lack of filtering in a classical transmission hologram is a result of its virtual thinness, that is, a short intra-emulsion path length and a low density fringe pattern. The end result is that the illumination beam encounters few fringes.

In the edge-lit case, based upon the longer path length within the emulsion and the relatively dense fringe structure, we would expect this edge-lit hologram to approach virtual thickness and to show some degree of filtering effect. This is consistent with experimental observation; the question of degree of filtering will be addressed below.

Finally, we can take a different look at relative spatial frequencies by observing the grating vectors for our examples. In the vector diagrams for our three examples, you will observe that the lengths of the grating vectors ***G***, which corresponds to spatial frequency have an approximate ratio of 4:2:1 for classical reflection, edge-lit transmission, and classical transmission respectively. This is perfectly consistent with our above calculations.

The filtering part of edge-lit holograms

The question of degree of filtering in the edge-lit region is one of keen interest. If strong Bragg effects are realized then full aperture transmission holograms might be achievable. Our example above with the 60° intra-emulsion reference and illumination beam angles does not produce holograms of sufficient virtual thickness to provide strong Bragg filtering.

Our example above and most experimentation to date have been done with silver halide emulsion. With silver halide, we were only able to open up an intermediary portion of the inaccessible zone, not fully penetrate it. This limited the virtual thickness of edge-lit holograms in two ways. First, it limits our intra-emulsion reference beam angle to less than 68° and thereby the intra-emulsion path length of the illumination beam. Secondly, achieving even steeper reference beams would produce higher spatial frequencies for our gratings.

Repeating our virtual thickness considerations with a photopolymer example, we can explore the potential for Bragg filtering in the extreme edge-lit region. Because photopolymer opens up the full extent of the inaccessible zone, we will use an illumination beam angle of 80° for our example. This will give us an intra-emulsion path length of 34 microns (6 / cos 80°).

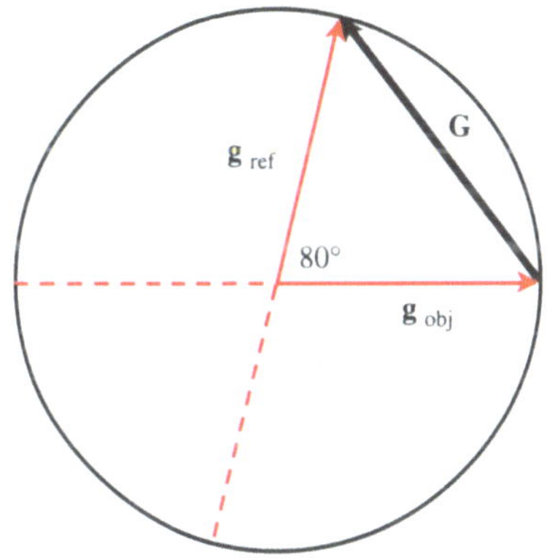

Extreme edge-lit transmission grating vector model for 80° intra-emulsion reference angle

Calculating the fringe spacing for photopolymer, we again use a 633 nanometer wavelength. But now we use a recording layer index of refraction of 1.49. The fringe spacing calculates to around 0.3 micron. Compare this to our intermediary region edge-lit example with silver halide emulsion; that example had a fringe spacing of 0.4 micron. Note that the increase in fringe spacing that would result from the lower index of refraction (resulting in a longer intra-emulsion wavelength) is overwhelmed by the effect of the increased reference beam angle that greatly reduces the spacing. Observe the grating vector; it has grown larger, signaling a denser spatial frequency.

With this very long path length within the emulsion and the rather dense spatial frequency, we can clearly describe this hologram, recorded in the extreme edge-lit region, as virtually thick.

Degree of filtering in the extreme edge-lit region

Path length within the emulsion and the spatial frequency are indicators of virtual thickness, because they suggest the number of fringes encountered by a ray of the illumination beam during a transit of the emulsion layer. Recall, the greater number of fringes encountered, the stronger the summation of constructively and destructively interfering rays; this serves as the basis of our Bragg-wavelength-selecting filtering effect. Let's go one layer deeper in considering the filtering effect in the extreme edge-lit region.

Our discussion of virtual thickness was based upon spatial frequency. But the number of fringes encountered within the emulsion is a product of both the spatial frequency and the angle of incidence of the illumination beam on the grating's fringes. The shallower the angle, the greater the distance between fringes encountered, and the fewer fringes encountered over a given path length. To see this, let's look at the number of fringes encountered by our example classical transmission and reflection holograms, which had equal intra-emulsion path lengths of 7 microns.

Observe the illustration of the transmission example of an illumination beam experiencing Bragg wavelength selection. For O_1 and O_2 to exit in phase and therefore interfere constructively, they must be in phase where line b meets both O_1 and O_2. For this to occur, the difference in the total path traveled by the diffracted rays must be equal to one wavelength. This is expressed as:

$$c - a = \lambda \tag{3}$$

where λ is the wavelength within the emulsion.

Looking at the geometry of the fringe structure, we see that:

$$a = c \times \cos\theta \tag{4}$$

where θ is the illumination beam angle within the emulsion: 25° in our example.

Combining equations we get:

Distance Traveled Between Adjacent Fringes

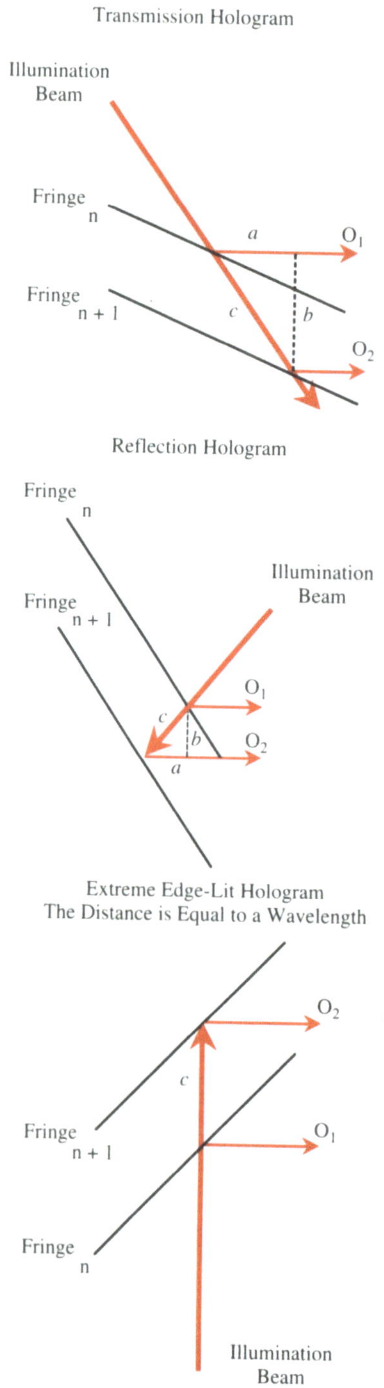

$$c = \frac{\lambda}{1 - \cos\theta} \tag{5}$$

Think of c as the distance traveled between fringe encounters. For our transmission example, c is 4.1 microns. Rays of the illumination beam, during their 7 microns emulsion crossing, will encounter only one or two fringes—the essence of virtual thinness!

When we consider our reflection hologram, we apply a similar logic, but note a key difference. Here for O_1 and O_2 to be in phase, the sum of paths c and a must equal one wavelength. Working through the same steps as in our transmission example, we arrive at:

$$c = \frac{\lambda}{1 + \cos\theta} \tag{6}$$

this gives us a distance between fringe encounters in our reflection example of about 0.2 micron. This makes perfect sense; the nearly orthogonal incidence of the illumination beam on the grating results in a transit between fringes nearly equal to the fringe spacing itself.

For our classical reflection hologram example, during its 7 micron emulsion crossing, the illumination beam will encounter approximately 35 fringes. Comparing this to the one or two fringes encountered in the classical transmission hologram example, we see the clear difference in their respective virtual thickness and virtual thinness, and recognize this as the basis of our characterization of distinction between transmission and reflection holograms.

So, now, what is the relative thickness or thinness of our extreme edge-lit transmission hologram, when we use a photopolymer emulsion ($n = 1.49$), and an intra-emulsion illumination beam angle of 80°? We apply the equation (5) above for transmission holograms to calculate the distance traveled between fringe encounters, and get a distance, c, of about 0.5 micron. Over the intra-emulsion path length for the illumination beam of 34 microns, a ray will encounter around 68 fringes! Clearly our extreme edge-lit transmission hologram has great virtual thickness.

Before stating a conclusion from our analysis of an extreme edge-lit hologram, let's look at one final point of interest. At the boundary of transmission and reflection holograms, where θ equals 90°, c equals λ using either the transmission or the reflection equation.

The hypothesis to be formulated from our theoretical examination of the extreme edge-lit region is that holograms in this region exhibit significant virtual thickness. We can anticipate that holograms recorded in this region will exhibit strong Bragg filtering, although this has yet to be established experimentally.

As a final consideration of anticipated filtering within the extreme edge-lit region, let's move the discussion from virtual thickness to actual thickness. Photopolymer emulsions which are the key to penetrating this extreme edge-lit region commonly have an emulsion thickness of around 12 microns—twice the thickness of typical silver halide emulsions. Doubling the actual thickness will double the intra-emulsion path length of the illumination beam, and double the number of fringes encountered. This reinforces our expectation that future experimentation in this extreme region will produce both transmission and reflection holograms with strong Bragg effects.

Those interested in a deeper mathematical treatment of thick holograms and volume effects are referred to the discussion in chapter 4 of Hariharan (1996).[vii]

Summary—penetration of the inaccessible zone

The inaccessible zone has been commonly penetrated to its intermediate limit of 68° intra-emulsion reference beam angles. In this intermediate region, some Bragg filtering effects are recognized, but not to the extent seen with classical reflection holograms.

The extreme region of the inaccessible zone is available through the use of recording media with indices of refraction lower than those of the glass substrates that support the emulsion. The significant virtual thickness of holograms recorded in the extreme region holds the promise of significant Bragg selection and the potential for full aperture transmission holograms.

This full penetration of the "inaccessible zone" alters our perspective of holograms. Where previously we thought of holograms as being classified into two types—transmission versus reflection—with very different optical properties, we can now consider holograms as a continuum. The location of any hologram on the continuum is determined by its grating structure, in correlation to its intra-emulsion reference beam angle. And the optical properties that we considered as distinctive to hologram type, can now be seen to be smoothly varying over the continuum in the transit from classical transmission to classical reflection geometries, and vice versa.

Conclusions

The application of the edge-lit geometry to holographic recording techniques is straightforward and adheres to the practices and mathematics of classical display holography. The most significant practical problem in recording edge-lit holograms is the formation of woodgrain fringes, and other spurious reflections that rob the recording layer of its diffraction efficiency. There are practical methods of mitigating this effect. This steep reference angle recording technique allows us to achieve intra-emulsion reference beam angles that had previously been inaccessible to holographers. This elimination of the inaccessible zone bridges the theoretical distinctions between reflection and transmission holograms, and especially with the particular benefits brought by photopolymer recording media, it is anticipated that the benefits of Bragg selection will be open to holograms in both transmission and reflection modes.

Readers interested in further study of this area may wish to consult additional writings by the author and his MIT colleagues.[viii,ix,x,xi,xii]

References

i. Leith, E. N., A. Kozma, J. Upatnieks, J. Marks, and N. Masseu (1966). "Holographic Data Storage in Three-Dimensional Media," *Applied Optics,* **5,** 8, pp. 1303-1311.

ii. Lin, L. (1970). "Edge Illuminated Hologram", *Journal of the Optical Society of America,* **60,** p. 714.

iii. Stetson, K. A. (1969). "An Analysis of the Properties of Total Internal Reflection Holograms," *Optik,* **29,** pp. 520-537.

iv. Nassenstein, H. (1969). "Interference, Diffraction and Holography with Surface Waves," *Optik,* **29,** pp. 597-607.

v. Upatnieks, J. (1992). "Edge-Illuminated Holograms," *Applied Optics,* **31,** pp. 1048-1052.

vi. Halle, M. W., S. A. Benton, M. A. Klug, and J. S. Underkoffler (1991). "The Ultragram: a Generalized Holographic Stereogram," *Proc. SPIE Practical Holography V,* **1461,** pp. 142-155.

vii. Hariharan, P. (1996). *Optical Holography: Principles, Techniques, and Applications,* Cambridge University Press, Cambridge, UK. Chapter 4.

viii. Birner, S. M. (1989). "Steep Reference Angle Holography: Analysis and Applications," S.M. Thesis, Department of Architecture, Massachusetts Institute of Technology, Cambridge, MA.

ix. Benton, S. A., S. M. Birner, and A. Shirakura, (1990). "Edge-Lit Rainbow Holograms," *Proc. SPIE Practical Holography IV,* **1212,** pp. 149-157.

x. Farmer, W. J., S. A. Benton, and M. A. Klug (1991). "Application of the Edge-Lit Format to Holographic Stereograms," *Proc. SPIE Practical Holography V,* **1461,** pp. 215-226.

xi. Farmer, W. J. (1991). "Edge-Lit Holographic Stereograms," S.M. Thesis, Program in Media Arts and Sciences, Massachusetts Institute of Technology, Cambridge, MA.

xii. Nesbitt, R. S. (1999). "Edgelit Holography: Extending Size and Color," S.M. Thesis, Program in Media Arts and Sciences, Massachusetts Institute of Technology, Cambridge, MA.

CHAPTER 19

Computational Display Holography

Wendy Plesniak, Ravikanth Pappu, John Underkoffler, Mark Lucente, and Pierre St.-Hilaire

Introduction

When Dennis Gabor invented holography in 1947, he was describing a new encoding and display method well in advance of the technology required for its execution. It was only in 1960, with Theodore Maiman's introduction of the pulsed ruby laser, and 1961, with Ali Javan's development of the continuous-wave Helium-Neon (He-Ne) laser, that the requisite source of coherent light finally became available, allowing the groundbreaking work in practical holography to be done by researchers like Emmett Leith and Juris Upatnieks.[i]

This early work inspired the idea of *computational holography*—of using numerical methods to simulate the physical processes underlying a real hologram's optical recording and reconstruction. The unavailability of spatial light modulation (SLM) devices or photographic emulsions meeting the daunting modulation and bandwidth requirements made writing and playback of such high-resolution computed patterns an initial impossibility; yet the optical mechanics to be mimicked via "fringe computation" were well understood from the start. And so, like optical holography awaiting the invention of the laser, computational holography remained for some time a discipline of theoretical promise but in practical limbo.

Clever techniques developed in the 1960s and onward applied esoteric variations on a Fourier theme to compute patterns that could be recorded using the limited media of the day and then reconstructed via Fraunhofer diffraction.[ii,iii,iv,v,vi] However, these structures were inappropriate for imaging dimensional, pictorial scenes with enormous information content, leaving the computation of synthetic holograms with the fidelity of their optical counterparts still a proposition fundamentally in search of a device.

The invention of a holographic video system by Benton *et al.* in the late 1980s[vii,viii] finally provided such a device: large, high-frequency interference patterns representing complicated three-dimensional scenes could be computed, written to the system and reconstructed for binocular viewing. The sudden availability of a general purpose "fringe output device," in turn, prompted rapid development of algorithms and approaches for computational display holography. With access to supercomputing resources and inspired by work of Leseberg (1987),[ix,x] Underkoffler in 1988 reported producing holographic images and short animated clips by simulating the interference of a monochromatic reference wave with light propagating from a computational object.[xi,xii] These first electroholographic images and movies were a historic step and demonstrated striking visual quality—bright, crisp, dimensional, and replete with many pictorial cues to depth and layout.

Electroholography's inception during a broad upswing of interest in *interactive systems* naturally influenced early directions; shortly after the first visually compelling images had been computed "off-line," image generation and update at rates close to real-time became a priority. Lucente (1993) reported the assembly of interference-modeled holograms using a table lookup scheme,[xiii] and extrapolated contemporary research in optically recorded holographic stereograms[xiv,xv] into the electroholographic domain.[xvi,xvii] Though the stereographic approach traded image quality for speed, this important work achieved landmark update rates and recast holography as an exciting and viable display technology for future interactive systems. Many subsequent contributions including algorithms for rapid hologram generation;[xviii,xix] methods for non-linear sampling,[xx] sample reduction[xxi] and holographic fringe compression;[xxii] hybrid computing methods;[xxiii,xxiv,xxv] experiments that included haptic interaction;[xxvi,xxvii] new display architectures that support full color and full parallax;[xxviii,xxix,xxx] and new technologies to support holography's required modulation, communication, and computation bandwidths[xxxi,xxxii,xxxiii] are propelling the field toward the mainstream.

In this chapter, we present a general overview of techniques, both historical and contemporary at the time of this writing, for computing display holograms. What follows intends to frame general approaches in the field and present specific examples pertinent to each; most of these examples were developed in Benton's laboratory at MIT over the decades he devoted to computational holography.

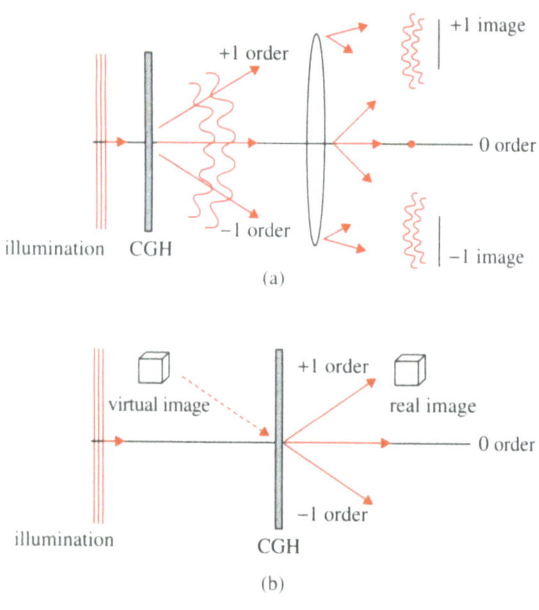

Comparison of Fourier (a) and Fresnel (b) playback geometries.

Fourier and Fresnel Holograms

Computed holograms traditionally fall into one of two classes: Fourier holograms, appropriate for far-field Fraunhofer diffraction, and Fresnel holograms, which produce images in the near field. The fundamental difference between these two diffraction models is the range of length scales in which they "operate." In particular, Fraunhofer diffraction requires that the monochromatic light source, the diffracting aperture, and the screen (or viewer) be sufficiently distant from each other that the paraxial approximation applies. The geometry of this requirement is captured by the Fresnel Number, defined by $F = a^2/L\lambda$, where a is the characteristic dimension of the diffracting aperture, L is the distance from the diffracting aperture to the viewer, and l is the wavelength of light illuminating the aperture. When $F \ll 1$, the optical system is said to be in the Fraunhofer zone; when $F \geq 1$, it is said to be in the Fresnel zone.

The mathematical framework that describes Fourier optics provides an accessible starting point for computing far-field holograms and underlies most early computing techniques. But the image-producing capability of such far-field holograms is not optimal for display holography; the images produced are flat, self-luminous, and clearly exhibit multiple orders (including the bright zero order) in the output. Further, they offer only modest parallax—in fact, there is very little "3-D" about them. Fresnel holograms, by contrast, offer greater capacity to produce realistic images for visual display; they can offer high angular and spatial resolution, can reconstruct deep, three-dimensional scenes, and can be generated so that only a single diffracted order is visible in the output. A comparison of the playback geometries of these two hologram classes is shown in the margin.

Computing Fourier Holograms

Historically (and still), generating a complex-valued holographic fringe pattern presents two difficult problems: first, simulating and computing the pattern with enough spatial resolution to produce the intended image, and second, writing the pattern to some material or device capable of recording its detail. Most early techniques computed holograms appropriate for Fraunhofer diffraction, where the reconstructed two-dimensional image $i(\mu, \upsilon)$ given by

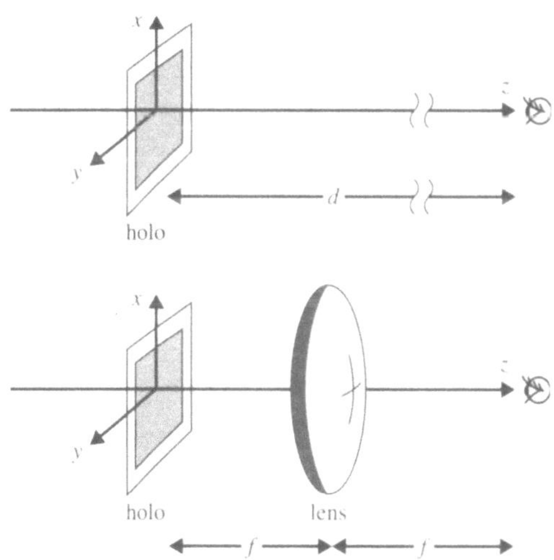

$$i(\mu,\upsilon) = \iint h(x,y)e^{j2\pi(x\mu+y\upsilon)}dxdy \tag{1}$$

is essentially the 2-D Fourier Transform of the hologram distribution, and with μ and υ in the reconstructed image corresponding to hologram coordinates x and y, scaled by a constant related to the reconstruction geometry. The holograms themselves were computed as inverse Fourier descriptions of the desired image amplitudes. The patterns were recorded using fairly low-resolution, grayscale or binary output devices, and then reconstructed in the manner shown in the illustration.

The computational *binary detour-phase hologram* developed by Brown and Lohmann (1969)[xxxiv] approximates the amplitude and phase of an object's inverse Fourier Transform with an $M \times N$ grid of binary-valued elements. Each element, corresponding to a single sample of the composite pattern, is an opaque cell of dimension $W \times H$ containing a rectangular aperture whose area, $w_{m,n} \times h_{m,n}$, is proportional to the magnitude of the Fourier coefficient it represents. The displacement $\Delta_{m,n}$ of the aperture's center from the cell's center encodes phase at that sample (see illustration in margin). At a plane in the far-field at z_0, the distribution $F(x, y)$ is given by the sum of diffracted fields $f_{m,n}(x, y, z_0)$ from each contributing aperture:

$$F(x,y) = \sum_{m=0}^{M-1} \sum_{n=0}^{N-1} f_{m,n}(x,y,z_0) \tag{2}$$

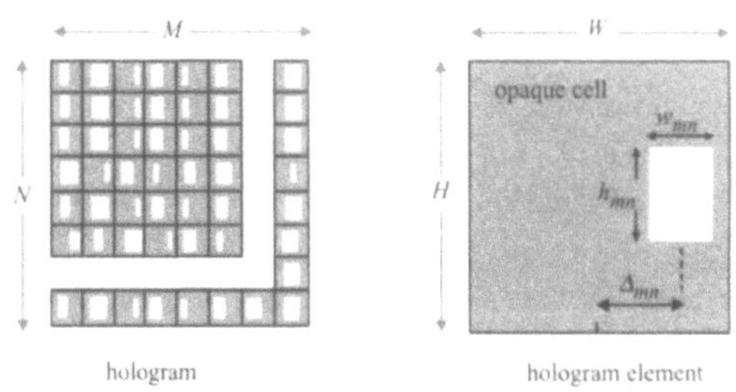

Binary detour-phase hologram (left) and a single cell (right).

If the hologram is reconstructed with collimated (plane) illumination, the $f_{m,n}$ can be described by

$$f_{m,n}(x,y,z_0) = E_0 w_{m,n} h_{m,n} \operatorname{sinc}\left(\frac{w_{m,n}x}{\lambda z_0}\right)\operatorname{sinc}\left(\frac{h_{m,n}y}{\lambda z_0}\right)$$
$$\times \exp\left[j\left(k_x + \frac{2\pi x}{\lambda z_0}\right)(mW + \Delta_{m,n})\right]\exp\left[j\left(k_y + \frac{2\pi y}{\lambda z_0}\right)(nH)\right] \tag{3}$$

where $k_x = 2\pi/W$ and $k_y = 2\pi/H$. In the far-field, where $w_{m,n}x$ and $h_{m,n}y$ are both $<< \lambda z_0$, this expression for $f_{m,n}$ can be given the following form:

$$f_{m,n}(x,y,z_0) = E_0 w_{m,n} h_{m,n} \exp\left[jk_x \left(mW + \Delta_{m,n} \right) + jk_y nH \right] \exp\left[j\left(x \frac{2\pi}{\lambda z_0} \left(mW + \Delta_{m,n} \right) \right) + j\left(y \frac{2\pi}{\lambda z_0} (nH) \right) \right] \quad (4)$$

Each aperture's area is related to the magnitude of its corresponding hologram sample by

$$w_{m,n} h_{m,n} = \left| h(mW, nH) \right| \quad (5)$$

and each aperture's offset is related to the sample's phase by

$$\Delta_{m,n} = \frac{W}{2\pi} \phi(mW, nH) \quad (6)$$

where $\Delta_{m,n} << \lambda z_0$. In the output plane z_0, the diffracted field from any one aperture is given by

$$f_{m,n}(x,y,z_0) = E_0 \left| h(mW,nH) \right| \exp\left[j\phi(mW,nH) \right] \exp\left[j \frac{2\pi}{\lambda z_0} (xmW + ynH) \right] \quad (7)$$

Superposing the diffracted fields from *all* apertures in the computed pattern produces the combined field:

$$\begin{aligned} F(x,y) &= E_0 \sum_{m=0}^{M-1} \sum_{n=0}^{N-1} \left| h(mW,nH) \right| \exp\left[j\phi(mW,nH) \right] \exp\left[j \frac{2\pi}{\lambda z_0} (xmW + ynH) \right] \\ &= E_0 \sum_{m=0}^{M-1} \sum_{n=0}^{N-1} h(mW,nH) \exp\left[j \frac{2\pi}{\lambda z_0} (xmW + ynH) \right] \end{aligned} \quad (8)$$

which expression describes the Discrete Fourier Transform of the hologram. Thus, if the hologram itself is computed as the inverse Fourier Transform of the desired final image, then the detour-phase encoding will produce the desired image upon replay.

Other methods similar to the binary detour-phase approach followed. Lee (1970)[xxxv] developed a method for representing a Fourier hologram by dividing each cell in the output plane into four different vertical subcells at laterally displaced locations. The subcells themselves represent phase angles of 0, 90, 180 and 270 degrees; in each hologram cell, an arbitrary phasor could be encoded with the assignment of opacity or semi-transparency to these subcells. Burkhardt (1970)[xxxvi] implemented an approach with only three apertures per cell, representing three mutually noncollinear phasors, and thus simplified Lee's method of encoding.

In holograms of this type, values are computed at sample locations, with the final patterns written on a printers or plotter (whose output mechanisms introduce a second level of "sampling") and then photographically reduced to obtain the high spatial frequencies required to diffractively produce images. The resulting images exhibit aliasing introduced by the printing process, which causes higher order images to reconstruct in lower diffracted orders. Other errors also diminish the quality of the final reconstructed image;[xxxvii] and the opacity of the overall mask reduces reconstructed image intensity.

Offering better image quality, the Kinoform[xxxviii] uses a similar encoding method but assumes unity magnitude of the Fourier coefficients (simulating a diffusely illuminated object) and modulates only the phase of these coefficients. Its computed grayscale pattern is then photographed and bleached to convert intensity variations into a phase relief pattern that matches the phases of the Fourier coefficients. An extension of the kinoform approach is the "referenceless on-axis complex hologram" (ROACH), which encodes both the amplitude and phase of the Fourier coefficients in separate layers of color film.[xxxix]

While detour-phase and kinoform-type holograms are useful as spatial filters in some imaging and image processing applications, they too produce effectively two-dimensional patterns far from the hologram plane. Attempts have been made to compute far-field holographic interference patterns of *three-dimensional* objects by "slicing" them in depth and stacking the slices' computed holograms, so that the far-field images of each slice form an image volume.[xl] Computing a "slice-stacked" hologram in this fashion is cumbersome, however, and the resulting images suggest heroism more than visual satisfaction.

Computing Fresnel Holograms

The difference in computational complexity between near-field and far-field holograms is substantial. Far-field holograms can easily be computed using 2-D Fourier transforms, a computationally expedient solution that can be accomplished using either fast algorithms or commercially available optimized hardware. Computing near-field holograms, which is mathematically equivalent to inverting a generalized scalar diffraction integral,[xli] admits no simple analytical approach. Instead, the vast majority of work on this topic investigates efficient numerical techniques for generating Fresnel fringe patterns.

Generally, computational approaches for Fresnel holograms fall into three categories: physically based techniques (interference modeling of scenes populated by holographic primitives such as points or lines, producing computed patterns similar in nature to optical holograms); holographic stereogram techniques (holographic encoding of 2-D parallax views, producing computed patterns similar in nature to optical holographic stereograms); and techniques which can be considered a hybrid of the previous two. The following sections describe some basic background on each (in the horizontal-parallax-only case) and provide a closer look at particular implementation techniques.

Full Parallax and Horizontal-Parallax-Only Holograms

Eliminating vertical parallax is a common practice in optical holography for reducing color blur in white light reconstruction. In computational holography, the elimination of vertical parallax has other benefits as well: it significantly reduces computational burden and can ease modulation, scanning, and other technical requirements on the display architecture as well. Of course, these benefits have tradeoffs: holograms without vertical parallax have constrained viewing

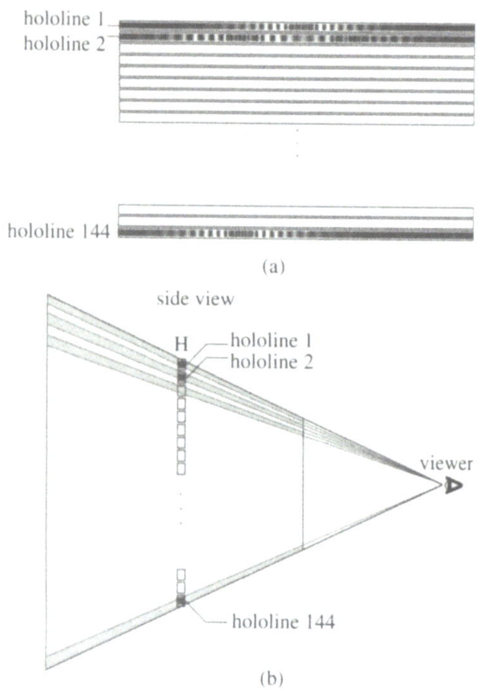

zones, astigmatic output, and diminished spatial resolution in the vertical direction.

Like their optically recorded counterparts, horizontal-parallax-only (HPO) computed holograms have diffractive power only in the horizontal (x) direction. They are comprised of a set of one-dimensional holograms arrayed in the vertical (y) direction ((a) in the figure), each a line hologram, or *holo-line* for convenience. Under collimated illumination along z, each holo-line reconstructs a planar "slice" of the image volume within its x-z plane, and all holo-lines collectively reconstruct the entire image volume. To optically direct the holo-lines' output planes toward the viewer's eyes so that the entire image volume is visible within a vertically constrained viewing zone, conditioning optics at the display output, such as a vertical diffuser in the plane of the hologram ((b) in the figure), may be used.

The techniques for computing holograms described here can be easily adapted to produce both full parallax and HPO display holograms; for the sake of simplicity, this chapter assumes the HPO case.

Physically Based Interference Modeling

Physically based techniques[xlii,xliii,xliv] represent three-dimensional scene geometry as a collection of optical primitives (often spherical or cylindrical emitters) that populate the edges or surfaces of a polyhedral model as shown in the margin. Their collective interference with a reference wave is computed at a set of discretized locations throughout the combined field to generate a hologram of that scene.

Holographic fringe patterns of three-dimensional pictorial scenes, even if "angularly modest," require representation at alarming spatial sampling rates. Computing full parallax holograms requires using the same high sampling rate both horizontally and vertically throughout the sampled plane, with the attendant N-squared escalation implied by allowing each primitive to contribute to the field accumulating at each sampled x,y location. Thus, as noted above, present practice often eliminates vertical parallax; such HPO schemes sample the holographic fringe pattern at an appropriately high frequency (*e.g.* 1000 mm^{-1}) in the horizontal direction, but at more conventional rates vertically (*e.g.* 2 mm^{-1}). At the same time, contribution to each one-dimensional holo-line is restricted to those "object points" that lie at the same vertical location (*i.e.*, in the horizontal plane containing that holo-line). The result, as with a "rainbow" or Benton hologram, is a reconstructed scene that exhibits visual parallax for horizontal movement of the viewer but none in response to vertical movement.

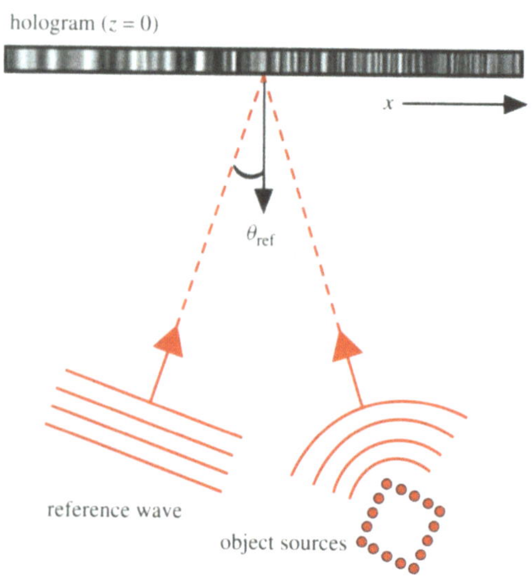

For HPO systems, the field radiated by a reference wave inclined at an angle θ_{ref} may be modeled as

$$E_{pl} = E_{pl_0} \exp\left[-j\frac{2\pi}{\lambda}(q_x x + q_z z) - j\phi_0\right] \tag{9}$$

where E_{pl0} is the wave amplitude, λ is the illumination wavelength, ϕ_0 is an arbitrary phase, and

$$q_x = \hat{q}\cdot\hat{x} = \sin\theta_{ref} \qquad q_z = \hat{q}\cdot\hat{z} = \cos\theta_{ref} \tag{10}$$

The field radiated by an individual spherical-emitting object source may be given as

$$E_{sph} = E_{sph_0} \exp\left[-j\frac{2\pi}{\lambda}[(x-x_0)^2 + (z-z_0)^2]^{1/2} - j\phi_0 \right] \quad (11)$$

with E_{sph0} as the wave amplitude, ϕ_0 as its associated initial phase (often randomized as a uniform deviate between 0 and 2π), and the location of the spherical emitter given by

$$\bar{r}_0 = (x_0, y, z_0) \quad (12)$$

These simplified models are derived from the field equations by ignoring both time dependence and propagation-dependent amplitude attenuation, and by assuming that all sources share an identical linear polarization.[xlv] Then, constructing an object wave as a sum of M spherical emitters and modulating it with an inclined plane wave produces a composite field given by

$$E_{tot} = \sum_{i=1}^{M} E_{sph_i}(x,z) + E_{pl}(x,z) \quad (13)$$

which can be evaluated at the hologram plane and used to produce an intensity pattern as follows:

$$I(x,z) \propto |E_{tot}(x,z)|^2 = E_{tot}(x,z)E^*_{tot}(x,z) \quad (14)$$

This intensity pattern is similar to that which film would (more continuously) record, if exposed at the same location to the same combined physical disturbance. A hologram can be thus be computed as the magnitude squared sum of object point wavefronts and the collimated reference wave, at sampled increments of Δx in the $z = 0$ plane.

Eliminating unwanted components of the diffracted field

An improved computational approach becomes apparent if the intensity pattern above is separated into its object and reference contributions; this reveals an implicit functional composition:

$$I(x,z) \propto |E_{tot}(x,z)|^2 = |E_{obj} + E_{ref}|^2 = \underbrace{|E_{obj}|^2}_{A} + \underbrace{|E_{ref}|^2}_{B} + \underbrace{2\,\mathrm{Re}\{E_{obj}E^*_{ref}\}}_{C} \quad (15)$$

Term A in Eq. (15) arises from the object points interfering with each other (and is thus called *object self-interference*). In traditional optical holography, the unwanted artifacts generated by this phenomenon can be spatially separated from the reconstructed image by using a reference beam angle several times larger than the angle subtended by the object at the hologram center. In computational holography, where a more shallow reference and illumination angle are used, this term often introduces unwanted image artifacts—but we have the option of not including it.

Term B in Eq. (15), called *reference bias*, represents a slowly spatially-varying bias that increases the intensity of the entire hologram. In traditional optical holography, the ratio between the object and reference beam intensities is carefully adjusted to optimize the modulation of the overall interference pattern, trading a low reference bias (and good fringe modulation) for weaker object self-interference fringes. In computational holography, again, we have the option of simply not including this bias term.

Finally, real-valued term C in Eq. (15) comprises the fringes required to reconstruct the desired image. This term fully describes the interference between the object point wavefronts and the reference wave, and without terms A and B, the calculation of these "useful" fringes can better fill the dynamic range available to the computed pattern.

The bipolar intensity method makes use of these observations;[xlvi] rather than computing the composite field from the complex wave interference, it computes only term C in the equation to produce the real-valued intensity distribution

$$I(x,z) \cong 2\,\mathrm{Re}\left\{E^*_{ref} E_{obj}\right\} = 2\,\mathrm{Re}\left\{E^*_{pl} \sum_i E_{sph_i}\right\} \tag{16}$$

This method not only simplifies computation of the intensity pattern, but eliminates object self-interference and reference wave bias. For an HPO computed hologram, this intensity pattern is computed for each holo-line by interfering the subset of object sources located within its xz-plane with a collimated reference wave.

The distance from a sample within a holo-line to its ith contributing spherical source is given by

$$r_i(x) = \left|\bar{r} - \bar{r}_{0_i}\right| = \left[(x - x_{0_i})^2 + z_{0_i}{}^2\right]^{1/2} \tag{17}$$

Using expressions for plane and spherical waves as given in (1) and (3) respectively, the interference pattern within any holo-line expands to

$$\begin{aligned} I(x,z) &= 2\,\mathrm{Re}\left\{\begin{array}{l} E_{pl_0} \exp(jkx\sin\theta_{ref} + j\varphi_0) \\ \qquad \times \sum_i E_{sph_{0_i}} \exp(-jkr_i(x) - j\phi_{0_i}) \end{array}\right\} \\ &= 2E_{pl_0} \sum_i E_{sph_{0_i}} \cos\left\{k[x\sin\theta_{ref} - r_i(x)] + \psi_{0_i}\right\} \end{aligned} \tag{18}$$

where $k = 2\pi/\lambda$, $z = 0$ in the hologram plane, $\hat{q}\cdot\hat{x} = \sin\theta_{ref}$, and ψ_0 describes the initial phases of the reference and i^{th} object waves bundled together. Finally, by setting E_{pl0} to eliminate any uniform bias, the final intensity pattern computed within each holo-line is given by

$$I(x,z=0) \cong \sum_i E_{sph_{0_i}} \cos\left\{k[x\sin\theta_{ref} - r_i(x)] + \psi_{0_i}\right\} \tag{19}$$

and the complete set of holo-lines collectively contains the HPO interference pattern generated by the entire scene.

Since populations of optical primitives—usually spherical emitting point sources—are used to represent underlying polyhedral geometry, each such source can inherit its associated polygon's actual or interpolated surface normal vector. Using this normal vector, an object source's amplitude $E_{sph_{0_i}}$ can be made to express the output of a computer graphics-style shading calculation, based on object surface parameters, texture or reflection maps, transparency or refractive properties, and specified light sources. To display lines and surfaces that appear solid and continuous, model geometry can be populated at densities that approach the arcminute resolution limit of human visual acuity[xlvii] (within the limit diffraction imposes on the size of a single point). Consequently, the images produced by interference modeling can have high angular and spatial resolution, can be strikingly realistic in appearance, can include rich pictorial information including occlusion,[xlviii] and can be free of visible diffraction artifacts.

Aliasing and over-modulation

Because computational holography is a sampled system—interference fringes are calculated at regular, discrete locations—it is necessary to prevent aliasing. The spatial frequency in the interference pattern produced by a plane wave and a single spherical emitter, for example, varies greatly across a typical sampled hologram's extent. It is thus possible for the spatial sampling rate of the system to be exceeded in some regions, which improper sampling, if ignored, leads to spatial aliasing and (upon optical reconstruction) image artifacts.

The most straightforward approach to preventing this undesired phenomenon is to locally disallow contributions from "object point" wavefronts at locations for which aliasing would otherwise occur. Treatment for the typical plane-wave-and-point-source case, whose intensity pattern is given by (19) above, proceeds by taking the spatial derivative of the cosine's argument to find an expression for the instantaneous, position-dependent spatial frequency:

$$\begin{aligned} f(x) &= \frac{1}{2\pi}\frac{d}{dx}k\left[x\sin\theta_{ref} - \sqrt{(x-x_0)^2 + z_0^{\,2}}\right] \\ &= \frac{1}{\lambda}\left[\sin\theta_{ref} - \frac{x-x_0}{\sqrt{(x-x_0)^2 + z_0^{\,2}}}\right] \end{aligned} \tag{20}$$

The range of "allowable" positions for which the spatial frequency is less than the maximum dictated by the system's sampling rate,

$$|f(x)| < f_{\max} \tag{21}$$

with the frequencies expressed as "cycles-per-unit-distance," is thus a simple window ($x_A < x < x_B$) outside which the accumulating field should not receive a contribution from the point source in question. The visual effect of bounding the field calculation in this spatial way is that individual three-dimensional elements of the image simply disappear beyond a certain angular vantage—that is, as the viewer moves her head before the reconstructed image, parts of it "wink out" past viewing limits to the right and left.

A second circumstance associated with the simulated recording of optical interference patterns concerns dynamic range and the digital representation of individual samples. A typical implementation of the field computation process might employ a pair of double-precision floating-point values (64 bits each) to represent the complex field accumulating at each location. After each of the participating primitives (point sources) has been allowed to make its contribution to the aggregate field, the resulting scalar fringe pattern's numerical values will be found to vary wildly, with detail spread across and concentrated in many orders of magnitude. However, the final repository for these calculated values is generally a frame buffer associated with the electro-optical output or reconstruction device, in which an individual fringe sample's representation is allotted only a single byte—eight bits. The crushing of these high-dynamic-range values into many fewer bits thus represents a potentially troublesome destruction of information, which can in turn manifest as image artifacts.

A similar issue obtains as hologram computation proceeds from line to line: a standard HPO approach dissects pictorial space vertically, by sequentially intersecting a stack of planes with the scene's three-dimensional objects and allowing only those optical primitives in a given plane to participate in the field calculation; each plane thus results in a single horizontal line of computed fringe pattern. Again, each such holo-line, represented initially as a sequence of high-precision numerical samples, must ultimately be normalized in order to "fit" in the lower-precision frame buffer. On the one hand, it might seem desirable to maximize diffraction efficiency for *each line* by allowing the normalized values to span the entire numerical range available at each buffer sample (say from 0 to 255) and thus, in turn, to use the full modulation depth provided by the reconstruction system. On the other hand, it may be visually prudent to normalize the entire "stack" of holo-lines simultaneously and with reference to the maximum value found in the full collection—in which case many individual holo-lines' fringes will be necessarily "underserved" by the numerical range of the frame buffer's samples.

The issue of modulation and dynamic range normalization is complex, and remains one of open research. Appropriate solutions must at least provide psychovisual equivalence: "corrected" results should produce images indistinguishable from (or superior to) "pre-correction images." One approach involves adaptively "jittering" the positions of individual object primitives by distances microscopic enough to incur no image difference but that result in an overall reduction of line-by-line fringe pattern dynamic range. An analogous technique that adjusts the arbitrary phase of each primitive (while leaving its position unchanged) also shows promise.

Accelerating computation

While a hologram computed via physically based modeling exhibits high resolution and accurate reconstruction throughout the scene volume, computation can be slow and the full computation must be done afresh whenever changes occur in the scene texture, shading, position, orientation, scale or geometry. Algorithms that make this computation more efficient include the use of difference calculations[xlix] and table lookup.[l]

In a table lookup method, a set of interference patterns, called *basis fringes*, are pre-computed, stored, and subsequently used to rapidly assemble the hologram. Basis fringes are generated by simulating the interference between a primitive (like a spherical emitting point) with a collimated reference wave; each fringe's primitive is located at the same *x-y* coordinate but at a different sampled distance from the hologram. Each of the resulting set of basis fringes is capable of reconstructing an image at the same *x-y* position but at a different z-depth, and the depth increment is chosen to sample that dimension sufficiently for a viewer. Using a collimated reference wave in the calculation enables a few simple rules to be employed for hologram assembly: to reconstruct a primitive's image in a different x-y location, the basis fringe is just translated an appropriate amount on the hologram; to reconstruct a primitive's image at a negative depth location, a basis fringe's samples can be mirrored on the hologram. In this manner, instances of the pre-computed fringes can be positioned and accumulated into the hologram as prescribed by the (x, y, z) locations of the individual primitives populating the scene, and the composite pattern can be normalized to fit the dynamic range of the display.

Incremental computing employs this kind of pre-computed table to render fast incremental changes to a holographic image.[li] The method is useful when smaller, localized changes to the image are required; appropriate local changes are made to the hologram instead of reassembling it entirely. In incremental computing, an initial hologram is generated by combining all basis fringes needed to reconstruct the combined field. An un-normalized version of the hologram is also maintained from which basis fringes may be incrementally subtracted and added: the images of object sources in the scene are *erased* by subtracting the basis fringes that represent them, and the image can be updated by incorporating new image-producing fringes. Only the parts of the hologram affected by scene changes are modified, and modifications are effected by simple operations.

Computer-Generated Stereogram Modeling

Computed holographic stereograms angularly multiplex 2-D computer-generated or optically captured parallax views of a scene. Computed holographic stereograms usually sample and display scene parallax more coarsely than traditional interference-simulated holograms would, and the same interference patterns can be used to encode any set of appropriately captured 2-D parallax views. These characteristics imply several notable advantages. First, capturing a scene optically or with computer graphics techniques and encoding them with a pre-computed pattern is generally more convenient and much faster than performing physically based modeling with the scene geometry. Second, with the benefit of modern computer graphic rendering and photographic techniques, carefully or subtly illuminated scenes replete with textures, reflections and material surface properties are much easier to incorporate. Third, coarsely sampling scene parallax helps to reduce the amount of information in each hologram, and both rendering hardware and image compression techniques can further reduce computation time and overall bandwidth requirements. Finally and perhaps most importantly, dissociating fundamental diffractive behavior from image content allows a single set of diffractive primitives to be pre-computed and repurposed to display any scene; to update a stereographic frame, only a set of parallax view pixels must be generated and used to modulate the pre-computed basis fringes.

Each pre-computed basis fringe comprises spectral content necessary to diffract input light in a different specific direction and uniformly through some small angular extent. In general, a set of N_b

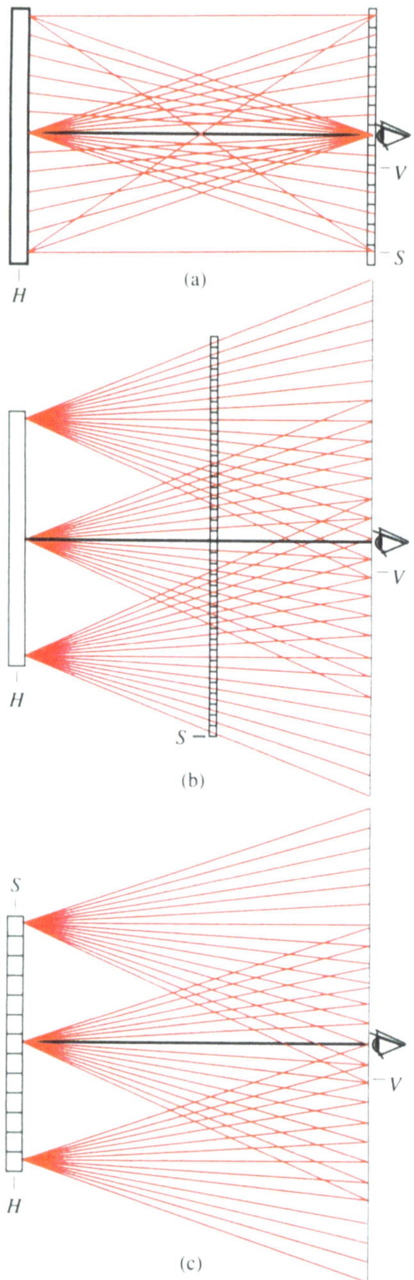

basis fringes can be used to angularly multiplex a set of N_p parallax views in a variety of ways: for instance, each view may be projected in a different and collimated direction; or views may be projected according to a more specialized prescription.

With reference to the parlance of the literature on optically recorded HPO holographic stereograms, it is useful to consider computed holographic stereograms as having a *hologram plane H* (the plane on which the hologram elements are written), a *slit plane S* (the exit pupil plane through which view pixels are projected), and a *viewer plane V* (the plane at which a human observer is located). Computer generated optically recorded holographic stereograms can arrange these planes in a variety of formats, as illustrated in the adjacent figure. However, in computed holography it is most common to co-locate the hologram plane and the slit plane as shown in (c), often with the number of parallax views equal to the number of basis fringes ($N_b = N_p$).

Conventional computer generated holographic stereograms

In HPO displays, the input parallax views and the final computed stereogram exhibit vertical resolution limited by the number of hololines N_v on the electroholographic display. In the horizontal dimension, however, conventional computed stereograms are comprised of an array of tiled chunks, each of which corresponds to a slit on the hologram plane and contains the same number of samples as a single basis fringe. A chunk, or hologram element (often called a *hogel*),[lii] is a linear combination of basis fringes, each weighted by an appropriate view pixel value. The hogel is designed to project a set of view pixels—each in a different direction—throughout the hogel's spatial extent. Resulting hogels are tiled on the hologram plane with their apertures abutting, and each is seen to "light up" with different pixel values when reconstructed and viewed from different angles within the viewzone.

To deliver the 2-D view pixels well, so that the image appears equally bright throughout the viewzone and exhibits minimal artifacts, the basis fringes are designed to satisfy a set of spatial and spectral constraints for a given reconstruction geometry. The spatial amplitude constraints ensure a fairly uniform level of diffraction throughout the hogel's extent and minimize hologram-plane artifacts. Generally, spatial phase has no direct constraint; however, constraining the endpoints of each basis fringe can serve to minimize inter-hogel phase discontinuities. Spectral constraints are most important and are chosen to perform specific diffractive tasks. For stereograms, each basis fringe diffracts light through an exit pupil at the viewzone distance, calling for a rectangular spectral amplitude of the appropriate width; if the basis fringes diffract light to a region that is too narrow, there will be gaps in the viewzone. Spectral phase (in most cases) is only indirectly constrained by the need to compute reasonably smooth spectral amplitude.

Basis fringes can be iteratively pre-computed using methods that attempt optimized satisfaction of the set of spatial and spectral constraints. Details of the design of constraints and these iterative nonlinear optimization methods are described by Lucente (1994).[liii] Methods include simulated annealing, genetic algorithms and the iterative-constraint methods used for phase retrieval. When applied to basis fringe pre-computation, this class of nonlinear optimization algorithms comprises several basic steps: (1) an initial random guess at a basis fringe (spatial phase); (2) random (or semi-targeted) alterations in the fringe; (3) a Fourier transform to analyze both spatial and spectral properties; (4) a quality function used evaluate the relative success; and (5) iteration until the basis fringe sufficiently satisfies the constraints. These algorithms are often slow, but are performed only once for a given holographic imaging system. In general, there is no closed-form algebraic method for calculating fringes that satisfy an arbitrary set of constraints.

Specifying stereogram parameters for the human viewer

In addition to adequately sampling a holographic fringe to avoid aliasing, setting stereogram amplitude and frequency sampling requirements to satisfy the spatial and angular resolving capabilities of the human viewer is also important in display holography. To generate output optimized for a human viewer, an HPO computed hologram must adequately sample (both spatially and spectrally) a continuous one-dimensional interference pattern $I(x, z, f)$ where z is the plane of the hologram, x denotes a location on the hologram, and f is the spatial frequency represented there. For holographic stereograms, which combine parallax views with an interference pattern that projects them, both the input parallax views and their angular projection are sampled. These sampling rates are not independent; thus, determining optimal spatial and angular sampling rates often involves some compromise between scene depth and resolution.[liv]

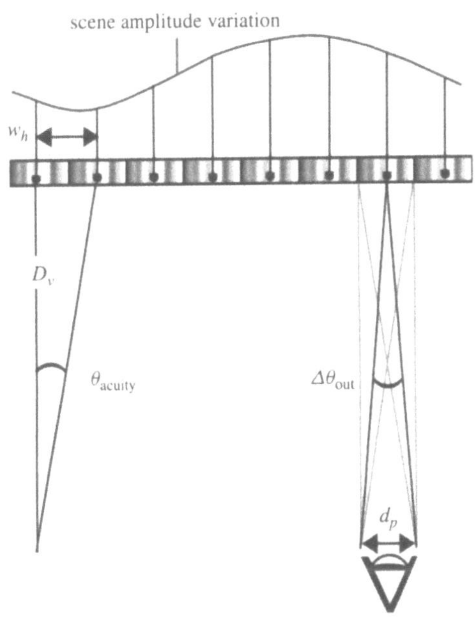

A simple geometric approach may be taken to determine reasonable stereogram sampling parameters. When the slit plane is located on the hologram plane, the parallax view sample spacing is equal to the hogel spacing w_h. Assuming that the most rapid perceptible amplitude variation within a parallax view is one minute of arc, the width of a view pixel, equal to the width w_h of a hogel in a conventional computed holographic stereogram, need be no larger than

$$w_h = p_{opt,x} = D_v \tan(1/60)^\circ \tag{22}$$

in order to be optimal for the human viewer positioned a distance D_v from the hologram. Since the sampling frequency must be more than twice the highest spatial frequency in the parallax views, they should be band limited using a filter with cutoff frequency $f_{max,\ amplitude}$

$$f_{\max,amplitude} \leq \frac{1}{2w_h} \tag{23}$$

This satisfies the Nyquist limit. Such band limiting of the parallax views may be accomplished during view capture by using a real or synthetic camera aperture at least $2w_h$ in diameter.[lv]

Within any hogel, the spectral sampling increment Δf used to generate the set of basis fringes must be chosen to encode the varying diffracted angular output of that region. We assume that a viewer's ability to perceive angularly varying phenomena (like changing occlusion relationships or specular reflections) is limited by the eye's pupil diameter d_p. To determine a Δf optimized for a human viewer, the grating equation may be used to relate angle of diffracted output θ_{out} to spatial frequency and the angle of incident illumination θ_{ill},

$$f\lambda = \sin\theta_{out} - \sin\theta_{ill} \tag{24}$$

For an input illumination $\theta_{ill} = 0$, and a maximum angular deviation $\Delta\theta_{out}$ given by

$$\Delta\theta_{out} = \tan^{-1}\left(\frac{d_p}{D_v}\right) \tag{25}$$

An optimal increment in spatial frequency $\Delta f_{opt,\ spectral}$ that will meet the spectral sampling requirements for a viewer positioned at D_v is given by

$$\Delta f_{opt,spectral} = \frac{d_p}{\lambda}\left(d_p^{\ 2} + D_v^{\ 2}\right)^{-\frac{1}{2}} \tag{26}$$

The number of basis fringes required to achieve this angular resolution across the stereogram's field of view θ_{view} is $\theta_{view}/\Delta\theta_{out}$. If each basis fringe is to angularly multiplex unique parallax information into the viewzone, $\Delta f_{opt,\ spectral}$ also implicitly defines an upper bound for spacing between capture cameras Δ_c; thus,

$$2w_h \le \Delta_c \le d_p \tag{27}$$

These "ballpark" spatial and spectral sampling requirements for the human viewer are illustrated in the margin. In addition, the hologram must contain spatial frequency content high enough to produce an angle of view that admits both an observer's eyes in the viewing zone (at minimum). These are some of the human factors benchmarks against which the modulation transfer function (MTF) of a computed stereogram can be evaluated.

Holographic stereogram parameters: partial coherence theory

The previous discussion uses a purely geometrical description to introduce the basic elements of holographic stereographic imaging. If we want to describe in detail the effects of recording geometry parameters on the final image, however, we need to take into account the wavelike nature of light. Such a "physical optics" treatment of holographic stereograms, based on partial coherence theory, is presented by St.-Hilaire (1994)[lvi] and (1995).[lvii] St.-Hilaire (1997) further extends the treatment to arbitrary stereogram geometries and viewer positions.[lviii]

Like the corresponding geometrical interpretation, a physical optics treatment of stereograms starts with the consideration that the planes of interest in the previous section (that is, the *hologram plane*, the *slit plane*, and the *viewer plane*) define a series of pupils through which the light propagates (the very last pupil being the observer's iris itself). It can be demonstrated that the discrete sampling of individual perspectives is equivalent to introducing a quadratic phase error across these pupils.[lix] This quadratic phase error corresponds to a defocusing error in the case of a conventional imaging system. Thus the discrete sampling operation results in a blurring akin to misfocus as the focal plane is moved away from the hologram plane. It should be noted that this effect is fundamentally different from the degradation observed when illuminating a conventional hologram with a spatially extended or non-monochromatic light source. Indeed, it constitutes the principal limitation of stereograms with respect to their phase coherent counterparts.

The correspondence between the phase error introduced by perspective sampling in a stereogram and misfocus in a conventional imaging system allows use of such standard analysis tools as the modulation transfer function (MTF) to study the behavior of stereograms. It is also useful in helping to derive the parameters that will maximize resolution, and to design appropriate filters to minimize sampling artifacts.[lx] In particular, it can be found that an optimum slit size results in a maximum resolution over the whole imaging volume. This situation is equivalent to adjusting the *f*-stop of a conventional camera to satisfy the hyperfocal condition.

Assembling conventional holographic stereograms

In the common implementation in which $N_b = N_p$, each hogel, with aperture size and spectral content satisfying the conditions above, is generated as a linear combination of pixel-weighted basis fringes: corresponding pixels from each of the N_p parallax views (*i.e.*, the first pixel from the first row in each image) forms a structure called a *pixel vector*. Each element of a pixel vector multiplies a basis fringe in the pre-computed set, and their sum is accumulated. The entire HPO hologram has N_v hologram lines containing a series of N_h hogels, each given by

$$\mathbf{h}_j = \sum_{i=1}^{N_p} p_{i,j}\mathbf{b}_i \quad (1 \le j \le N_v N_h) \tag{28}$$

where $p_{i,j}$ are pixel weights expressed by the pixel vector and $\mathbf{b}_i$ are basis fringes.

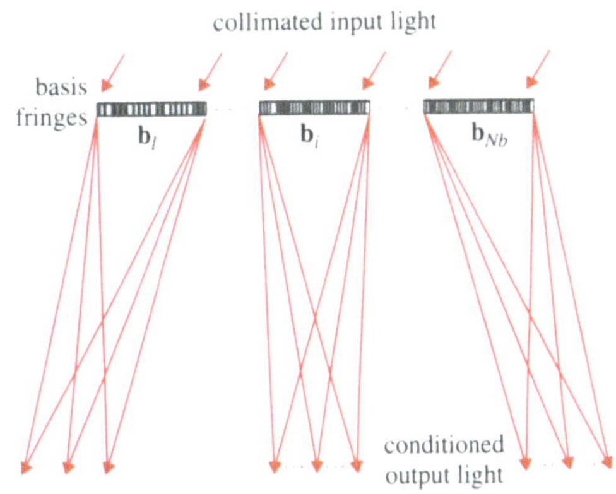

In early work at MIT,[lxi] the number of basis fringes was usually made equal to the number of parallax views, and each view was projected from the hologram plane in a different direction and in a collimated fashion as illustrated in the margin. $N_b = N_p = 32$ basis fringes were used, each with $B = 1024$ samples. N_p monochromatic parallax views were generated as input to stereogram computation, each of dimension $w_i = 256$ by $h_i = 144$ pixels. Pixel vectors formed from these parallax views were used to modulate the set of basis fringes, creating the 256 × 144 hogels, each roughly 0.6 mm wide, which tiled the computed holographic stereogram.

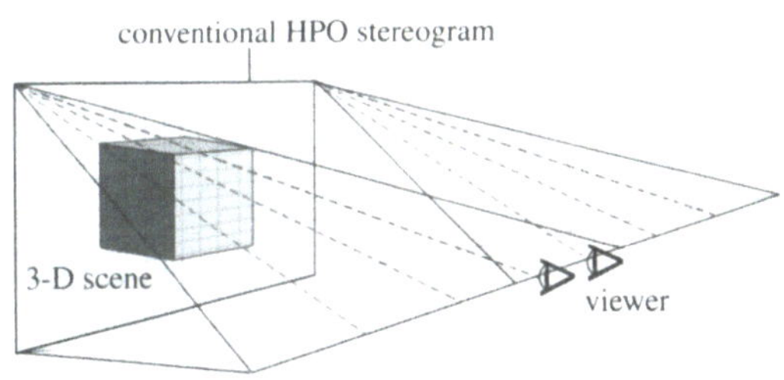

This work produced images with landmark speed from a wide variety of photographic and computer rendered input, and increased excitement about the viability of holographic video systems. However, the technique also exhibited a number of visual shortcomings that did little to justify the use of holography over, say, more commonplace stereoscopic techniques in spatial display work. Images had noticeably lower spatial resolution than those produced by interference modeling, and had spatial and angular resolution similar to other pixel-based stereographic displays.

Since hogel aperture width is physically related to the diffracted resolution of its output, a compromise that minimizes hogel aperture size while maximizing angular resolution during the synthesis of basis fringes is required, as noted previously. This basic tradeoff leaves its mark on image quality. The technique can also introduce visual artifacts. Phase discontinuities at the hard boundaries between adjacent hogels cause diffraction artifacts in the reconstructed light field, which manifest as visually distracting dark vertical bands in the image. Further, given the astigmatic optical projection of the component parallax views and the HPO nature of the display, the parallax view capture geometry should match the hologram viewing geometry by exhibiting orthographic character in the horizontal direction but perspective character vertically, as shown here in the sketch. Yet standard perspective rendering is usually employed rather than this unconventional capture geometry, resulting in output that appears hyperstereoscopic and anamorphically distorted.

Nonetheless, the synergy between computer graphics rendering, digital capture techniques, and holographic stereogram computing remains potentially great. The basic multiply-and-accumulate (MAC) operations at the heart of stereogram computation are fast and well matched to hardware implementation—suggesting that a platform for generating holographic stereograms could be efficient enough for use in real-time applications.[lxii]

Diffraction-Specific Modeling: a Hybrid Technique

Computing conventional holographic stereograms is in fact a special case of a more general approach called "diffraction-specific computing."[lxiii] This approach applies to forms of computed holography ranging from conventional holographic stereograms to physically-based interference modeled holograms, and may thus be regarded as a hybrid approach. The diffraction-specific approach considers only the reconstruction step in holography rather than the generation of fringes through simulation of optical interference. It produces a spatially and spectrally sampled treatment of a holographic fringe and therefore of its associated diffracted wavefront.

Diffraction-specific fringe computation has the following four features:

1. *Spatial discretization:* The fringe is treated as a regular array of functional holographic elements (hogels). For example, in HPO holograms, a holo-line is a catenation of evenly spaced hogels with center-to-center spacing w_h, each comprising on the order of hundreds of samples.
2. *Spectral discretization:* A *hogel vector* is a sampled spectral representation of a hogel. Each component represents the spectral energy within a small range of spatial frequencies, each distributed throughout the useful spectrum.
3. *Basis fringes:* Pre-computed basis fringes combine to convert hogel vectors into physically useful fringes. Each basis fringe represents a portion of the fringe spectrum and is pre-computed with appropriate sample spacings.
4. *Rapid linear superposition:* A hogel fringe is computed through the rapid real-valued linear superposition (modulation) of the pre-computed basis fringes, as specified by hogel vectors.

Applied to the computation of conventional holographic stereograms, hogel vectors are normalized pixel vectors (as described previously); basis fringes are different for each hogel location, and must be pre-computed to diffract light to the array of exit pupils at the specified viewzone distance. The general approach, however, is analogous to the conventional stereogram approach described previously, which places the viewzone at infinity (using no exit pupils); this simplifies fringe computations in many ways. First, the same set of basis fringes can be used for all hogels. Second, and most profound from the perspective of sampling theory, this is equivalent to spectral sampling of the diffracted hologram wavefront.

This general diffraction-specific approach allows for the direct sampling of hogel spectra by generating orthographic views during the rendering step. These views can be assembled (as previously described) into pixel vectors, normalized, and used as hogel vectors to compute hogel fringes. Each hogel is computed based on this sampled representation of the spectrum required to diffract light in specific amounts for each of the discretized directions. An efficient application of diffraction-specific computation applies the standard rules of sampling theory: given sufficient levels of spatial and spectral sampling, a diffracted wavefront can be reproduced to generate the desired 3-D image.

Hogel vectors can be generated by normalizing pixel vectors, analogously to the conventional stereogram approach described above. However, diffraction-specific computing can also generate hogel vectors using the (x, y, z) locations and amplitudes of object *elements* within a scene. In this approach, a *diffraction table* is used to describe the mapping of (x, y, z) locations to specific hogel-vector contributions. An image element at some (x, y, z) location in the image volume can be represented in a sampled manner as components of a particular set of hogel vector contributions. A diffraction table maps this sampled relationship. Each entry of the diffraction table is an amplitude factor, which can be multiplied by the desired element magnitude and then summed with other contributions to calculate the total hogel-vector array for a specific 3-D scene. Diffraction table entries are pre-computed for all possible (x, y and z) values of image element; pre-computation involves spatially and spectrally sampling (with the same spacings used to pre-compute the associated basis fringes) the theoretical spectral content of a diffracted wavefront. The spectrum is related to an image element through optical propagation. For example, for an image point, the desired diffracted spectral content can be shown to be a roughly linear distribution of plane-wave spectra.[lxiv]

During computation for a particular 3-D object scene, the (x, y, z) location of each image element serves as an index into the diffraction table, retrieving a set of mapped magnitudes which are then scaled by the element's magnitude and summed to calculate the total hogel-vector array. For image points, the magnitude is the square root of the intensity calculated from the 3-D scene and lighting information. To compute fringes, the hogel vectors are used to modulate pre-computed basis fringes and accumulate specific superpositions for each hogel. Depending on the particular basis fringes used and the object elements represented in the diffraction table, the wavefront diffracted by the resulting fringes may range from being similar to one projected by a conventional holographic stereogram, to being similar to one projected by a physically based interference

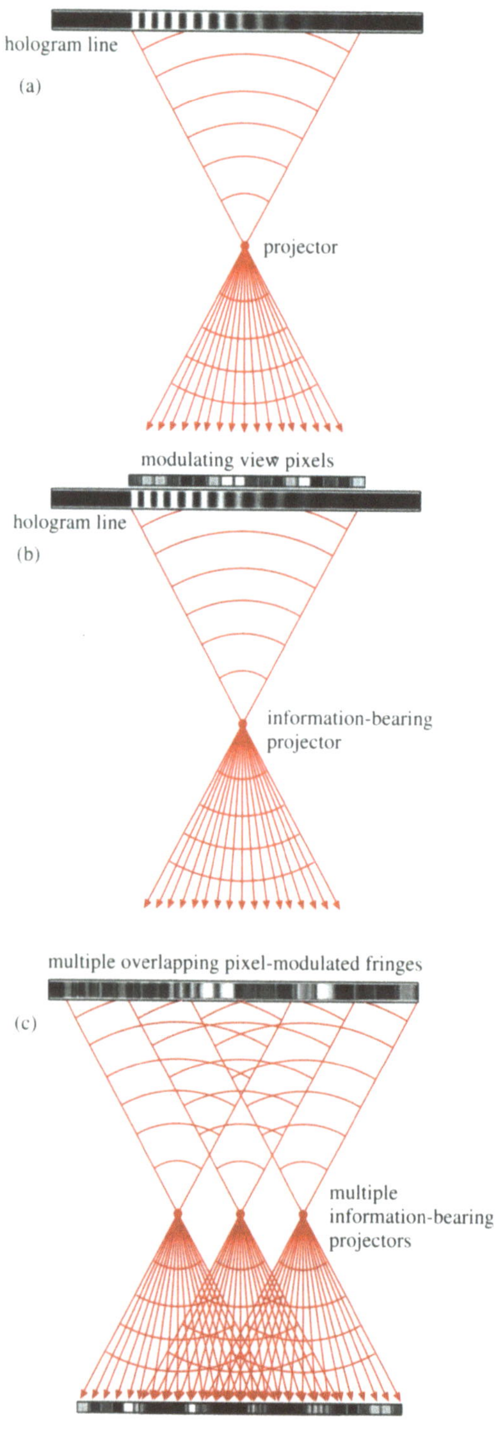

modeled hologram.

Diffraction-specific fringe computation can be viewed as a form of bandwidth compression. Because a set of hogels represents a spatially sampled fringe, each hogel vector contains the information representing one hogel (*i.e.*, its descretized spectrum). A 3-D scene encoded as sampled diffraction specifications (*i.e.*, an array of hogel vectors) requires less bandwidth because hogel vectors are more compact than physically modeled fringes. This "compression ratio" results in a smaller fringe representation and a proportionally faster computation speed; it also reduces image resolution by increasing image blur—a tolerable trade-off in some imaging situations. Diffraction-specific techniques reduce the implementation of the most computationally intensive step—hogel-vector decoding—to a large number of basis fringe modulations, which are simply MAC operations.

The diffraction-specific method has been used thus far to generate images for a variety of electro-holographic display architectures. One implementation drives a display in a Fourier geometry, *i.e.*, incorporates a Fourier-transform lens after the modulator. Another implementation shows good results for full parallax 3-D images.[lxv] This work also exploits the simplicity of hogel-vector decoding by implementing the multiply/accumulate operations on a variety of hardware, including clusters of GPUs, FPGAs (field-programmable gate arrays) and ASICs (application-specific integrated circuits).[lxvi,lxvii]

A Related Hybrid Technique: Reconfigurable Image Projection (RIP) Holograms

Other hybrid methods have been used to generate images with quality that approaches or matches that of interference modeled holograms;[lxviii,lxix] these implementations often construct the hologram plane using abutting hogels. While this tiled arrangement of hologram elements can prove useful for some parallelized computational pipelines and display architectures, it is not strictly necessary. Particularly when the slits are located away from the hologram plane, abutting hogels can be a limiting construct.

Reconfigurable Image Projection (RIP) holograms[lxx,lxxi] allow individual basis fringes to be windowed to minimize diffraction artifacts at their boundaries, modulated by view pixels, and overlapped on the hologram plane. The superposed hogels collectively produce a slit plane (or more complicated projection surfaces) in space, off the hologram plane. This approach is flexible and offers many advantages: it combines the tremendous speed, efficiency, and flexibility of stereogram computing, with image quality that can match the metric accuracy and realism of interference-modeled holograms. The technique supports embodiments ranging from stereograms with single or multiple slit planes or variously-shaped slit-surfaces, to physically-based holograms where slits correspond to the object primitives themselves.

RIP holograms project one or more series of parallax views of a 3-D scene through one or more holographically reconstructed projection (slit) surfaces. Projection surfaces are populated by a collection of holographic primitives reconstructed some *(x, y, z)*-distance offset from the hologram surface. Each holographic primitive behaves as a projector that can relay parallax view information to a "sweep" of locations in the viewzone. Holographic primitives may be points, lines, microfacets or other primitives with spatial and projective characteristics that vary according to their modeling.

A holographic primitive is encoded within the hologram by an instance of a pre-computed basis fringe. Depending on the desired diffractive behavior, basis fringes may be generated by simulating interference between an object wave with the primitive's desired projective behavior and a reference wave that matches the reconstructing wave, or by some other technique. If a simple isotropic spherical emitter is interfered with a plane wave reference for example, its reconstructed holographic primitive will radiate uniformly through some angle of view ((a) in the figure). The spatial and angular resolutions of a projection surface populated by these holographic primitives can match the surface population density and diffractive resolution found in interference-modeled holograms.

As in conventional stereogram computing, basis fringes can be modulated by a parallax view pixel vector describing one or a set of view pixels' values across their angle of view; and thus modulated basis fringes may also be considered hogels. Depending on the method of view generation or capture, the view pixel vector may be assembled from a "slice" through the volume of captured parallax views, from a more complicated indexing scheme, or directly by a specialized renderer,[lxxii] eliminating altogether the need to pre-render and index views.

The resulting hogel reconstructs an information-bearing holographic primitive that relays the view pixels back out along the direction of original capture ((b) in the figure). Hogels are subsequently accumulated into the final RIP hologram ((c) in the figure), and so their manner of assembly and angular multiplexing behavior are much like those of conventional stereograms. Upon reconstruction, a static or moving observer within the viewzone sees with each eye an appropriate, disparate view of the reconstructed scene. This technique borrows from both interference-modeled holograms and conventional computed stereograms to provide a high-quality and configurable light field projector.

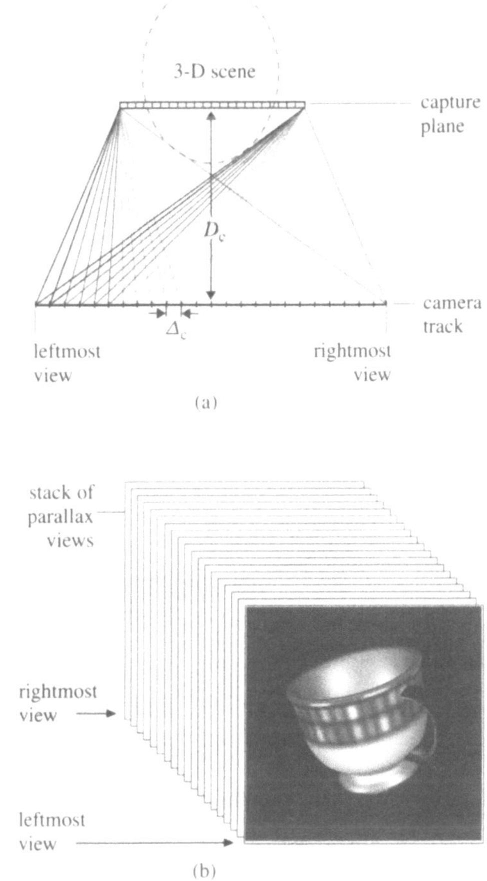

(a) Rendering a set of parallax views using a capture geometry that corresponds to the hologram projection geometry, and (b) the resulting stack of horizontal parallax views.

In the simplest case, each in the set of HPO basis fringes $f_i(x)$ used for RIP hologram computing can be derived from modeling the interference of an inclined plane wave and a spherical wave whose source is located at $x = 0$ and distance $z = d_{o_i}$ from the hologram plane. The same prescription given in Equation (19) may be used with amplitude $E_{sph0i} = 1.0$ and initial phase $\psi_{0_i} = 0.0$ to describe the signal-bearing part of this interference pattern:

$$f_i(x) = \cos\frac{2\pi}{\lambda}\left(r_i(x) - x\sin\theta_{ref}\right) \; for \quad \left(d_{o_i}\tan\theta_{out,\max}\right) \le x \le \left(d_{o_i}\tan\theta_{ill}\right) \tag{29}$$

where $r_i(x)$ is given in (6) for $z = d_{oi}$. The expression is computed with $\theta_{ill} = \theta_{ref}$, and for values of x that allow the basis fringe's spatial frequency to vary from 0 to some f_{max}. To prevent visible diffractive artifacts generated by abrupt phase discontinuities at its edges, each basis fringe $f_i(x)$ may be windowed by a filter with passband width W_i

$$W_i = d_{o_i}\left(\tan\theta_{ill} + \tan\theta_{out,\max}\right) \tag{30}$$

The ith windowed basis fringe, $f_{win_i}(x)$, will reconstruct an image of a point at $z = d_{o_i}$ from the hologram, projecting into the viewzone through an angle

$$\theta_{win_i} = \theta_{out,\max} + \theta_{ill} \tag{31}$$

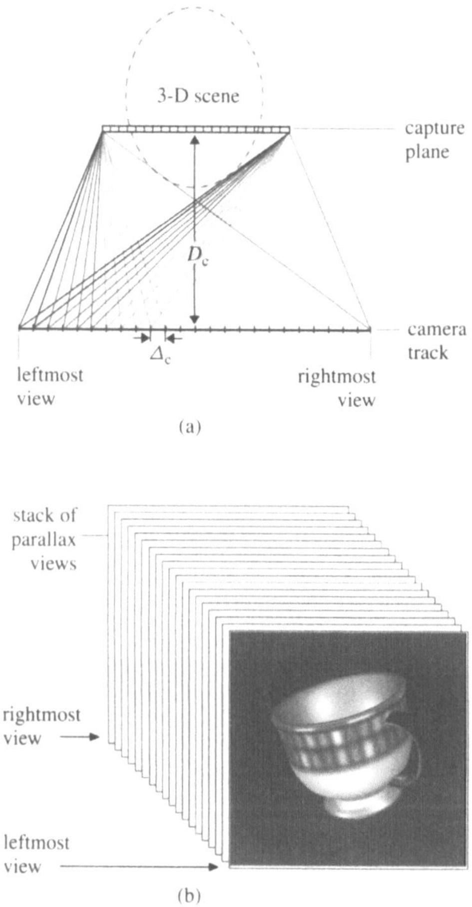

RIP hologram assembly: (a) indexing parallax views and (b) accumulating view-modulated basis fringes into the hologram

This image, the resulting holographic primitive, acts as a simple light projector, and a densely populated collection of such holographic primitives comprises the projection plane or surface. If the projection surface is planar and located at $z = d_o$, translated instances of a single windowed basis fringe, accumulated into the hologram, can be used to populate holographic primitives over one entire projection plane. If the desired projection surface is tipped with respect to the hologram plane, is multi-planar or is non-planar, instances from a set of pre-computed basis fringes will be required. Basis fringes that model holographic primitives with other spatial and projective characteristics may be used as well.

For the simplest case, in which a single projection plane is produced, a sweep of parallax views can be rendered using a capture geometry that corresponds to the RIP projection geometry (one type is illustrated in the margin). These parallax views can be used to modulate windowed basis fringes, which are, in turn, accumulated into a RIP hologram. In this process, the basis fringe corresponding to each of the holographic primitives on the projection plane is modulated by an appropriate view pixel vector **p** which is extracted from the parallax view volume of dimension M pixels wide by N pixels high by K parallax views. The method of selecting the appropriate pixels to modulate each basis fringe depends on way the parallax view volume is formatted, and thus on the view capture or generation method that created it. If the parallax view volume with $M = N_h$ and $N = N_v$ created by a shearing-recentering perspective camera is considered (see illustration, next page), the number of samples n within any pixel vector **p** to be extracted from the volume is determined as a function of the projection and capture geometries:

$$n = \frac{D_c}{\Delta_c}(\tan\theta_{out,\max} + \tan\theta_{ill}) \tag{32}$$

where D_c is the perpendicular distance from the capture track to the shear plane, and Δ_c is the capture camera increment. To modulate a basis fringe corresponding to holographic primitive location (i, j) for

$$\begin{aligned} 0 \le i \le N_h - 1 \\ 0 \le j \le N_v - 1 \end{aligned} \tag{33}$$

The corresponding (i, j) location must be indexed in the parallax view volume to retrieve the appropriate n view pixels between k_{min} and k_{max} (where $k = 0$ corresponds to the left-most captured view and $k = K-1$ corresponds to the right-most captured view). For any (i, j), k_{min} and k_{max} are given by

$$k_{\min} = round\left[\left(\frac{K-n}{N_h}\right)i\right]$$
$$k_{\max} = round\left[\left(\frac{K-n}{N_h}\right)i\right]+n \qquad (34)$$

and their values undergo discrete jumps as i increases.

Once an appropriate **p** has been extracted from the view volume, its values are interpolated to fill a vector **p'** with the same number of samples as the basis fringe it will multiply. As shown in the margin, the hogel that will reconstruct the primitive at (i, j) is created by modulating the windowed basis fringe with **p'**. To create a single flat projection plane, the hogel is positioned in the hologram at location $(i \times \Delta x, j \times \Delta y)$, where Δx is the spacing between holographic primitives and Δy is the holo-line spacing, and accumulated into the pattern H_{RIP}

$$H_{RIP} = \sum_{j=1}^{Nv}\sum_{i=1}^{M} f_{win_{i\times\Delta x,\, j\times\Delta y}}\,\mathbf{p}_{i,j}{}' \qquad (35)$$

Once an entire pattern is finally assembled in this MAC fashion, it can be normalized to fit the bit-depth of the display system. When reconstructed, each hogel projects the parallax information it encodes, and a viewer can observe appropriate scene parallax as she moves through the viewzone. The technique improves upon the image quality of conventional stereograms while affording similar efficient computation. By eliminating abutting hogel apertures on the hologram plane, this approach allows scene capture and projection to be tuned together, according to the sampling requirements of the scene and the constraints of a given display architecture.

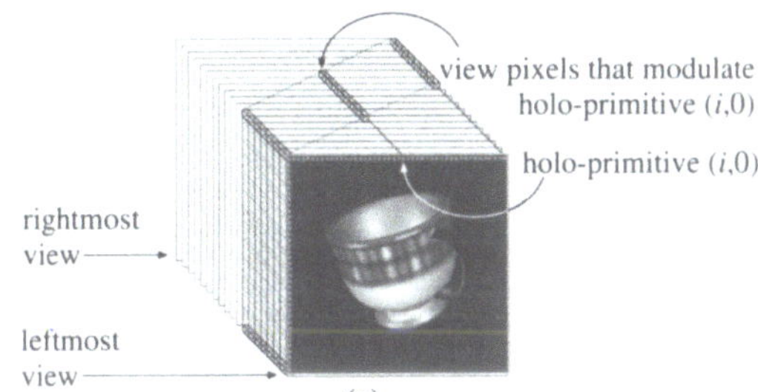

(a)

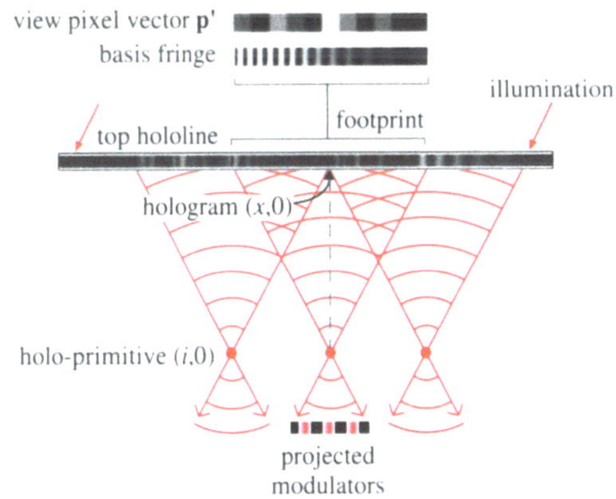

(b)

A continuum between stereograms and volumetric displays

This flexible hybrid approach can be used to generate output along a continuum from stereograms to true volumetric displays. To illustrate this continuum—the particulars of parallax view generation aside—first consider the case ((a) in the figure on the next page) in which the projection surface describes a single plane, which can be produced using multiple instances of a single basis fringe. Each holographic primitive populating the plane projects parallax views acquired using the same projection geometry (via computer graphics or optical capture). This process of view capture and display is stereographic in nature. Next, one can imagine the projection surface being shaped as a hemi-cylinder or hemisphere, wrapped around the scene's perimeter ((b) in the figure). Instances from a larger set of basis fringes are required to populate the surface in depth, and view capture and indexing must proceed in a different but easily programmable fashion; the stereogram-style assembly remains similar in principle. A more specialized projection surface may be shaped to "shrink-wrap" the scene so that holographic primitives populate object surfaces ((c) in the figure), or to densely populate the object volumes themselves (in principle, the same input as interference modeling would require). One way to accomplish this representation would populate the display volume with many projection planes, finely

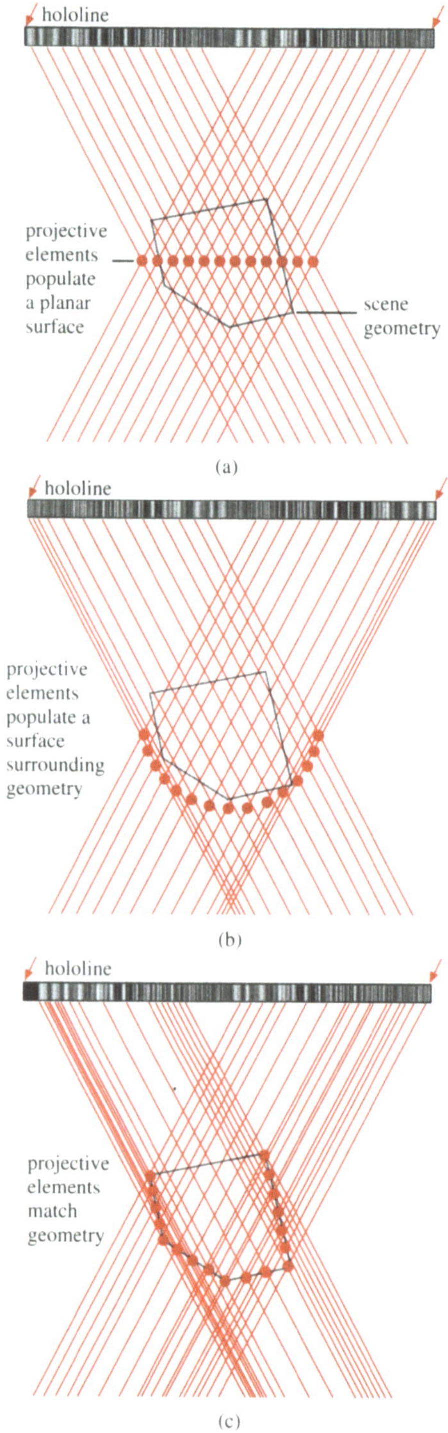

stacked in depth; such a configuration is volumetric in nature and could effectively display transparent or refractive volumes and/or opaque and occluding surfaces. Flexible hybrid techniques like this one can be adapted to drive many different kinds of electroholographic displays.

Toward Interactive, High-Quality Displays

The allure of holographic imaging remains absolute: it interpolates technical eventuality and popular imagination. The first was established with Gabor's discovery that a record of optical interference might be invertible through optical diffraction; the second has been fed on a healthy diet of science fiction TV and cinema depicting foregone dimensional imaging.

This compound inspiration underlies much of the work described here, which promises the addition of motion and interactivity to the present reach of holography. Yet much more discovery stands between field's current state and the ability to compute and display large scale and moving holographic projections real enough to fool human eyes.

At present, electroholography's principal challenges still persist; high-speed computation and high-bandwidth modulation of visible light still constrain the kinds of holograms that can be created and the nature of the images they project. Advances throughout a broad set of technologies and fields of research—materials science, SLM devices, computer hardware, light field representation and encoding, and computational algorithms—will offer new possibilities for creating moving, responsive, full color spatial images of realistic and expressive content.

References

i. Leith, E. N. and J. Upatnieks (1962). "Reconstructed Wavefronts and Communication Theory," *J. Opt. Soc. Amer.*, **52,** pp. 1123-30 (1962).

ii. Brown, B. R. and A. W. Lohmann (1966). "Complex Spatial Filtering with Binary Masks," *Applied Optics,* **5,** p. 967.

iii. Brown, B. R. and A. W. Lohmann (1969). "Computer Generated Binary Holograms," *IBM Journal of Research and Development,* **13,** pp. 160-167.

iv. Lee, W. H. (1970). "Sampled Fourier Transform Hologram Generated by Computer," *Applied Optics,* **9,** pp. 639-643.

v. Burckhardt, C. B. (1970). "A Simplification of Lee's Method of Generating Holograms by Computer," *Applied Optics,* **9,** pp. 1949-1951.

vi. Waters, J. P. (1968). "Three-dimensional Fourier Transform Method for Synthesizing Binary Holograms," *Journal of the Optical Society of America,* **58,** pp. 1284-1288.

vii. Kollin, J. S., S. A. Benton, and M. L. Jepsen (1989). "Real-Time Display of 3-D Computed Holograms by Scanning the Image of an Acousto-Optic Modulator," *Proc. SPIE Holographic Optics II: Principles and Applications,* **1136,** pp. 178-185.

viii. St.-Hilaire, P., S. A Benton, M. Lucente, M. L. Jepsen, J. Kollin, and H. Yoshikawa (1990). "Electronic Display System for Computational Holography," *Proc. SPIE Practical Holography IV,* **1212,** pp. 174-182.

ix. Leseberg, D. (1987). "Computer Generated Holograms: Cylindrical, Conical and Helical Waves," *Applied Optics,* **26,** 20, pp. 4385-4390.

x. Leseberg, D. and C. Frere (1988). "Computer-Generated Holograms of 3-D Objects Composed of Tilted Planar Segments," *Applied Optics*, **27,** 14, pp. 3020-3024.

xi. Underkoffler, J. S. (1988). "Development of Parallel Processing Algorithms for Real-Time Computed Holography," SB Thesis, Department of Electric Engineering and Computer Science, Massachusetts Institute of Technology, Cambridge MA.

xii. Underkoffler, J. S. (1991). "Toward Accurate Computation of Optically Reconstructed Holograms," SM Thesis, Media Arts and Sciences Section, Massachusetts Institute of Technology, Cambridge MA.

xiii. Lucente, M. (1993). "Interactive Computation of Holograms Using a Look-Up Table", *J. Electron. Imaging*, **2,** 1, pp. 28-34.

xiv. Benton, S. A. (1983). "Survey of Holographic Stereograms," *Proc. SPIE Processing and Display of Three-Dimensional Data,* **367,** pp. 15-19.

xv. Halle, M. W., S. A. Benton, M. A. Klug, and J. S. Underkoffler (1991). "The Ultragram: a Generalized Holographic Stereogram," *Proc. SPIE Practical Holography V,* **1461,** pp. 142-155.

xvi. Lucente, M. and T. Galyean (1995). "Rendering Interactive Holographic Images," *Proc. ACM SIGGRAPH'95,* pp. 387-394.

xvii. Lucente, M. (1996a). "Computational Holographic Bandwidth Compression," *IBM Systems Journal,* **35,** pp. 349-365.

xviii. Yoshikawa, H., S. Iwase, and T. Oneda (2000). "Fast Computation of Fresnel Holograms Employing Difference," *Proc. SPIE Practical Holography XIV and Holographic Materials VI,* **3956,** pp. 331-335.

xix. Plesniak, W. (2003). "Incremental Update of Computer Generated Holograms," *Journal of Optical Engineering,* **42,** 6, pp. 1560-1571.

xx. Pappu, R. S. (1996). "Nonuniformly Sampled Computer-Generated Holograms," *Optical Engineering,* **35,** 6, pp. 1538-1544.

xxi. Pappu, R. S. (1995). "Minimum Information Holograms," S.M. Thesis, Program in Media Arts and Sciences, Massachusetts Institute of Technology, Cambridge MA.

xxii. Lucente, M. (1996b). "Holographic Bandwidth Compression Using Spatial Subsampling," *Optical Engineering,* **35,** 6, pp. 1529-1537.

xxiii. Lucente, M. (1994). "Diffraction-Specific Fringe Computation for Electro-Holography," Ph.D Thesis, Electrical Engineering and Computer Science Department, Massachusetts Institute of Technology, Cambridge MA.

xxiv. Cameron, C. D., D. A. Payne, M. Stanley, and C. W. Slinger (2000). "Computational Challenges of Emerging Novel True 3D Holographic Displays," *Proc. SPIE Critical Technologies for the Future of Computing,* **4109,** pp. 129-140.

xxv. Yoshikawa, H. and H. Kameyama (1995). "Integral Holography," *Proc. SPIE Practical Holography IX,* **2406,** pp. 226-234.

xxvi. Pappu, R. and W. J. Plesniak (1998). "Haptic Interaction with Holographic Video Images," *Proc. SPIE Practical Holography XII,* **3293,** pp. 38-45.

xxvii. Plesniak, W. J., R. S. Pappu, and S. A. Benton (2003). "Haptic Holography: a Primitive Computational Plastic," *Proc. of the IEEE,* **91,** 9, pp. 1443-1456.

xxviii. Slinger, C. W., R. W. Bannister, C. D. Cameron, S. D. Coomber, I. Cresswell, P. M. Hallett, J. R. Hughes, V. Hui, J. C. Jones, R. Miller, V. Minter, D. A. Payne, D. C. Scattergood, D. T. Sheering, M. A. Smith, and M. Stanley (2001). "Progress and Prospects for Practical Electroholographic Display Systems," *Proc. SPIE Practical Holography XV and Holographic Materials VII,* **4296,** pp. 18-32.

xxix. Slinger, C. W., C. D. Cameron, S. D. Coomber, R. J. Miller, D. A. Payne, A. P. Smith, M. G. Smith, M. Stanley, and P. J. Watson (2004). "Recent Developments in Computer-Generated Holography: Toward a Practical Electroholography System for Interactive 3D Visualization," *Proc. SPIE Practical Holography XVIII,* **5290,** pp. 27-41.

xxx. Slinger, C. W., C. Cameron, and M. Stanley (2005). "Computer-Generated Holography as a Generic Display Technology," *IEEE Computer,* **38,** 8, pp. 46-53.

xxxi. Petz, C. and M. Magnor (2003). "Fast Hologram Synthesis for 3D Geometry Models using Graphics Hardware," *Proc. SPIE Practical Holography XVII and Holographic Materials IX*, **5005,** pp. 266-275.
xxxii. Quentmeyer, T. (2004). "Delivering Real-Time Holographic Video Content with Off-the-Shelf PC Hardware," SM Thesis, Department of Electrical and Computer Engineering and Computer Science, Massachusetts Institute of Technology, Cambridge MA.
xxxiii. Bove, Jr., V. M., W. J. Plesniak, T. Quentmeyer, and J. Barabas (2005). "Real-Time Holographic Video Images with Commodity PC Hardware," *Proc. SPIE Stereoscopic Displays and Applications*, **5664A,** 2005.
xxxiv. Brown, B. R. and A. W. Lohmann (1969). *Op. cit.*
xxxv. Lee, W. H. (1970). *Op. cit.*
xxxvi. Burckhardt, C. B. (1970). *Op. cit.*
xxxvii. Tricoles, G. (1987). "Computer Generated Holograms: an Historical Review," *Applied Optics*, **26,** 20, pp. 4351-4360.
xxxviii. Lesem, L. B., P. M. Hirsch, and J. A. Jordan (1969). "The Kinoform: A New Wavefront Reconstruction Device," *IBM Journal of Research and Development*, **13,** pp. 150-155.
xxxix. Chu, D. C., J. R. Fienup, and J. W. Goodman (1973). "Multi-Emulsion, On-axis, Computer Generated Hologram," *Applied Optics*, **12,** pp. 1386-1388.
xl. Waters, J. P. (1968). *Op. cit.*
xli. See for example the discussion in Chapter 8 of Born, M., and E. Wolf (1999). *Principles of Optics: Electromagnetic Theory of Propagation, Interference, and Diffraction of Light*, 7th Ed., Cambridge University Press, Cambridge, UK.
xlii. Underkoffler, J. S. (1988). *Op. cit.*
xliii. Underkoffler, J. S. (1991). *Op. cit.*
xliv. Underkoffler, J. S. (1997). "Occlusion Processing and Smooth Surface Shading for Fully Computed Synthetic Holography," *Proc. SPIE Practical Holography XI and Holographic Materials III*, **3011,** pp. 19-30.
xlv. Underkoffler, J. S. (1991). *Op. cit.*
xlvi. Lucente, M. (1992). "Optimization of Hologram Computation for Real-Time Display," *Proc. SPIE Practical Holography VI*, **1667,** pp. 32-43.
xlvii. Campbell, F. W., and D. G. Green (1965). "Optical and Retinal Factors Affecting Visual Resolution," *J. Physiol.*, **181,** pp. 576-593.
xlviii. Underkoffler, J. S. (1997). *Op. cit.*
xlix. Yoshikawa, H., S. Iwase, and T. Oneda (2000). *Op. cit.*
l. Lucente, M. (1993). *Op. cit.*
li. Plesniak, W. (2003). *Op. cit.*
lii. Lucente, M. and T. Galyean (1995). *Op. cit.*
liii. Lucente, M. (1994). *Op. cit.*
liv. St.-Hilaire, P. (1994). "Modulation Transfer Function and Optimum Sampling of Holographic Stereograms," *Applied Optics*, **33,** 5, pp. 768-774.
lv. Halle, M. W. (1994). "Holographic Stereograms as Discrete Imaging Systems," *Proc. SPIE Practical Holography VIII*, **2176,** pp. 73-84.
lvi. St.-Hilaire, P. (1994). *Op. cit.*
lvii. St.-Hilaire, P. (1995). "Modulation Transfer Function of Holographic Stereograms," *Proc. SPIE Applications of Optical Holography*, **2577,** pp. 41-49.
lviii. St.-Hilaire, P. (1997). "Optimum Sampling Parameters for Generalized Holographic Stereograms," *Proc. SPIE Practical Holography XI*, **3011,** pp. 96-104.
lix. St.-Hilaire, P. (1994). *Op. cit.*
lx. St.-Hilaire, P. (1997). *Op. cit.*
lxi. Lucente, M. (1993). *Op. cit.*
lxii. Watlington, J., M. Lucente, C. J. Sparrell, V. M. Bove, Jr., and I. Tamitani (1995). "A Hardware Architecture for Rapid Generation of Electroholographic Fringe Patterns," *Proc. SPIE Practical Holography IX*, **2406,** pp. 172-183.
lxiii. Lucente, M. (1994). *Op. cit.*
lxiv. Lucente, M. (1996). *Op. cit.*
lxv. Slinger, C. W., *et al.* (2001). *Op. cit.*
lxvi. Slinger, C. W., *et al.* (2004). *Op. cit.*

lxvii. Slinger, C. W., *et al.* (2005). *Op. cit.*

lxviii. Cameron, C. D., *et al.* (2000). *Op. cit.*

lxix. Yoshikawa, H. and H. Kameyama (1995). *Op. cit.*

lxx. Plesniak, W., M. Halle, S. D. Pieper, W. Wells III, M. Jakab, D. S. Meier, S. A. Benton, C. R. G. Guttmann, and R. Kikinis (2003). "Holographic Video Display of Time-Series Volumetric Medical Data," *Proc. IEEE Visualization 2003*, pp. 589-593.

lxxi. Plesniak, W., M. Halle, V. M. Bove, Jr., J. Barabas, and R. Pappu (2006). "Reconfigurable Image Projection Holograms," *Optical Engineering*, **45,** 11, 115801.

lxxii. Halle, M. and A. Kropp, (1997). "Fast Computer Graphics Rendering for Full Parallax Spatial Displays," *Proc. SPIE Practical Holography XI and Holographic Materials III*, **3011,** pp. 105-112.

CHAPTER 20

Holographic Stereograms and Printing

Michael Klug and Mark Holzbach

Holographic Stereograms

Parallax panoramagrams and integral photographs significantly predate the advent of modern holography as three-dimensional imaging media.[i] "Integral photograph" or "Fly's eye lens array photograph" are terms describing 3-D displays consisting of a photographic medium and a spherical or cylindrical lenslet array placed in close proximity to the medium surface.[ii] In integral photography, each small lens captures the scene from a slightly different perspective than its neighbor, and a large array of these perspective views is recorded in correspondingly small areas across the film. Viewing the photo with the lenslets in place results in a simultaneous reconstruction of the array of discrete perspectives, enabling the viewer to perceive a 3-D representation due to parallax displacements for each eye.

Invented in 1947, and made practical with the advent of the laser in the early 1960s, holography replaced the need for discrete perspective recording by providing a means to record and reconstruct actual wavefronts of light reflecting off of the surface of an object, by encoding those wavefronts through the physical phenomenon of interference.[iii] The perspective information captured in the holographic process is nearly continuous, offering much higher fidelity 3-D reproduction capability than that of integral photography. Holography has limited utility, however, because the object or scene must remain utterly motionless during the recording process, lest the interference pattern be blurred and unresolvable. In an effort to find a way to record holograms of living subjects and objects that cannot remain motionless for the duration of an exposure, hybrid approaches have been conceived, coupling the off-line parallax capture of integral photography with the encoding and optical capabilities inherent in holography. This hybrid has been termed "holographic stereography."

A number of early "holographic stereogram" techniques were proposed and demonstrated, notably those of Pole, DeBitetto, and Redman.[iv,v,vi] In the simplest form, a holographic stereogram is created by sequentially projecting perspective views onto the recording plate with laser light, and recording a separate hologram of each, as illustrated in the figure in the margin. The resultant multi-exposure hologram (termed "angularly multiplexed") successfully reproduces a 3-D image, but the reduced diffraction efficiency resulting from many sequential exposures of the photosensitive medium produces dim results. In other schemes, parallax information is captured, and then recorded on a segmented holographic emulsion through a combination of laser-light projection techniques and masking of the holographic plate—so-called "spatially-multiplexed" approaches. In DeBitetto's method, diagrammed in the next figure, a plate is recorded with vertical strip holograms, each containing a different discrete perspective of the scene. The composite hologram serves as a kind of memory window analog, through which the viewer can see different perspectives with each eye. All of these holographic stereo-

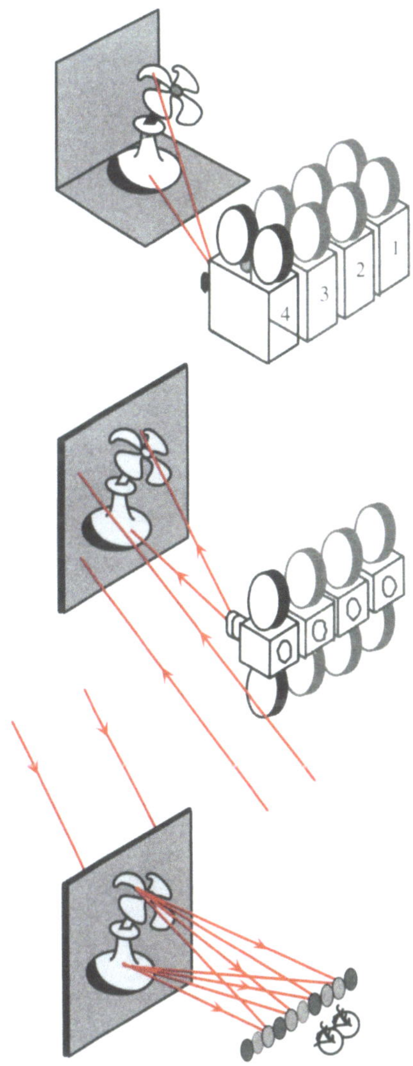

Simple holographic stereogram (after Redman (1969))

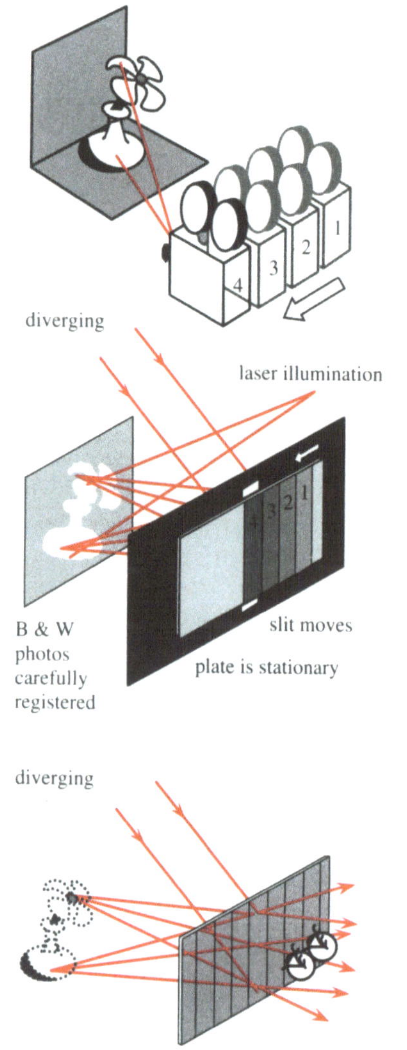

Stereogram composite H1 (after DeBitetto (1969)

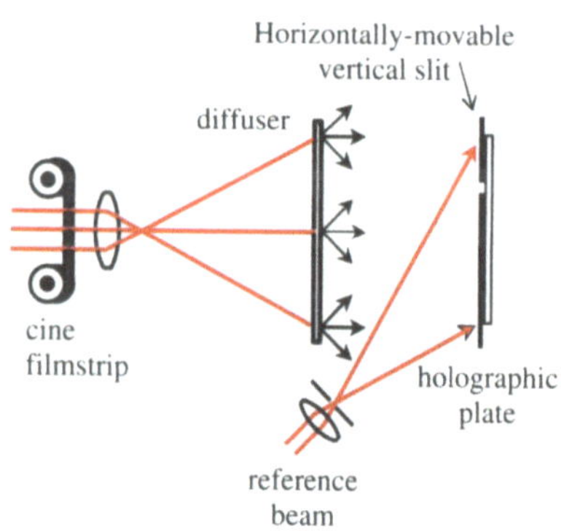

gram methods result in the elimination of phase continuity from point-to-point on the hologram due to the sampling of perspectives and the sequential recording of that information. In this way the coherence of the phase front from point to point on the holographic plane is sacrificed in favor of the reduction of information and the ability to record a broader range of scenes than is possible with classic holography.

In many of the initial holographic stereogram methods, the resulting image must be viewed with laser illumination due to blurring caused by chromatic dispersion of the remote image plane of the projection screen. In addition, the "image-plane" of spatially-multiplexed stereograms lies in the virtual image space beyond the surface of the holographic window, in contrast with the angularly-multiplexed case in which that plane is coincident with the surface of the window itself. A second holographic recording step (the "transfer step") can be employed to produce an "image-plane" hologram derived from the multi-perspective bearing spatially-multiplexed master plate. The second step merges the diffusion plane (that of the original ground glass projection screen) with the hologram plane, minimizing chromatic dispersion blurring effects. Also, most holographic stereogram techniques employ horizontal parallax-only ("HPO") recording, eliminating vertical parallax in order to reduce information content and minimize recording time. Another feature of HPO holographic stereograms is the minimization of dispersion blurring, which tends to manifest itself most onerously in the direction aligned with the reference beam inclination, due to the geometry of the interference pattern. In a process using a movable slit (see diagram), two-step holographic stereograms can be made to produce full color reproduction, using the rainbow hologram mixing methods pioneered by Benton in the 1970s. Additionally, by recording the master hologram at the "achromatic angle"—an angle roughly matching the dispersion distribution of the grating when recorded and illuminated off-axis, a "black and white" transfer hologram can be made.[vii]

Full-color holographic stereograms that provide a wide viewing angle in which the color remains relatively consistent are produced by recording image-plane transfers in a "reflection" geometry. In this case, illustrated on the next page, the reference beam is directed to the recording plane from the side opposite that of the master plate. A volumetric fringe pattern is formed, having a significantly higher spatial frequency than that of the transmission, "rainbow" geometry, and also exhibiting Bragg-selectivity, thus reducing the bandwidth of diffracted light upon reconstruction. The resulting hologram is illuminated from the same side from which it is viewed, usually with a light source that is positioned above the head of the viewer. The image usually has the coloration of the laser used for the recording, and maintains color consistency over a broad viewing angle. Manipulation of the recording layer thickness prior to or post recording can result in the reproduction of colors other than that used for recording. In silver-halides, the effect is achieved through diffusion-swelling of the emulsion with fairly inert large molecules in solution. Alternatively, Walker's (1989) "in-situ" approach, using various concentrations of water in an Isopropanol bath, enables practical 3-color separation recording in the transfer step without the usual negative side effects of the large-molecule approach.[viii] In this way, full color holographic stereogram transfers are made using a single long-wavelength laser. A similar approach, this one using post-recording diffusion swelling, was developed using dry-process photopolymer recording materials developed by Dupont in the 1990s.[ix]

The mechanical systems for producing holographic stereograms are relatively complex. First, perspective information of the scene or object must be collected or generated. Until the mid-1990s, the perspectives were recorded on film, one frame for each, using a calibrated camera system. The recording geometry of the individual perspectives was set to match that of the holographic stereogram recording geometry and, more importantly, the viewing geometry of the final hologram. A "re-centered shear-camera" geometry was adapted in order to minimize distortions and create the most accurate, realistic-looking scenes.[x] The film strips containing perspective frames would be inserted into a laser projector on the optical bench, usually using pin-registration to ensure positional accuracy. The recording process would entail exposing each image to the holographic plate, followed by advancing the frame and moving the mask (or plate) between exposures. It is necessary to keep vibrations to a minimum in this process, while at the same time minimize the time for recording the hundreds of views comprising the holographic stereogram plate. With the advent of high quality electronic spatial light modulators, such as liquid-crystal displays, the film strips and the mechanical apparatus associated with their projection and indexing could be eliminated, simplifying the process.

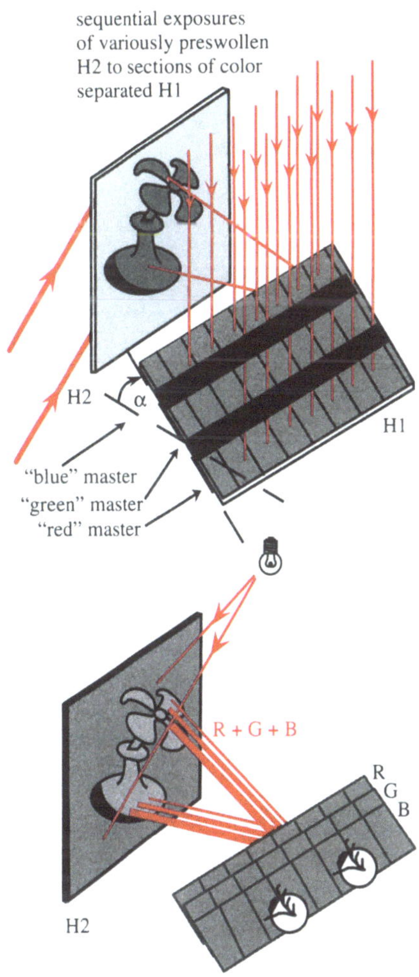

Multi-color reflection holographic stereogram

One-Step Approaches

Despite the successes of two-step holographic stereograms in production of full color and even large scale images the technique is cumbersome, requiring multiple large, stable recording setups, and thus does not lend itself to wide-spread practical utility as a visualization medium.[xi] One-step approaches were developed that enabled direct production of white-light viewable holograms, such as Lloyd Cross' Multiplex™ of the early 1970s. These were also HPO displays, made using diffusion-less anamorphic projection systems. In the Multiplex™ case, a large cylindrical lens is used in place of the projection screen. Images projected onto the lens focus at a different plane (ideally, at the viewer's position) in the vertical direction, and focus at the hologram film plane in the horizontal (cylindrical lens power) direction. The resulting hologram, though highly chromatically-dispersive, displays a sharp image because of the astigmatic displacement of the two focal planes. Benton's refinement of the technique in the late 1970s and early 1980s enabled the imagery to pass through the hologram plane with minimal distortion, creating a true image-plane one-step HPO holographic stereogram.[xii]

The advent of modern computer graphics made it possible to create holographic stereograms of non-physical subject matter since the early 1980s. Raytracing methods were developed that enabled direct computation of imagery for non-standard optical projection systems like those required for one-step holographic stereogram recording. In addition, image processing algorithms, derived from models of the optical recording system and applied to perspective imagery made it possible to anticipate distortions and pre-distort component imagery to preclude their effects. In 1986, work by Teitel and Benton combined computer-graphic ray-tracing and an anamorphic optical recording system, resulting in the first successful production of an HPO holographic stereogram that traversed the film plane using computer-graphic imagery.[xiii] Holzbach (1986) adapted similar principles to perspective computer-generated imagery produced with a standard camera model, and then post-processed to produce the pre-distortions necessary to create an undistorted holographic image.[xiv] Sometimes called "slice and dice" by the creators, the method called for subdivision of the original perspective views,

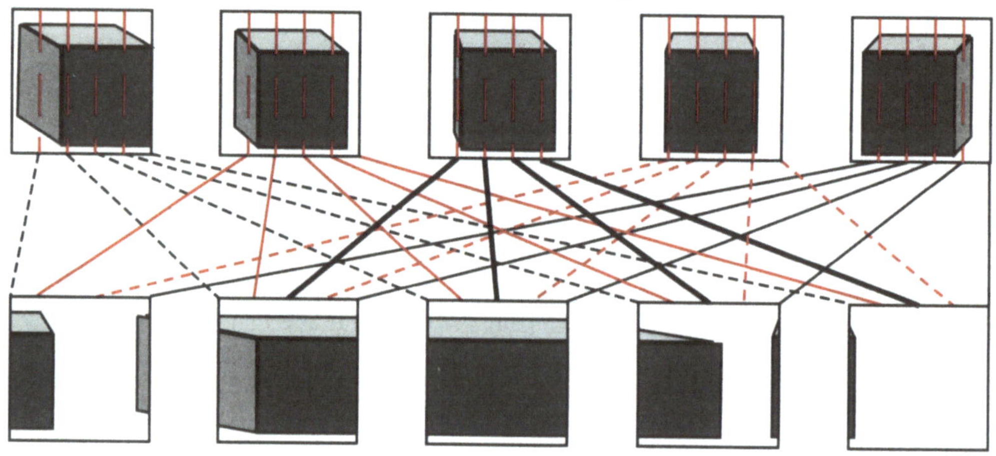

followed by re-assembly of the components into new hologram views, as depicted here.

Further sophisticated modeling of the recording geometries contributed to the development of the Alcove Hologram in 1986—a format that produced an image that floated in the center of a hemicylinder.[xv] Due to the distance from the film plane to the subject (generally about 30 cm), the wide 180-degree horizontal viewing angle, and the translucence of the cylinder bearing the film, the result was an eerily convincing aerial 3-D image. The original Alcove was made with a diffusing projection screen, and required laser illumination in a unique but awkward display system in order to avoid image blur due to dispersion. Further work on a white light-illuminated reflection-format of the Alcove was successfully demonstrated in 1987.[xvi]

HPO one-step holographic stereogram development progressed in the development of the Ultragram—a collection of generalized optical recording and corresponding computer-graphic techniques that enabled production of white-light viewable holographic stereograms in a variety of formats.[xvii] A large-scale approach, analogous to the Alcove, was used to produce meter-square scale holograms with minimal distortion, using an optical system that re-positioned the reference beam for each component hologram recording. The resulting images, illuminated with a filtered arc-lamp, resulted in the first life-size holographic stereograms of computer graphic subjects such as automobile components. Chromatic computer graphic pre-distortion, coupled with manipulation of the recording film to produce RGB images, resulted in the production of the first full-color one-step holographic stereograms.[xviii]

Holographic Printing

Despite advances made in producing full color computer-generated holographic stereograms, the focus on horizontal-parallax only, and the need for pre- or post-exposure emulsion manipulation in order to reproduce full color has limited the utility of the medium. The desire for more realism in holographic imagery has driven the reconsideration of full-parallax and the subsequent simplification of recording methods. This, in turn, has resulted in a hybridization of holographic stereograms, integral photographs, and modern printing techniques called "holographic printing." Consideration of this technique can be approached from a "printer-centric" vantage point that provides an understanding of the optics and component data generation techniques without some of the complexities of historical methods already mentioned.

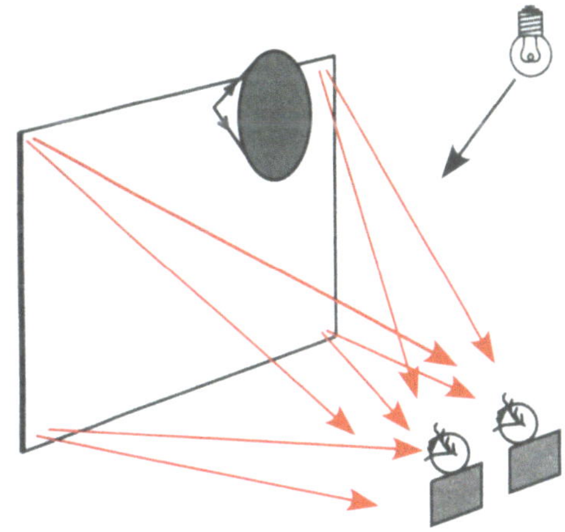

2-D planar display: Isotropic Lambertian reflection (no angular intensity variation); modest information density (1□)

Holographic elements

A photograph or picture can be thought of as a two-dimensional representation of a scene, projected onto a single plane. When one views a picture, each eye receives the same intensity information from light emanating or reflecting from each picture element ("pixel") on the surface bearing the picture. The isotropism of the display results in the same value of information distributed to all points in space. The image is devoid of parallax and most of the depth cues needed for 3-D perception. This is why, when one views a photograph or a drawing on a piece of paper, one generally perceives a two-dimensional representation of the subject on a flat surface. As illustrated in the margin, mechanisms with which the planar medium can be made to distribute different values to different points in space enable a three-dimensional representation of the scene, if that information is parallactically-distributed to the viewer's position. This is the effect that is achieved in integral photography, in which information in each small area on the recording medium is anisotropically distributed into the viewer's space by passing through a lenslet placed in front of it. Holographic printing is an approach in which the refractive lenslet is replaced by a diffractive analog that also simultaneously encodes the image information in the form of a complex interference pattern.

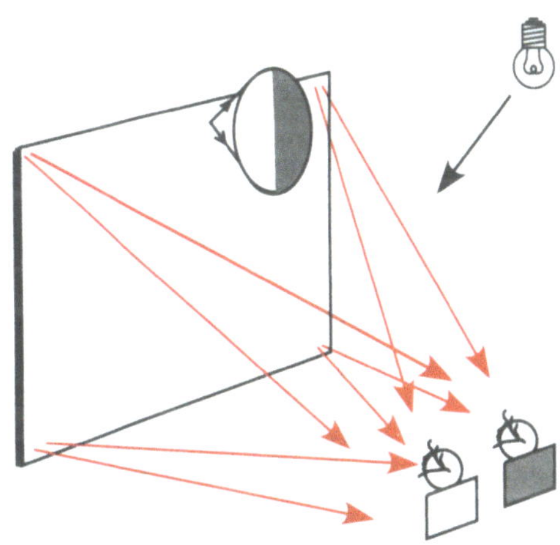

3-D planar display: Anisotropic specular reflection (angular multiplex of perspectives); high information density (2□–10^6□)

When two mutually-coherent laser beams collide on a piece of photosensitive film, a grating is formed and recorded at that spot. When one of those same beams subsequently illuminates that spot, the second beam will be reconstructed with an intensity and **k**-vector similar to the recording conditions. This phenomenon is governed by the grating equation, the most fundamental mathematical relationship in holography. Upon illumination, light is only visible emanating from the spot on the film from one particular angle—the angle of

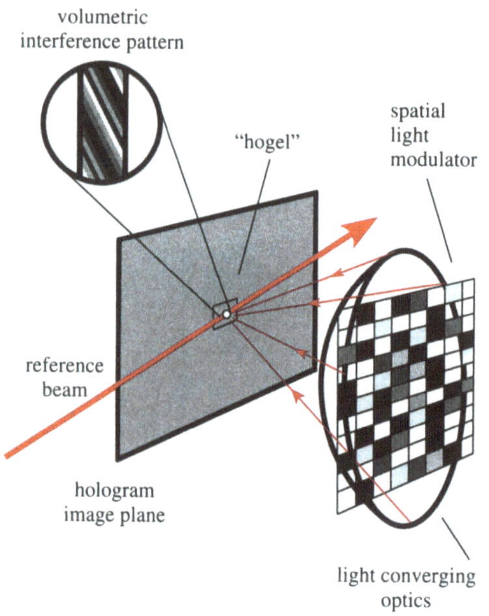

the original second beam (the "data beam"). If, instead of a single beam with a single intensity, the data beam is comprised of a large number of beams, all of differing intensities, the resulting *holographic element* (or "hogel") faithfully reproduces a distribution of rays, each reconstructed with relatively varying intensities, when illuminated by the original first beam. The adjoining figure schematically illustrates this case. Since human eyes are separated in space, a viewer gazing toward such a point will receive different information in each eye, resulting from the anisotropic distribution of light from the small point on the film. An array of such points on a piece of film, recorded in such a way as to provide a completely different picture of a scene to each eye, can result in viewer perception of the scene in 3-D, due to parallax and disparity effects.

It is possible to produce an array of varying-intensity data-beams by masking and converging a broad-cross-section laser beam using a lens or mirror. At the focal point of that optic, a Fourier transform hologram may be recorded. Reconstruction with a plane wave replica of the reference beam will faithfully reproduce an infinity-focused image of the mask. The highest diffraction efficiency (and therefore, brightest data beam reconstruction) occurs when the beam ratios for the data beam and reference beam are nearly equal. In the case of Fourier transform holography, however, data beam intensity at the film plane is dominated by the low frequency contributions in the transform distribution. If the reference beam intensity is matched to the highest data beam intensity, the higher frequency components form relatively low efficiency gratings, due to the large disparity in beam ratios. In addition, since the Fourier transform itself has theoretically infinite spatial distribution in the Fourier plane, and since the reference beam, for reasons of practicality, must be truncated in distribution, high frequencies are clipped due to lack of reference beam at the extremities of the frequency range.

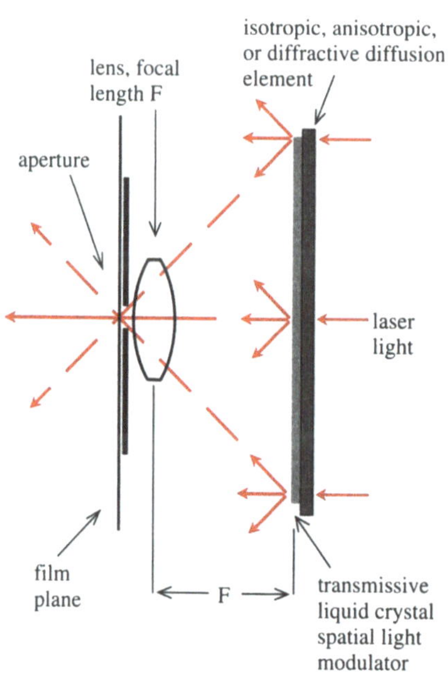

A number of techniques may be employed to represent all spatial frequency information in a well-defined, limited area on the film. These methods are diagrammed in the margin.

In one approach the hogel view is projected onto an isotropic diffuser, and the recording film is masked except for a single hogel-portion aperture. A lens is placed between the diffuser and the aperture, with the diffuser at its back focal plane. The lens serves to image the diffuser at an infinite distance, collimating the rays from individual elements on the diffuser surface. The phase information across the aperture mask is randomized, and the hogel is uniformly illuminated with the data beam information. Though effective, this approach is optically inefficient unless an anisotropic holographic diffuser is employed. A uniform distribution of light within the aperture mask results, as does a representation of most spatial frequencies in the mask plane, and a 1:1 ratio can be achieved between the reference beam and all components of the data beam, resulting in a holographic recording with high diffraction efficiency.

Alternatively, a "pseudorandom" or "band-limited" phase mask is employed in place of the diffuser in order to produce a specific intensity distribution at the conjugate focal plane of the lens coinciding with the recording film plane.[xix] The distribution is commonly that of a "top hat", in order to facilitate abutment of neighboring exposures and uniform distribution of exposure across the film plane. Gaussian intensity distributions may also be employed along with overlapping hogel exposures to smooth the effects of hogel-to-hogel boundaries. Band-limited phase mask elements eliminate the need for a physical aperture at the film plane, since all data beam energy is concentrated in the desired hogel area.

In a third method, a one-to-one image relay optical system that employs a diffuser or micro lenslet array at the input plane, and an aperture at the Fourier plane is placed upstream of the final hogel-view collimating lens. The collimating lens produces an image of the Fourier plane at its output, and the replication of the field at that plane produced by the lenslet array or diffuser is replicated in the hogel. This approach tends to be less efficient, but equally as effective as the direct approach, with the added flexibility of adjustment of the size of the aperture, and, thus, the size of the exposed area.

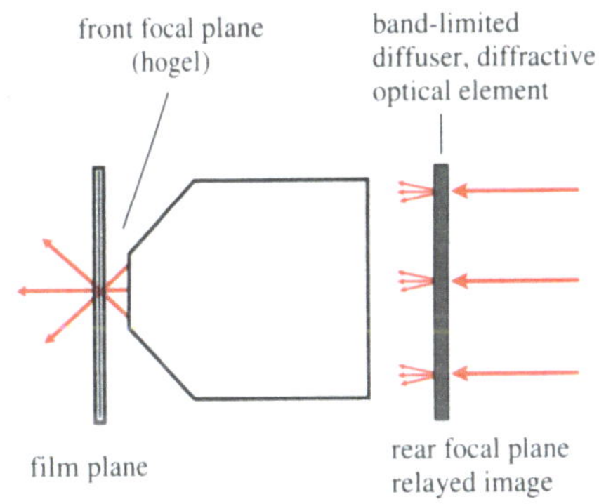

Multi-element Fourier transform lens

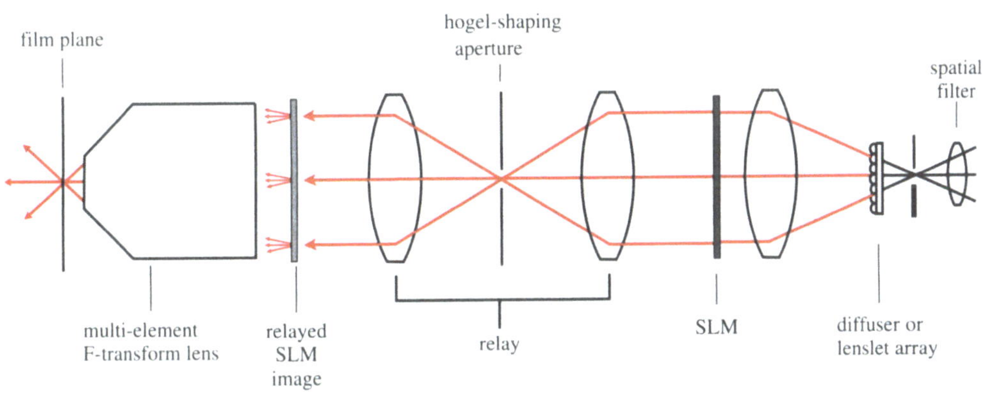

The exposed area at the film plane may be referred to as a holographic element, or *hogel*. Successive exposures in an array on the film produces a composite display that, for each hogel, independently controls the intensity and directional distribution of the light falling upon it. The resultant diffracted "lightfield" can produce three-dimensional images, or can be used as a holographic optical element-equivalent of a lens or mirror. Since phase information is randomized both within each hogel's boundaries, and from hogel-to-hogel, performance of the system is far from diffraction-limited. However, for the purpose of producing effective three-dimensional displays, it is more than adequate. Hogel-based "digital holograms" have been produced with hogel sizes from 100 microns to 3 millimeters, depending on the desired image scale and resolution of input data.

Hogel mask image generation: direct

Among the computer graphic rendering techniques that became practical and popular during this period is a technique known as *ray tracing*.[xx] The basic concept behind this technique is straightforward, it is a mathematical approximation of geometrical optics performed by tracing at least one light ray from an eyepoint through each target image pixel, and through a mathematical 3-D scene description that can contain surfaces that are reflective, refractive, transparent, opaque and shadow-casting, or any mixture/combination of the above. Compared to other computer graphic rendering techniques, ray tracing is capable of yielding the most photorealistically accurate simulation of a scene. However, raytracing requires significantly more computation, usually using general purpose central processing unit (CPU) hardware, and thus can be an order of magnitude or more slower than less-accurate techniques implemented in special purpose graphics processing unit (GPU) hardware. Teitel's previously mentioned work on traversing the hologram image plane using computer graphic imagery made use of custom-developed anamorphic raytracing algorithms.

Full-parallax systems that produce holograms that can be viewed with minimal distortion at any distance were achieved in the 1990s.[xxi,xxii] In the mid 1980s, reasonably priced small liquid crystal screen video displays became commercially available and could serve as an electronic film transparency equivalent, thus removing the need for intermediate 2-D cinefilm. Another important advantage afforded by the liquid crystal screen was that the 2-D imagery could be changed without any mechanical motion: a solid-state device had replaced the film-transport's many moving parts, and thus dramatically increased the systems' stability and reliability. In 1990, Yamaguchi used a raytracing technique similar to that of Teitel, coupled with a spherical-lens and liquid-crystal screen-based recording system to "print" simple full-parallax holographic stereograms.[xxiii]

Most uses for computer graphics do not actually require physical accuracy, and so computer graphic techniques that favor computational speed over the accuracy of ray tracing have always been most common. Efficient rendering techniques generally known as *scan line* techniques have become codified into standards such as openGL and DirectX and accelerated in GPU hardware under popular operating systems including Microsoft Windows, Apple MacOS, and Linux. Despite their limited accuracy, the realism of these techniques often seems to rival the realism of ray tracing, incorporating things like environmental *reflection mapping*, shadows and sophisticated diffuse *global illumination* lighting approximated using "baked in" texture maps draped on object surfaces.

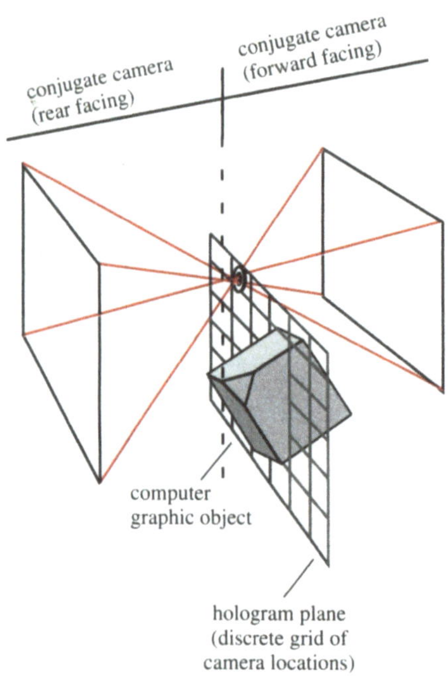

Although anamorphic (*e.g.*, cylindrical lens) optical effects are not possible with conventional scan line rendering techniques, Halle and Kropp (1997) described how the techniques could be adapted in order to simulate a large spherical lens with its focal plane at the holographic display surface.[xxiv] They understood the utility of this adaptation for fast and efficient rendering for full-parallax holograms, and called their invention the "double frustum camera" because it required one camera frustum to image the volume behind the holographic display surface, and another to image the volume in front. The images created by the two frusta are simply composited together with the elements in front potentially occluding those in back. The figure graphically explains the concept of the double frustum hogel-view camera.

Halle and Kropp considered the problem of how to deal with the desirability for imagery directly on or near the display surface, the position of a well-known mathematical singularity, the conventional eyepoint in the rendering calculation. In conventional rendering a "near clipping plane" is imposed before the eyepoint singularity, and imagery is excluded from that location. Their proposed solution was to align the near clipping planes of the front and back frusta before compositing. Although their solution still avoids the mathematical singularity, it does not avoid discontinuous representation for objects points that lie between the eyepoint and the near clipping plane of either frustum. There is also a mathematical precision problem for surface normal calculations near the eyepoint that can result in noticeable shading and reflection mapping errors. Improved solutions to this problem, including multi-sample rendering and polygonal subdivision near the eyepoint, were proposed by Holzbach and Chen in 1999.[xxv]

Indirect image generation

Although ray tracing and scan line rendering techniques require a 3-D description of the object and scene, including geometry, lights, and material qualities, another form of rendering which does not is called *image-based* rendering or *light field* rendering. It has been a major subject of interest within the computer-graphic community since 1996.[xxvi] These computer graphic techniques borrow from years of research in the computer vision community and their earliest application to holographic display was at MIT in 1986.[xxvii] They can accept sets of 2-D imagery of an object or scene taken from a variety of different perspectives as input, and create new synthetic views by resampling or remapping the input as output. This remapping may require either interpolation or extrapolation from the image samples. There often is a large degree of redundant information between neighboring perspectives (also sometimes referred to as image *coherency*) and some of the research in this area has to do with effective data structures including compression to allow interaction. The image in the margin shows the process of lightfield remapping used for producing hogel views for holographic printing.

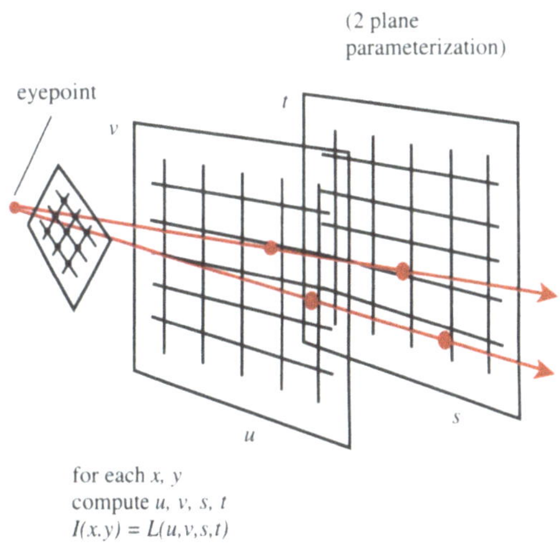

One attraction of image-based data and rendering techniques for holographic display, is that they can be used to manipulate complex (often real-world) objects and scenes independently of their complexity. When real-world imagery is used, it also eliminates the need for time-consuming traditional 3-D computer graphic geometric modeling, lighting, shading, and texturing. Full-parallax holographic display datasets are often batch-rendered using conventional cameras (e.g. from a synthetic computer graphic camera array or from a real-world camera array) and then batch-transformed into the hogel views. The ideal batch pipeline would be implemented as an uncompressed single *block transform* operation in high-speed volatile memory. Unfortunately, full-parallax datasets in the hundreds of gigabytes to single-digital terabytes are common, sizes for which volatile memory is still far from being a practical solution.

Holographic printer implementation

Production of practical holographic printers has required the maturation of component technologies, including spatial-light modulators and actively-stabilized and solid-state lasers, resulting in smaller, faster, and more reliable printing. Also, photopolymer-based recording materials have replaced silver-halide emulsion, simplifying the production process by eliminating the need for wet chemical development. These materials have also improved quality by producing less scatter and exhibiting less humidity sensitivity. Higher speed computer graphic rendering algorithms and graphics-dedicated hardware like that pioneered at Silicon Graphics Corporation in the 1980s and 1990s and the now-ubiquitous graphics processing unit ("GPU") have also contributed to simplifying and accelerating the holographic printing process. Using 1997 levels of technology, including actively-controlled continuous-wave lasers, liquid-crystal and digital micromirror spatial light modulators, and rapid computer graphic generation and software-controlled automation, the recording times for a typical holographic "print" ranged from 24 to over 100 hours, depending on hogel and image size. This production time represents an order of magnitude improvement in holographic image production rate as compared with prior two-step HPO approaches, with the added benefit of full parallax.

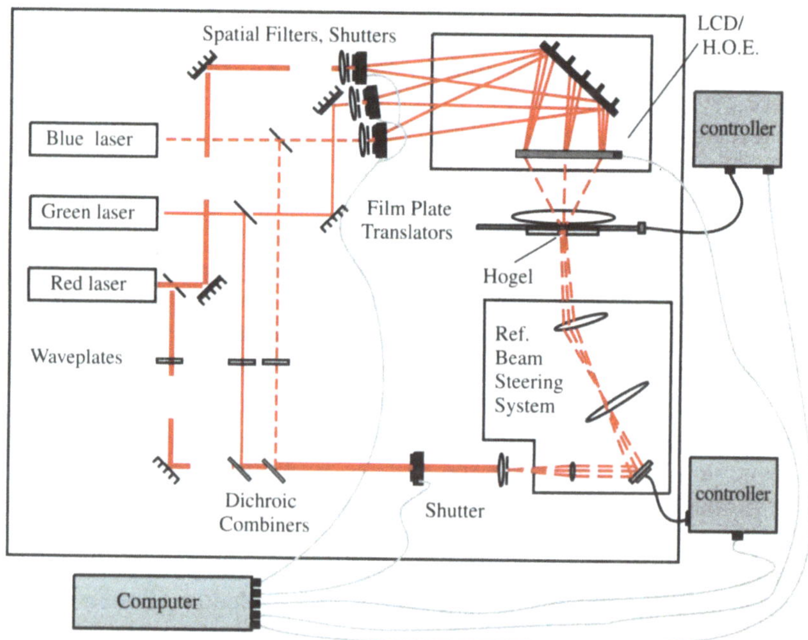

Functional holographic printing systems have been constructed by Klug, *et al.*, in 1997, and Smith in 2000.[xxviii,xxix] In a typical system, diagrammed above, commercially-available lasers with good color rendering-capable wavelengths were employed, including Krypton-ion (647nm red), frequency-doubled Nd:YAG (532nm green) and frequency-doubled ND:YLF (457nm blue). The system incorporates a spatial light modulator with a resolution of 1280 × 1024 RGB pixels, and prints hogels directly onto Dupont OmniDex 801 dry-process photopolymer film. A simple baking step after exposure is followed by a lamination step to produce the finished image. In order to produce images for undistorted point-source illumination, the printer incorporates a variable reference beam angle system, shown at the top of the next page. The reference beam is directed through an optical system designed to enable manipulation of the impingement angle of the beam on the media so that any hogel may be recorded with a different reference beam angle as compared with its neighbor. This "beam-steering" methodology enables illumination of the complete matrix of holographic exposures from a single point source at a specific position in space, predicted by the spatial point of coincidence of the collective hogel reference beams. This feature also facilitates the reconstruction of each hogel's image with maximum efficiency, by Bragg angle matching from hogel to hogel between recording and replay.

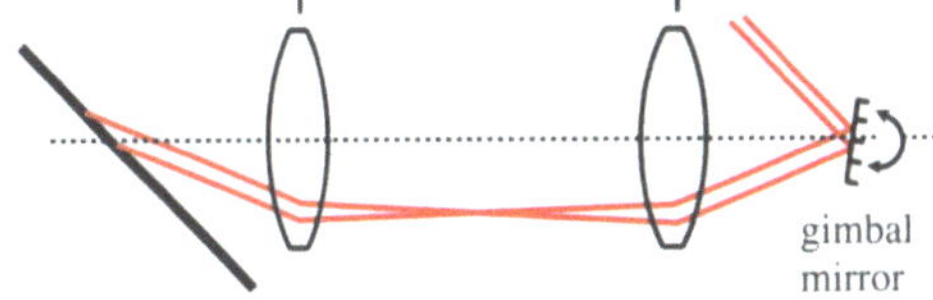

Scale, resolution, and format

Holographic printing scalability may be achieved, despite film width limitations, through tiling of unit holographic prints. In monolithic holographic exposures, this is usually not feasible due to quasi-Gaussian diffraction efficiency falloff from center to edge of the hologram. Since each hogel is has the same potential diffraction efficiency, there is no such falloff in brightness at the edges of each hogel array, rendering the process adaptable to tiling. Hogel image rendering software incorporates routines that automatically subdivide the image plane within the scene, and partition the image tile components and reference beam directions appropriately for printing. After recording, the film is laminated to a scratch-resistant transparent PET layer, and this composite is then laminated to a tile-registered rigid or flexible substrate. Good results have been obtained laminating to aluminum honeycomb composite materials, which retain excellent rigidity, with a significantly reduced thickness to weight ratio as compared to glass or acrylic. As illustrated in the margin, the tiles are fitted to a backing plate, and can be adjusted using three-point threaded mounts. The entire image may then be illuminated with a single point source to reconstruct the image. This process was first demonstrated in 1998 in the production of the largest single holographic image known to-date: a 1.8 m by 5.4 m image of a full-size automobile. The image comprises 27 tiles, each 60 by 60 cm, with 2.5 mm square hogels, and required approximately 3 weeks of total printing time.

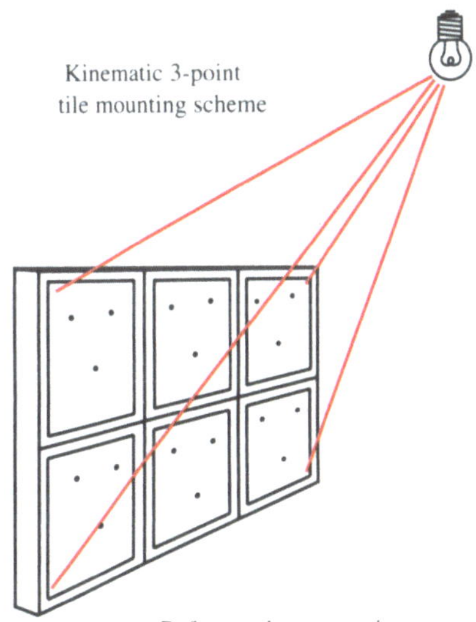

An important consideration in the production of holographic prints involves that of sampling and resolution. Hogel size (and density) determines the resolution of a print of given size. SLM pixel count determines maximum theoretical perspective resolution for the scene, with a printer optical system of a given *f*-number. Both parameters are affected by other effects inherent in the recording process, including diffraction, scatter, and noise. Typical spatial light modulators have resolutions as high as SXGA-standard (1280 × 1024 pixels). For a solid cone angle, θ, reproducible by a hogel made with such a modulator, the perspective density of the display can be approximated to be $1000/\theta$. For an effective *f*/0.5 optical system, producing a 90-degree hogel field of view, angular resolution is approximately 10 samples/degree. Reduction of the spatial light modulator image resolution can result in inadequate sampling, an effect manifested as an interview aliasing artifact in deep portions of the image volume. These effects are predicted and documented in the literature.[xxx]

Hogel resolution, sometimes referred to as "image-plane" resolution is variable with hogel size and shape, as well as inter-hogel spacing. Common resolutions for holographic prints include 2 mm, 1 mm, and 0.5 mm top-hat distributions, and 100–200μ Gaussian distributions. Hogel size may vary, depending on final image size (larger images can generally support larger hogels), viewing distance, content data resolution, and application. Due to the time required for recording each hogel using CW-laser-based printing systems, the cost of increasing resolution by decreasing hogel size must be carefully balanced with the display size and the quality requirements for the particular image and application.

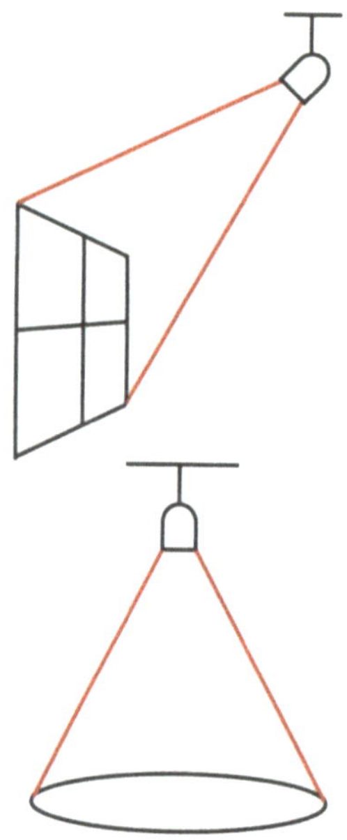

Two illumination and orientation formats for holographic printer output have been demonstrated. A vertically oriented format, much like that used for standard, off-axis reflection holograms, requires a point light source oriented 45 degrees from the normal of the film plane. Sometimes called "wall-mode," this format is useful for elevation or three-quarter views of the scene or subject, or for portraits. An alternative, horizontally oriented format, sometimes called "disk" or "table-top mode" is illuminated from a point source placed normal to the hologram surface. This format allows viewing from 360 degrees annularly, as well as from above the depicted scene. This is useful for plan views, presentation of terrain, or depiction of subjects where a full 360 degree "God's eye view" is useful or necessary. In each case, the reference beam in the printer must be oriented properly in order to produce the correct illumination conditions. The horizontal format possesses the additional feature of being able to be correctly viewed with the hologram placed on a turntable. The two formats are illustrated here.

Improving printing speed

Continuous-wave laser-based holographic printing systems provide high quality images, but still retain vibration sensitivities that limit the speed and adaptability of the technique to broad commercial application. This is due to the fact that while each hogel exposure may be relatively rapid (on the order of 10–50 milliseconds in a typical system), this exposure window is still subject to detrimental vibration effects that result in dim or blank hogels. In a CW laser-based holographic printer, a hogel is exposed by activating mechanical or electro-optical shutters in the beams, the film is indexed to the next hogel position, and a "settling time" follows in order to allow vibrations to damp out before the next exposure. The equipment used for film indexing, beam routing, and data conveyance must be vibrationally stable, usually requiring massive, cumbersome components. Such encumbrances do not easily lend themselves to mass-production and broad application. The printing speed of such systems is also limited by the laser power (governing the exposure time for each hogel). Typical hogel printing cycle times for a CW-laser based system are approximately .75–1.0 seconds, resulting in production times on the order of 72 hours for 60 by 60 cm holograms with 1 mm hogels. Additionally, image quality in such systems can be affected by environmental conditions that may vary randomly over such a long recording period.

Holographic printing speed may be increased by replacing the continuous-wave laser system with a pulse laser system. In this case, the laser provides all of the required energy for each hogel exposure in a single pulse of short duration, typically less than 50 nanoseconds. Relative to the CW case, this represents six orders of magnitude in exposure time reduction and implies that a pulse laser-based hologram printer can print images much more rapidly than a CW-laser based system. This is because the tolerance for vibration or translational displacement of the recording film during such a short time interval is reduced, given the resolution of the interference pattern being recorded in the hogel and the level of tolerable blur in that interference pattern before the blur effect is noticeable to the human eye. Through experiment and calculation, it has been shown that relative motion between the recording beams and the film of up to 2 meters/second does not produce discernable hogel brightness degradation. Update rates for the images displayed on the SLM, and the maximum pulse repetition rate for the laser determine printing speed limitations up to this rate. These may be delimited by fundamental refresh rates for the SLM, the bandwidth of the transmission channel from the image data source to the SLM, and physical limitations in the laser mechanism itself. Typical commercial pulse laser systems can operate at repetition rates ranging from 30Hz to multiple kHz for the pulse durations and energies required. For a printing system operating at 60 hogels per second, the production time for the benchmark 60 by 60 cm hologram with 1 mm hogels is approximately 2 hours.

Conclusions

Holographic printing represents an amalgamation of 3-D hardcopy concepts proposed, developed, and refined over the last one hundred years. The evolution of this medium has spanned numerous technical challenges, and refinement continues today, with efforts that concentrate on further size and cost reductions, speed color, and viewing angle improvements, and integration into initial practical applications. In addition, standards for source data must also be developed in order to provide a simple means for preparing content imagery for predictable output quality. As the medium of holographic printing becomes adopted into regular use in niche applications, understanding of the broad capabilities and strengths of this new medium will feed its further development and growth.

References

i. Okoshi, T. (1976). *Three-Dimensional Imaging Techniques*, Academic Press, New York.

ii. Lippmann, G. (1908). "Épreuves Réversibles. Photographies Intégrales," *Comptes Rendus*, **146**, pp. 446-451.

iii. Gabor, D. (1948). "A New Microscopic Principle," *Nature*, 161, pp. 777-778.

iv. Pole, R. V. (1967). "3-D Imagery and Holograms of Objects Illuminated in White Light," *Applied Physics Letters*, **10**, pp. 20-22.

v. DeBitetto, D. J. (1969). "Holographic Panoramic Stereograms Synthesized from White Light Recordings," *Applied Optics*, **8**, pp. 1740-1741.

vi. Redman, J. D. (1968). "The Three Dimensional Reconstruction of People and Outdoor Scenes Using Holographic Multiplexing," *Proc. SPIE*, **15**, pp. 117-122.

vii. Benton, S. A. (1976). "Three-Dimensional Holographic Displays," *Proc. Electro-Optical Syst. Des. Conf. 1976*, pp. 481-485.

viii. Walker, J. L. and S. A. Benton, (1989). "In-Situ Swelling for Holographic Color Control," *Proc. SPIE Practical Holography III,* **1051**, pp. 192-199.
ix. Hubel, P. and M. Klug, (1992). "Color Holography using Multiple Layers of DuPont Photopolymer," *Proc. SPIE Practical Holography VI,* **1667**, pp. 215-224.
x. Molteni, W. (1991). "Shear Lens Photography for Holographic Stereograms," *Proc. SPIE Practical Holography V,* **1461**, pp. 132-141.
xi. Outwater, C. and C. Newswanger (1985). "Large Format Holographic Stereograms and their Applications," *Proc. SPIE Applications of Holography,* **523**, pp. 26-32.
xii. Benton, S. A. (1978). "Distortions in Cylindrical Holographic Stereogram Images," *J. Opt. Sci. Am.,* **68**, p. 1440.
xiii. Teitel, M. (1986). "Anamorphic Imaging for Synthetic Alcove Holographic Stereograms," SM Thesis, Department of Architecture, Massachusetts Institute of Technology, Cambridge, MA.
xiv. Holzbach, M. (1986). "Three Dimensional Image Processing for Synthetic Holographic Stereograms," SM Thesis, Department of Architecture, Massachusetts Institute of Technology, Cambridge, MA.
xv. Benton, S. A. (1987). "'Alcove' Holograms for Computer-Aided Design," *Proc. SPIE True 3-D Imaging Technologies and Display Technologies,* **761**, pp. 53-61.
xvi. Krantz, E. (1987). "Optics for Reflection Holographic Stereogram Systems," SM Thesis, Department of Architecture, Massachusetts Institute of Technology, Cambridge, MA.
xvii. Halle, M., S. A. Benton, M. Klug, and J. Underkoffler (1991). "The Ultragram: A Generalized Holographic Stereogram," *Proc. SPIE Practical Holography V,* **1461**, pp. 142-155.
xviii. Klug, M., M. Halle, and P. Hubel (1992). "Full Color Ultragrams," *Proc. SPIE Practical Holography VI,* **1667**, pp. 110-119.
xix. Klug, M., A. Holzbach, and A. Ferdman (2001). U.S. Patent 6,330,088, "Method and Apparatus for Recording One-Step, Full-Color, Full-Parallax, Holographic Stereograms."
xx. Whitted, T. (1980). "An Improved Illumination Model for Shaded Display," *Comm. of the ACM,* **23**, 6, pp. 343-349.
xxi. Klug, M., *et al.* (2001). *Op. cit.*
xxii. Klug, M. (2001). "Scalable Digital Holographic Displays," *Proc. PICS 2001: Image Processing, Image Quality, Image Capture Systems Conference,* **4**, pp. 26-32.
xxiii. Yamaguchi, M., N. Ohyama, and T. Honda (1990). "Holographic 3-D Printer," *Proc. SPIE Practical Holography IV,* **1212**, pp. 84-92.
xxiv. Halle, M., and A. Kropp (1997). "Fast Computer Graphics Rendering for Full Parallax Spatial Displays," *Proc. SPIE Practical Holography XI and Holographic Materials III,* **3011**, pp. 105-112.
xxv. Holzbach, M. E., and D. T. Chen (2002). U.S. Patent 6,366,370, "Rendering Methods for Full Parallax Autostereoscopic Displays."
xxvi. Levoy, M., and P. Hanrahan (1996). "Light Field Rendering." *Proc. ACM SIGGRAPH '96,* pp. 31-42.
xxvii. Holzbach, M. (1986). *Op. cit.*
xxviii. Klug, M. (2001). *Op. cit.*
xxix. Smith, S. (2000). Private communication.
xxx. Halle, M. (1994). "Holographic Stereograms as Discrete Imaging Systems," *Proc. SPIE Practical Holography VIII,* **2176**, pp. 73-84.

CHAPTER 21

Holographic Television

The Holy Grail

If a hologram is the most complete possible 2-D record of a still 3-D scene, then a dynamically updatable electronic hologram, or "holo-video," should be the ultimate medium for reproducing moving 3-D scenes. Now that we know the optical characteristics of holograms, how to illuminate them, and how to generate holographic images computationally, we possess most of the background we need to build such a thing. We might finally be able to see a walking, talking 3-D image of Princess Leia (even if she has to stand behind—or straddle—some sort of window). In order to accomplish that goal, we'll have to replace the holographic plate with some device that is quickly electronically rewritable, and "as good as" the static holograms we've been discussing—so we need to figure out what "as good as" means, and if we can't find a device like that, we'll have to invent some other tricks so that we can employ something that's only almost good enough.

There's also a nomenclature issue we need to take care of. There have been a number of systems developed that were called by their inventors or marketers "holographic TV," even though they weren't truly holographic as we have come to know and love the term. When people who *really* know what holography is (a group which includes the readers of this book, if you've stayed with us this far) use the term "holo-video," we're referring to a system that operates at video rates (*i.e.* the image can be changed fast enough that it appears to move smoothly, though we might be willing to stretch things a bit and accept a device that can be updated only a few times a second as long as the images themselves don't flicker annoyingly), and that uses diffraction to reconstruct wavefronts, just like the static holograms we've been discussing. "Volumetric" displays, which emit or reflect spots of light at points throughout some region of three-space, don't count—such displays also generally don't handle occlusion cues to depth properly, and the viewer can see through a front surface to a back surface (which is fine for data visualization but doesn't give a realistic-looking rendering of a real scene).

Space-Bandwidth Product

Readers probably have some familiarity with the liquid-crystal display (LCD) panels used for televisions, computer screens and such things. It seems as if it should be easy to put a pixel pattern into one of them representing a computed hologram, illuminate it suitably, and see a holographic image that can move as fast as we can compute (or pull from memory) new collections of fringes.[i] In a context like this, the LCD panel (or any other electronic device that acts in a manner analogous to a slide in an old-fashioned slide projector) is often called a *spatial light modulator* (SLM) and more specifically an electrically addressed SLM (EASLM). SLMs can also have their pixels addressed optically (in which case they are called OASLMs, which we'll discuss in a later section) or using electron beams

(EBSLMs, which have been used to display simple holograms as well).[ii] An overview of various SLM technologies is provided by Goodman (2005).[iii]

The most significant question we have to ask regarding whether such a panel would be able to show a useful hologram is what the maximum diffraction angle available is (since for 3-D perception, both the viewer's eyes have to fit into the view zone, and ideally the viewer can move around a bit and see even more of the scene). Recall our grating equation (6) from Chapter 5:

$$\sin\theta_{\text{out},m} = m\,\lambda\, f + \sin\theta_{\text{in}} \tag{1}$$

In the simplest case, we'll use a sinusoidal grating so we get only a first diffracted order ($|m|$ is one). As of the date of publication, a typical high-quality direct-view LCD panel has a pixel spacing ("pitch") of about 0.25 mm, which translates into 4000 pixels/meter, and since the highest spatial frequency sinusoid we can show takes two pixels/cycle (one pixel at the peak and one at the trough of the sinusoid), spatial frequency f becomes 2000 cycles/m. If we then illuminate this panel with our familiar 633 nm laser (at an input angle of zero, for simplicity), the largest diffraction angle the panel can give us is about 0.0725°. It doesn't look as if that approach is going to get us a very useful or satisfying holographic image. Nevertheless, researchers have applied LCD panels in this fashion to show that the basic idea works, even if the available devices weren't quite up to the job.[iv]

A variety of microdisplay technologies have been developed for use in video projectors, head-mounted displays, or camera viewfinders—LCD, liquid crystal on silicon (LCOS), and digital micromirror (DMD, DLP) among them—which have the same number of pixels as a direct-view panel but in a much smaller area and thus can create much greater diffraction angles owing to the correspondingly higher maximum spatial frequency.[v,vi] At the same time, these diffraction angles come from a holographic "window" (to use our term from the beginning of the book) that is too small to be of use, unless we are building holographic TV sets for insects.

The take-away from this discussion is that if we want a holographic image that's big and has a lot of look-around, there seems to be no way to escape having a huge number of pixels in our electronic replacement for the traditional hologram, since we need the display device to be large and we need the pixels to be close together (this notion is sometimes referred to as *space-bandwidth product,* and is calculated as the dimensionless product of the maximum spatial frequency in cycles/meter and the width of the diffractive device in meters). We can trade off one parameter against the other to some degree—if it's easier to make a device have a lot of pixels than it is to make the pixels small, we can add *demagnifying* lenses to the display and increase the view angle in return for making the hologram smaller—but optics won't let us increase both size and angle. Similar space-bandwidth product considerations will also apply if we want to make a holo-video camera by replacing the light-sensitive plate in a holographic exposure setup with an electronic image sensor. There are also *temporal* bandwidth requirements: in either the display or the capture case, we need a device with a refresh rate in at least the tens of Hz, so we can create the appearance of smooth motion.

As a rough guide, for a display that most viewers would regard as television-like in size and with something in the range of 30° to 60° of look-around, we would need to have in the neighborhood of one to ten million pixels per scan line (try the math yourself—measure the width of your TV screen in meters and multiply that by twice the spatial frequency you get from the above equation (1) to determinc the number of pixels needed per scan line to get that sort of angle). We'd also need the same order of magnitude number of scan lines, so we're talking about between 10^{12} and 10^{14} total pixels in the image before we're done. As one surprised MIT freshman put it, "That's, like, a million TV screens!" The appeal of horizontal-parallax-only (HPO) imagery becomes clear, as in the HPO case we still would need the high pixel count per scan line, but only the same number of scan lines as in a TV picture (so we're down to "only" around a thousand TV screens' worth of pixels).

It's not too easy to make display devices that have a million pixels per scan line, whether they have a million scan lines or only a thousand, so in all likelihood we'll have to use one or both the techniques of *tiling* (using an array of devices) or *scanning* (recycling repeatedly a smaller number of electronic pixels to paint the entire image).

Scophony-Style Displays and Scanning

The first couple of pioneering holo-video systems built at the MIT Media Laboratory in the early 1990s ("Mark I" and "Mark II") were variants of a diffractive (though 2-D) television display from the 1930s called the Scophony system. It's worth exploring this old system a bit, both because it is a wonderful early use of electronically controlled diffraction and because it works according to an extremely simple principle that still looks like one of the more promising ways to get a high enough space-bandwidth product for our purposes.

Acousto-optic modulators

In 1921 Brillouin[vii] predicted that compression plane waves of ultrasound traveling through a liquid would locally change the index of refraction, creating a grating pattern that could diffract light. In 1932, two different teams[viii,ix] observed the phenomenon, and within a few years a working television display system ("Scophony") was designed around this principle.[x] A piezoelectric quartz crystal is placed at one end of a trough filled with a liquid and when electrically stimulated oscillates to create a compression wave. Electrically changing the amplitude of the wave (proportionally to the brightness of a video signal) changes the intensity of the diffracted first order of a beam of light, which, in turn, is scanned by rotating mirrors to form an image. Transparent solids also exhibit the same *acousto-optic* phenomenon, and the resulting device is often called an *acousto-optic modulator (AOM)* or a *Bragg cell.* The wavelength of the grating is inversely proportional to the speed with which sound travels through the liquid or solid used.

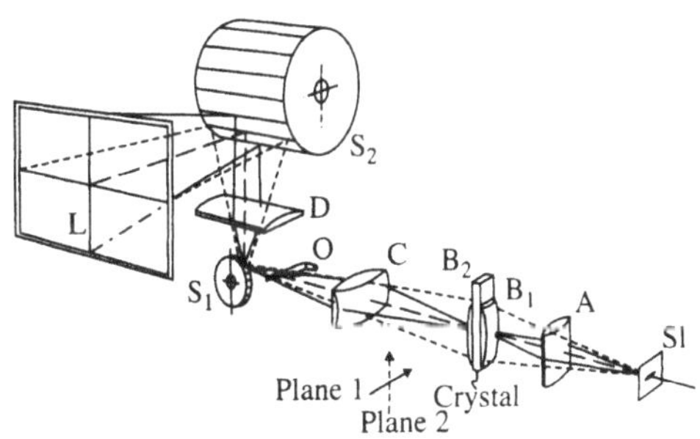

Scophony television display, 1930s. S_l anc are rotating scanning mirrors. (Reprir from reference [10] with permission).

This all seems so delightfully simple and useful that there must be a reason we don't see more of it today! The big catch is that the grating pattern is *traveling* with the speed of sound in the liquid, and in order to get a stationary image we have to reflect the diffracted output of the AOM from a mirror that is moving very fast in the opposite direction. It is typical (as in the Scophony system, illustrated in the margin) to make this mirror in the form of a faceted rotating drum, which may have to turn at tens of thousands of revolutions per minute and is a major limiting factor in scaling these sorts of displays to a large size or a high resolution. The AOM and horizontal mirror create one line of the image at a time, and another much slower moving mirror provides vertical scanning.

The MIT Mark I display

Instead of taking a fixed-frequency sinusoid and changing just its amplitude (and thus the amount of light diffracted at a fixed angle), imagine changing the frequency as well (and thus the angle of the diffracted first order). And further consider driving the AOM with a sum of many gratings at different frequencies. Then we can think of the AOM as a one-dimensional hologram, as in one line of an HPO holographic image.

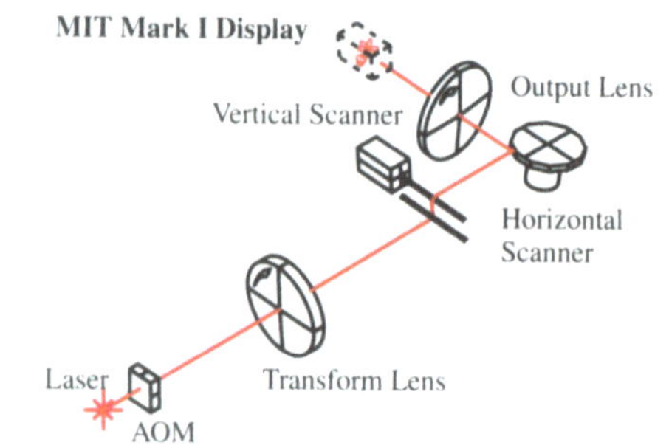

This is the basic idea behind the first MIT holo-video display (illustrated here).[xi] The AOM consists of a crystal of tellurium dioxide with an ultrasonic transducer at one end, and is driven by a fringe pattern stored in a computer frame buffer; as the useful electrical operating range of the AOM is between 50–100 MHz, the video signal is multiplied by a sinusoidal carrier at 100 MHz and low-pass filtered to retain only the lower sideband. The AOM is illuminated by a 10 mW He-Ne laser, and its output is scanned vertically with a mirror attached to a galvanometric scanner. A spinning polygonal mirror makes the image stationary in the horizontal direction and creates a "virtual AOM" that is one holo-line in length, longer than the actual AOM. From the perspective of the viewer (though too fast for the viewer to see) the effect is that of a long holo-line that is scanned by an aperture the length of the actual TeO_2 crystal. Because the AOM can provide only about 3 degrees of diffraction, it is optically demagnified to give a 15 degree viewing angle. The resulting images are able to fill about a 25mm × 25mm × 25mm view volume. Imagery for the Mark I display was computed on a massively parallel supercomputer (a Thinking Machines CM2); the display received a 32768 × 192 raster image from the frame buffer.

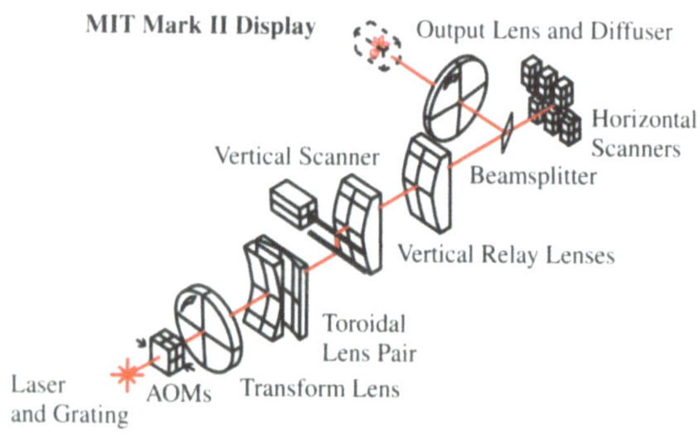

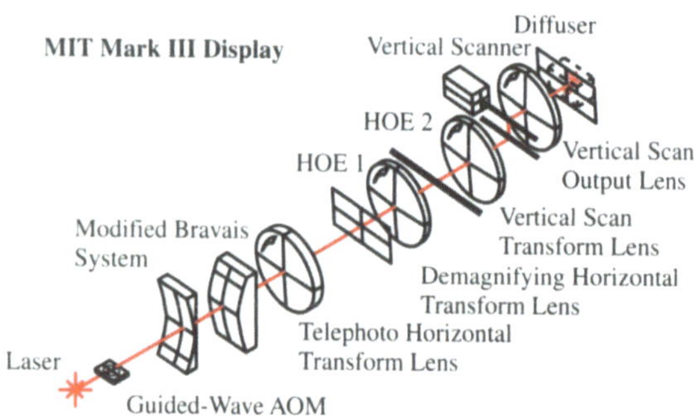

Color images were demonstrated on this display in 1992, by installing a three-channel AOM, and illuminating the channels with red, green, and blue laser light (and connecting the inputs to the R/G/B video outputs of the computer).[xii] Readers who have been following the math in this book closely will likely have figured out that the color images won't overlap exactly, and it was necessary to compensate for that effect by adding a holographic grating after the output of the AOM. As we noted in Chapter 5, the green and the blue don't diffract as "radically" as the red, and thus the available view angle for full-color images is limited by the diffraction angle of the blue channel.

The MIT Mark II display

It was apparent to those who worked with Mark I that scaling the system up so that it made a big enough image to get both eyes into the view zone with some amount of look-around required increasing the space-bandwidth product of the diffractive part of the display, and also required a horizontal mirror that was bigger than would be practical with a spinning mirrored drum.

The Mark II display, designed by St. Hilaire,[xiii,xiv] brought two major improvements over Mark I. The first was the use of an 18-channel TeO_2 AOM, such that 18 scan lines could be output in parallel (with the vertical mirror moving in 18-line steps instead of a continuous sweep); the second was the replacement of the large horizontal polygonal mirror with a linear array of small galvanometric scanners. The goal of this modular architecture was to be able to scale up the image in size and resolution by adding more AOM channels and more horizontal mirrors. The raster image driving the display is 18 channels of 262144 × 8, and was generated by a compact dataflow supercomputer called Cheops developed at the MIT Media Laboratory,[xv] and later by PC graphics cards (see below).

Images from Mark II have a 30° view angle and a volume of 150mm × 75mm × 150mm (width × height × depth). This is not only large enough to permit both eyes to be in the view zone, but to permit experiments with such concepts as haptic (force-feedback) interfaces,[xvi] though with only 144 vertical scan lines it still isn't really "television" quality.

The MIT Mark III display

Besides making images that were smaller and lower-resolution than desired, Mark II's architecture is physically large (almost the size of a dining table top) and relatively expensive. The scalable nature of Mark II pointed the way toward increasing its image size and resolution, but a direct scale-up (adding more AOM channels and more horizontal mirrors) would have further increased the cost and size.

The idea behind the Mark III display, whose development started in 2005, was to try to improve the image quality and at the same time reduce the size and cost of the display to the point that it would be practical to think of a holo-video "monitor" as a PC peripheral. In order to do this, it will be necessary to increase the space-bandwidth product yet again, and somehow to eliminate the moving horizontal mirrors (it would be nice to get rid of the vertical scanning mirror, too, as it's the last vestige of early-1900s "mechanical" television, but as it's relatively slow-moving—one facet per video frame—it is much less of an engineering or cost challenge than the horizontal mirrors).

A PC holo-video screen isn't going to be of much use unless the PC is able to drive it. There are two dimensions to this issue: generating the video signals (which have a lot more pixels per scan line than those for normal screens) and generating the diffraction patterns from a 3-D model or a set of stereogram views (as discussed in the preceding chapter). Consideration of the rapid increase in the computational power of graphics chips used in PCs and game machines led Bove's group and others to investigate whether these chips would be able to handle the computational and video-driving load of holo-video displays.[xvii,xviii] Although these chips aren't intended to support scan lines of hundreds of thousands of pixels, they do allow software to set the horizontal blanking interval between lines to a very small value such that the display can treat a set of scan lines as if they were a single holo-line; since the video buffer is being filled with diffraction patterns rather than normal pictures, occasional "holes" in the pattern if the blanking can't be reduced all the way to zero won't significantly affect what the viewer sees. Because these chips can do multiply-accumulate operations very efficiently, it proved to be possible to generate fringes for Mark II using three dual-output PC graphics cards (in order to get the needed 18 video signals) faster than with the previous special-purpose hardware.

The AOM in the Scophony-style displays above is sometimes called a *bulk-wave* device, as both the sound waves and the light diffracted by them are able to travel freely throughout the bulk of the material. In another class of devices that convert electrical signals to acoustic signals, *surface-acoustic wave* (SAW) devices, the sound travels on and just below the surface. If one of these devices also contains an optical waveguide such that light waves travel in a very thin (a few wavelengths) layer just below the surface of the substrate, the surface acoustic waves can diffract the light in the waveguide, and the result is a *guided-wave acousto-optic device.* [xix] Such a guided-wave device must be made of material that has appropriate piezoelectric and optical characteristics; lithium niobate ($LiNbO_3$) is often used. A diffusion process is applied which changes the refractive index in a suitably shallow area below the surface and then metal electrodes of suitable size for the acoustic wavelength are placed on the surface to form a transducer such that the acoustic waves travel across the light beam traveling through the waveguide and diffract it. Through the use of multiple different-sized transducers it is possible to make wideband devices with bandwidths in the GHz; hence guided-wave devices have great potential to serve in a holo-video display, though to date they have mostly been used in nondisplay applications like signal processing and optical switching.

Proklov, *et al.* (1992) [xx] demonstrated that these interesting devices can be made to deflect light in two perpendicular directions, if an additional transducer is oriented such that the acoustic waves travel toward the incoming light. Changing the frequency of the signal to this transducer causes the exiting first-order diffracted beam to move up and down, in addition to the side-to-side movement from the transducers that send acoustic signals across the waveguide. Although the vertical deflection angle available to date is a lot less than the horizontal deflection, and not enough to provide the vertical scanning for a video display, Smalley, Smithwick and Bove (2007)[xxi] at the MIT Media Laboratory decided to try using the vertical effect for another purpose: eliminating the need for the fast horizontal mirror that limits Scophony-type displays.

The available bandwidth of a wideband $LiNbO_3$ guided-wave device is so high that one of them can more than replace the stack of AOMs in Mark II. In having only a single light modulator, the MIT Mark III holo-video display is architecturally similar to its predecessor Mark I, with one big trick: as the fringe pattern moves across the device to create a holo-line, the vertical input (a sine wave) of the guided-wave device increases in frequency such that the line descends as it travels. A special optical element–either a helical mirror or a holographic optical element of similar properties–converts the vertical motion into a horizontal motion that counteracts the travel of the fringe pattern and results in a stationary holo-line from the viewer's perspective. This idea is in some respects a continuous version of a stepped mirror that was used by Son *et al.* in conjunction with a pulsed laser in an HPO display.[xxii]

A remaining bit of business that it appears must be managed is how to generate a single video signal that has a gigahertz or more of bandwidth! Recall that Mark II didn't have that problem as each AOM could be driven by a separate video signal. It fortunately works out that one can do a similar thing with Mark III, such that it's not necessary to create (by making an ultra-high-speed frame buffer) a single video signal at all—a set of transducers on the guided-wave device are designed to have passbands the same width as the bandwidth of one video output from a graphics chip, with each one's center frequency shifted up by that same amount above the preceding. Then each separate video signal is upconverted in frequency (by single-sideband modulation) to fit into its corresponding transducer's passband. Note that this means that rendering for this display is a bit unusual, as each video output is not a different line as in Mark II, but a different range of angles for the same line.

Tiling and the QinetiQ Display

Earlier we referred to a tiled approach to achieving a high space-bandwidth product (indeed, the 18-channel AOM in MIT's Mark II is an example of tiling). While it would be possible to make a large holo-video display with a large view angle by tiling a massive number of microdisplay chips, the wiring and video-generation requirements for such a display are unpleasant to contemplate. A small version of such a system with six microdisplays has been developed by a group at the University of Hyogo.[xxiii] A holographic display system developed by Slinger, *et al.* at QinetiQ[xxiv] achieves an extremely high space-bandwidth product by doing the tiling in part optically. Many microdisplay devices have only binary (on or off) pixels, but also have very high temporal refresh rates (in the kHz), so that they can give the appearance of continuous-tone images by turning pixels on and off rapidly with a varying duty cycle depending on the gray level. If a binary grating can be made to produce a tolerably good holographic image, the high refresh rate that these EASLMs use to create gray scale can be reallocated to image more than one region of a hologram. The QinetiQ system uses the EASLM to write a fringe pattern onto a region of another device called an optically addressed SLM (OASLM), which is essentially an LCD panel with a photosensitive layer in place of the addressing wires and transistors of an EASLM; the OASLM can "remember" a fringe pattern imaged onto part of it by the EASLM. A lens and shutter arrangement allows QinetiQ's 1024 × 1024 EASLM to build up a 5120 × 5120 fringe pattern onto the OASLM at video rates. QinetiQ's modular system then allows stacking the EASLM-and-OASLM modules like bricks to get an even higher space-bandwidth product. QinetiQ has demon-

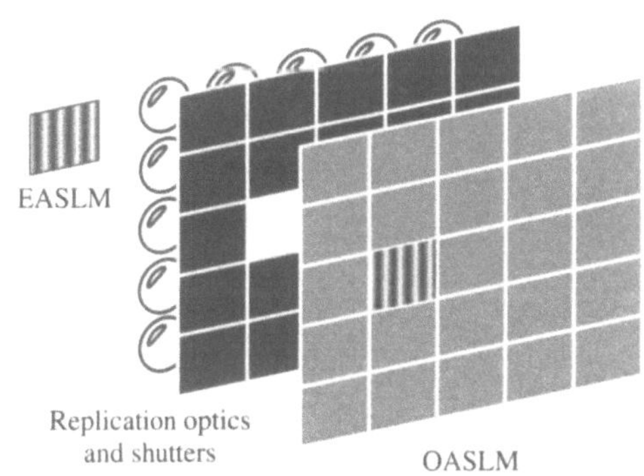

strated a group of four modules side-by-side with a resulting fringe pattern of 5120 × 20480, including both monochrome and frame-sequential color images (where the SLM is illuminated, in turn, by red, green, and blue light sources).

Electronic Capture

In preceding chapters we've discussed how to generate imagery for a holo-video display from a 3-D model or from an array of parallax 2-D views. But what about capturing a hologram of a real scene electronically? Leaving aside issues of the huge amount of data involved (a good 3-D model of a scene is a far more compact representation of it than the fringe pattern of a hologram of the scene, so if we're trying to do real-time TV broadcasting or streaming over a limited-capacity channel it might be more efficient to transmit a model and render the hologram fringes at the receiving end), we have precisely the same space-bandwidth product considerations as with the display. If we could get it to work, though, we could record or transmit the scene in real time.

Probably the first electronic capture and transmission of a hologram was done at Enloe *et al.* (1966) at Bell Labs in the 1960s, and involved a vidicon tube camera; in lieu of a suitable real-time display device, the video was sent to a cathode-ray tube whose image was photographed to become a holographic transparency that was illuminated by a laser.[xxv] Sato *et al.* (1992), mentioned above with respect to LCD panels, also used a vidicon camera for capture.[xxvi] Schnars and Jüptner (1994)[xxvii] recorded a real-scene hologram with a solid-state sensor, placing a charge-coupled device (CCD) in the taking setup for an off-axis Leith and Upatnieks hologram. Recall from Chapter 10 that such holograms typically have fringes of around a micron in size; current CCD and CMOS (complementary metal-oxide semiconductor) image sensors have pixel sizes in the range of about 3–20 microns, so in order to be able to resolve the fringe pattern the angle between the object and reference beams has to be reduced to only a few degrees so that the fringes get larger. A result of the small angle is small separation of the zero-order beam and the first-order image; thus this idea really works only for small objects far away from the sensor. If large-area sensors with a lot of small pixels become available, this limitation will of course relax. It is also possible to post-process a small-angle or even in-line hologram to separate the zero-order beam, if multiple images representing different phases of the reference beam can be captured; see for example Sato (2007).[xxviii] Poon *et al.* (2005)[xxix] acquired HPO holograms in real time by scanning an object with a plane wave and a spherical wave and measuring the local interference between them; the output of such a system would be a good match to Scophony-style holo-video displays.

Consider that a holo-video capture device is a funny sort of "camera," as it has no lens! It therefore also has none of the optical effects that camera lenses create, like limited depth of field. Where there are small moving objects at lots of different depths, holographic capture can be useful despite the limitations of current technology. One area with precisely those requirements is the study of small marine organisms such as plankton, and a team at the University of Aberdeen has built an electronic holo-camera—both in-line and off-axis—for underwater applications.[xxx]

Scene illumination poses a challenge. As with still-life holograms, the scene and the camera can't move more than a few tens of nanometers during the exposure time for a frame, or the interference fringes will vanish. Such a requirement is hardly compatible with the notion of a hand-held camera (or even a tripod-mounted camera) recording moving real-world scenes, unless the exposure time is extremely brief. Since electronic image sensors are not sensitive enough to do available-light exposures of only a few nanoseconds, it's necessary to use something like a pulsed laser to emit a short flash of light for each frame (akin to an electronic flash in conventional photography, but coherent).

Conclusions

Holographic television displays in the early 21st century are about where conventional television was in the 1930s (and not just because some of the development is based on 1930s inventions!). The fundamental principles are understood by researchers (and by readers of this book), demonstration systems exist (though not a lot of people have seen them), and widespread use will likely require only technological refinements of the "making things smaller, faster, and cheaper" variety rather than big breakthroughs. Holographic image sensors lag behind displays in development, but as more displays are created they will likely provide a big incentive to increase the pace of sensor research.

With the closing of this chapter, we also close our journey through the holographic-imaging realm. The authors now look to our readers to develop the next set of advances, such that everyone can better enjoy "The Window View Upon Reality."

References

i. Depending on how an LCD panel is built, it can work as either an intensity modulator (with polarizers installed) or a phase modulator (with no polarizers).

ii. Poon, T.-C., B. W. Schilling, M. H. Wu, K. Shinoda, and Y. Suzuki (1993). "Real-Time Two-Dimensional Holographic Imaging by Using an Electron-Beam-Addressed Spatial Light Modulator," *Optics Letters,* **18**, 1, pp. 63-65.

iii. Goodman, J. W. (2005). *Introduction to Fourier Optics,* Roberts & Co., Englewood, CO, chapter 7.

iv. Sato, K., K. Higuchi, and H. Katsuma (1992). "Holographic Television by Liquid-Crystal Device," *Proc. SPIE Practical Holography VI,* **1667**, pp. 19-31. Rather than working with synthetic imagery, these authors captured fringe patterns on a television camera.

v. Hermerschmidt, A., G. Wernicke, S. Krüger, A. Langner, H. Gruber, and M. Durr (2006). "High Resolution Coherent Optical Reconstruction of Digital Holograms and Their Applications," *Proc. SPIE Intl. Conf. on Holography, Optical Recording, and Proc. of Info.*, **6252**, p. 62521J.

vi. Huebschman, M. L., B . Munjuluri, and H. R. Garner (2003). "Dynamic Holographic 3-D Image Projection," *Optics Express,* **11**, 5, pp. 437-445.

vii. Brillouin, L. (1921). *Ann. De Physique,* **17**, p. 103.

viii. Debye, P. and F., W. Sears (1932). *Proc. Nat. Acad. Sci.,* **18**, p. 409.

ix. Lucas, P. and P. Biquard (1932). *J. Phys. Radium,* **3**, p. 464.

x. Lee, H. W. (1938). "The Scophony Television Receiver," *Nature*, **142**, 3584, pp. 59-62.

xi. St.-Hilaire, P., S. A Benton, M. Lucente, M. L. Jepsen, J. Kollin, and H. Yoshikawa (1990). "Electronic Display System for Computational Holography," *Proc. SPIE Practical Holography IV*, **1212**, pp. 174-182.

xii. St.-Hilaire, P., S. A. Benton, M. Lucente, and P. M. Hubel (1992). "Color Images with the MIT Holographic Video Display," SPIE Vol. 1667, *Proc. SPIE Practical Holography VI*, **1667**, pp. 73-84.

xiii. St.-Hilaire, P. (1994). *Scalable Optical Architectures for Electronic Holography*, PhD dissertation, MIT, Cambridge MA.

xiv. St.-Hilaire, P., S. A Benton, M. Lucente, J. D. Sutter and W. J. Plesniak (1993). "Advances in Holographic Video," *Proc. SPIE Practical Holography VII*, **1914**, pp. 188-196.

xv. Watlington, J. A., M. Lucente, C. J. Sparrell, V. M. Bove, Jr., and I. Tamitani (1995). "A Hardware Architecture for Rapid Generation of Electro-Holographic Fringe Patterns,'" *Proc. SPIE Practical Holography IX*, **2406**, pp. 172-183.

xvi. Plesniak. W. J., R. S. Pappu, and S. A. Benton (2003). "Haptic Holography: a Primitive Computational Plastic," *Proc. of the IEEE*, **91**, 9, pp. 1443-1456.

xvii. Bove, Jr., V. M., W. J. Plesniak, T. Quentmeyer, and J. Barabas (2005). "Real-Time Holographic Video Images with Commodity PC Hardware," *Proc. SPIE Stereoscopic Displays and Applications*, **5664A**, pp. 255-262. This reference discusses real-time implementations of both a stereogram algorithm and the Reconfigurable Image Projection algorithm on a standard graphics card.

xviii. Petz, C. and M. Magnor (2003). "Fast Hologram Synthesis for 3D Geometry Models Using Graphics Hardware," *Proc. SPIE Practical Holography XVII and Holographic Materials IX*, **5005**, pp. 266-275.

xix. Tsai, C. S. (ed.) (1990). *Guided-Wave Acousto-Optics*, Berlin: Springer-Verlag.

xx. Proklov, V. V. and E. M. Korablev (1992). "Multichannel Waveguide Devices Using Collinear Acoustooptic Interaction," *Proc. IEEE 1992 Ultrasonics Symposium,"* pp. 173-178. This paper also references an earlier (1981) paper in Russian reporting on the authors' work in this area. See also Tsai, C. S., Q. Li, and C. L. Chang (1998). "Guided-Wave Two-Dimensional Acousto-Optic Scanner Using Proton-Exchanged Lithium Niobate Waveguide," *Fiber and Integrated Optics*, **17**, pp. 157-166.

xxi. Smalley, D., Q. Smithwick, and V. M. Bove, Jr. (2007). "Holographic Video Display Based on Guided-Wave Acousto-Optic Devices," *Proc. SPIE Practical Holography XXI*, **6488**, p. 64880L.

xxii. Son, J.-Y., S. A Shestak, S.-K. Lee, and H.-W. Jeon (1996). "Pulsed Laser Holographic Video," Proc. SPIE Practical Holography, **2652**, pp. 24-28.

xxiii. Sato, K., A. Sugita, M. Morimoto, and K. Fujii (2006). "Reconstruction of Color Images at High Quality by a Holographic Display," *Proc. SPIE Practical Holography XX*, **6136**, p. 61360V.

xxiv. Slinger, C., C. Cameron, S. Coomber, R. Miller, D. Payne, A. Smith, M. Smith, M. Stanley, and P. Watson (2004). "Recent Developments in Computer-Generated Holography: Toward a Practical Electroholography System for Interactive 3D Visualisation," *Proc. SPIE Practical Holography XVIII*, **5290**, pp. 27-41. See also Slinger, C., C. Cameron, and M. Stanley (2005). "Computer-Generated Holography as a Generic Display Technology," *IEEE Computer*, **38**, 8, pp. 46-53.

xxv. Enloe, L. H., J. A. Murphy, and C. B. Rubinstein (1966). "Hologram Transmission Via Television," *Bell Syst. Tech. J.*, **45**, 2, pp. 335-339.

xxvi. Sato, K. *et al.* (1992) *op. cit.*

xxvii. Schnars, U. and W. Jüptner, (1994). "Direct Recording of Holograms by a CCD Target and Numerical Reconstruction," *Applied Optics*, **33**, pp. 179-181.

xxviii. Sato, K. (2007). "Simultaneous Recording of Practical 3D Color Images by Phase-Shifting In-Line Holography," *Proc. SPIE Practical Holography XXI*, **6488**, p. 64880V.

xxix. Poon, T.-C., T. Akin, G. Indebetouw, and T. Kim, (2005). "Horizontal-Parallax-Only Electronic Holography," *Optics Express*, **13**, 7, pp. 2427-2432.

xxx. Watson, J., M. A. Player, H. Y. Sun, D. C. Hendry, and H. P Dong (2004). "eHoloCam–An Electronic Holographic Camera for Subsea Analysis," *Proc. Oceans '04/MTS/IEEE Techno-Ocean '04,* v. 3, pp. 1248-1254. See also Dong, H., C. Khong, M. A. Player, M. Solan, and J. Watson (2004). "Algorithms and Applications for Electronically Recorded Holography," *Proc. SPIE Sixth Intl. Conf. on Correlation Optics,* **5477**, pp. 354-365.

Index

www.ingramcontent.com/pod-product-compliance
Lightning Source LLC
LaVergne TN
LVHW071243210926
848486LV00007B/328